Angels in the Gate

New York City and the
General Slocum Disaster

Karen T. Lamberton

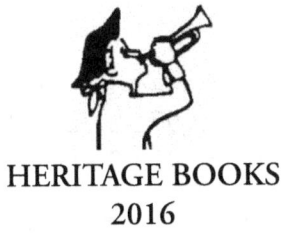

HERITAGE BOOKS
2016

HERITAGE BOOKS
AN IMPRINT OF HERITAGE BOOKS, INC.

Books, CDs, and more—Worldwide

For our listing of thousands of titles see our website
at
www.HeritageBooks.com

Published 2016 by
HERITAGE BOOKS, INC.
Publishing Division
5810 Ruatan Street
Berwyn Heights, Md. 20740

Copyright © 2006 Karen T. Lamberton

All rights reserved. No part of this book may be reproduced or transmitted in any form or by any means, electronic or mechanical, including photocopying, recording or by any information storage and retrieval system without written permission from the author, except for the inclusion of brief quotations in a review.

International Standard Book Numbers
Paperbound: 978-0-7884-3827-1
Clothbound: 978-0-7884-6439-3

TABLE OF CONTENTS

Table of Contents	i
Illustrations	iv
Preface	vii
Introduction	xi
Notes on Names Listed Throughout the Text	
Family Relationships	
Notes on *Kleindeutschland*, the German Community of New York City, 1904	1
A Gay Excursion to Hell	8
Our Families' Stories	16
Andrew Andrews	17
Henry Borsum	20
Albertina Burkhardt	21
Catherine (Ulmeyer) Gallagher Connelly	22
Michael Calderone	24
Frank Daniels	24
Barbara (Rayle) Dillon	24
Luigi & Adelena DeLuccia	25
James Duane	27
Harry G. Firneisen	28
Edwin Fitch	29
Michael J. Fitzgerald	30
Edward Flanagan	31
Albert F. Frese	32
Louise Gailing	33
Emily Halley	36
David Harris (nee Rosenholz)	38
Archibald Alexander Hill	39
Meta Jaeger	40
Mary (Hicke) Kershaw	40
John W. Klenck	41
Peter Knell	42
Elizabeth Kuhn	44
Anna Charlotte Lange	45
Adella (Liebenow) Wotherspoon	47
Gustav Adolf Lutz	49
Patrick J. Lynch	51

Henry Neats Mallabar	53
John McAllister	54
Michael McGrann	55
Franz William Moller	57
Charles & Louisa Motzer	58
Edward Mueller	60
Agnes Mundle	61
The Muth Families	62
John T. O'Connor	65
Eileen "Nellie" O'Donnell	66
William John O'Gorman	68
The Osmers Families	70
Pauline Peutz	72
Wilhelmina "Minnie" Rauch	73
The Rheinfrank Families	75
Henry Seifert	76
John Augustus Scheuing	77
Frances Ethel Stecker	79
Johann Dietrich Stelling	80
William B. & Sophie (Hansen) Tetamore	82
Dora (Laik) Krauss Thoma	84
Sophie Ulrich	86
Capt. William H. Van Schaick	87
James Celestine Ward	91
Edward N. Weaver	93
Rush Adelbert Webster	94
The Weis Families	96
Frederick & Sophie Zipse	98
The Bandelow Boy	100
Mourning & Memorial	**101**
In Mourning	102
The Memorial Meeting at St. Marks Evangelical Lutheran Church	105
The Service for Mrs. Anna E. & Miss Gertrude Haas	110
Sympathies Received	112
In the Schools	116
Pastor Haas Sermon	117
Services for the Unknown	118
The Most Awful Grave	119
The General Slocum Disaster Victims' Monument	119
Other Memorials	122
A Cross Too Great to Bear	123

Relief 124
    Mayor's Committee on Relief 124
    The General Slocum Survivors Association 132
    A Word About the Appendices That Follow 134

Appendix 136
    Passing Through Hell Gate 139
    The Work of Heroes 151
    Roll of Honors 185
    Legal Repercussions of the Disaster 194
    Bibliography 249
    Other Sources for Research 254
    Lists of Persons aboard the *General Slocum* June 15, 1904 264
    Validated list of Victims / Survivors / Missing Persons 266
    Non-Validated list of Victims / Survivors / 356
        Missing Persons

ILLUSTRATIONS

Illustrations from Documents in the Author's Collection:

Page	5	*Tompkins Square Park Outdoor Gymnasium, The Playground City*, by G. W. Harris, *The American Monthly Review of Reviews*, (1904)
	84	*Captain William H. Van Schaick, New York's Awful Steamboat Horror*, H. D. Northrup, National Publishing Company, (1904).
	101	*Every House in Mourning, New York's Awful Steamboat Horror*, H. D. Northrup, National Publishing Company, (1904).
	102	*Quest of the Dead, Munsey's Magazine*, (June 25, 1904).
	117	*Pastor Haas, New York's Awful Steamboat Horror*, H. D. Northrup, National Publishing Company, (1904).
	118	*One of the Hundreds of Funerals Enters Lutheran Cemetery, New York's Awful Steamboat Horror*, H. D. Northrup, National Publishing Company, (1904).
	119	*Trench was Dug for the Sixty One, New York's Awful Steamboat Horror*, H. D. Northrup, National Publishing Company, (1904).
	152	*Patrol, Massasoit, Zophar Mills, and a Hartford Tug, The Evening Mail Illustrated Saturday Magazine*, (June 18, 1904).
	157	*Franklin Edson, Riverside Hospital, Cosmopolitan Magazine*, (1892)
	168	*The Riverside Hospital, Cosmopolitan Magazine*, (1892).
	175	*Ghastly Sights*, scanned image of original news article from unknown Schenectady, New York newspaper, (1904).
	176	*John M. Rice, Diver, New York's Awful Steamboat Horror*, H. D. Northrup, National Publishing Company, (1904).
	182	*Three Heroes, New York's Awful Excursion Boat Horror*, John Wesley Hanson, (1904).
	201	*Volunteer Rescuers Cluster Around the* Wreck, *Munsey's Magazine*, (December 1904).
	214	*Coroner O'Gorman, New York's Awful Steamboat Horror*, H. D. Northrup, National Publishing Company, (1904)
	216	*Rotten Life Preserver stamped June 1891, New York's Awful Steamboat Horror*, H. D. Northrup, National Publishing Company, (1904).
	259-263	*Compilation of Photos of Victims and Survivors*; in *The World*, newspaper; republished by H. D. Northrup, *New York's Awful Steamboat Horror*, National Publishing Company, (1904). and by John Wesley Hanson, *New York's Awful Excursion Boat Horror*, (1904).

Images used with the permission of the Greater Astoria Historical Society:

	8	Newspaper ad for an excursion on the General Slocum
	145	Map of Hellgate and the Brothers Islands

Images used with the permission of the ECLU Archives, Wagner College, Staten Island, New York:

	110	Gertrude Haas, *Der Lutherische Herold*, New York City (June 25, 1904).
	110	Mrs.Anna Haas, *Der Lutherische Herold*, New York City (June 25, 1904).
	111	St. Marks Sunday School Teachers (composite picture), *Der Lutherische Herold*, New York City (June 25, 1904).
	112	Telegram from President T. Roosevelt

113	Telegram from the German Ambassador at Washington, D.C., Baron Sternberg
114	Telegram from the Kaiserin of Germany
114	Letter from Archbishop John M. Farley
115	Letter from Mayor McClellan
115	Letter from the Salvation Army
149	*General Slocum, Der Lutherische Herold*, New York City (June 25, 1904).

Images from the Collection of Al Trojanowicz, Fireboat Historian, Queens, New York:

154	*The Zophar Mills*

Images used with the permission of:

63	*The Muth Children*, Original in the collection of Mrs. Walter Bollinger
188	*Letter to Rush Webster from Otto Meinhardt;* Collection of Sharon Webster Harvey
188	*Letter to Rush Webster from Representative Goulden;* Collection of Sharon Webster Harvey

Post Cards from the Author's Collection:
Numbers refer to the original publisher's catalog number for the card.

Title Pg	*The General Slocum Rounding the Battery*, #1000, Success Postal Card Company, Publisher, New York
xvi	*Ellis Island*, publisher unknown
xvi	*Statue of Liberty*, publisher unknown
7	*Williamsburg Bridge*, publisher unknown
10	*Knickerbocker Steamboat company Trade Card* c. 1880, publisher Unknown
113	*The Kaiser & Kaiserin*, #3230, Alterocca –Terni, Publisher unknown.
151	*The Grand Republic*, publisher unknown
158	*Ferry Boat Bronx*, publisher unknown
160	*Tugboat on the Hudson*, #448, published by the Hagemeister Co., New York
171	*NorthBrother Light, East River, N.Y.;* #28117 the Hugh C. Leighton Co., Portland, Me.
172	*Bellevue Hospital, New York;* #12857 Phostint card, Detroit Publishing Co.
174	*Lincoln Hospital*, publisher unknown
183	*"L" Station -110th St., New York*, #100, Arthur Martin, Publisher, New York
219	*Sing Sing*, publisher unknown
231	*The Whitehall Building*, publisher unknown

Photographs from the Collection of Francis J. Duffy:

22	Catherine Gallagher Connelly
47	Adella (Liebenow) Wotherspoon
119	Ribbon from the Dedication of the General Slocum Monument in Lutheran Cemetery, Queens, N.Y.
122	Tompkins Square Monument Prior to Restoration

Photographs from the James P. Lamberton Collection:

120	The General Slocum Disaster Monument
121	Kathinka M. Stoss Side Statue –Angel with Child

121	Kathinka M. Stoss Side Statue –Angel of Resurrection
136	Plaque on the Face of the Slocum Monument, Lutheran Cemetery

Photographs, from the Author's Collection:
Front Cover	*Detail of Tompkins Square Monument*
xi	Dad and his Father on an Excursion Boat
15	The *General Slocum* in her Glory Days
122	General Slocum Monument in Tompkins Square Park,(2003)
135	Centennial Wreath at North Brother Island, (2004)
177	A Diver Ascends (stereo photo card), H.C. White Co., North Bennington, Vt.
220	Theodore Roosevelt Political Campaign Button
220	William Howard Taft Political Campaign Button

Photographs of Medals in the Collection of Ed Sere
185	Department of the Treasury, Silver Life Saving Medal
186	U. S. Volunteer Life Saving Service, Bronze Life Saving Medal of Honor

PREFACE

The disastrous fire aboard the steamboat *General Slocum* devastated the largest German ethnic population in the United States in a matter of two hours. It was the greatest disaster in the first 350 years of New York's history. Happily oblivious to any threat of danger, almost two thousand people[1], mostly mothers and children sailed from the foot of 3rd Street on the East River for a picnic on Long Island Sound. Fathers kissed their wives and children goodbye on their way to work. Children ran ahead to the docks with their friends. As the neighborhood emptied out, laughter and happy greetings filled the air and a band played German tunes. Their pleasure craft, the *General Slocum,* was the largest and fastest steamboat in the harbor. They would be at the picnic grounds well before noon, and back by bedtime. As the ship left the pier, the neighborhood fell silent and empty. And so it remained.

We still seek lost family members and the answer to "Why?" Some had recently arrived as emigrants, many were carving their niche in the American Dream, and some had finally made it into the growing middle class. All bequeathed us, by their lives before the *Slocum*, and their courage afterwards, a legacy which remains larger than life. They still remind us that there is light at the end of an endless night; that hope survives in the worst of times; that life wins out in the end.

As the 100th Anniversary of the tragedy approaches, the need for facts, if not answers remains. Many authors have attempted to make the tragedy of the *General Slocum* either a sensational novel of historical fiction or a footnote of statistics within a broader topic. Our objective is twofold; first, to create a living memorial to those brave progenitors; second, to create a body of information which may serve as a reference on the tragedy. In order to commemorate the victims, heroes, caregivers, and good Samaritans, their stories, as handed down in their families, have been collected and are presented in each descendant's own words. Some contributors had photos of their ancestors, and have included them on their story page. To further help the reader, the author has included family trees for each contributor and ancestor on the family's page. This information is intended to assist genealogists in determining the possible connections between themselves and the contributing families.

As these events were investigated and individuals researched, the project seemed at times to become something like digging a stone out of the garden which, as you go deeper, becomes bigger and bigger, until it is a boulder too large to move. Sometimes people of high importance to the disaster have been impossible to locate, in other cases the material was not only voluminous but contradictory. In fairness to other researchers, it was decided to include as many names and ways in which various people were implicated in the events of the week of June 15, 1904 as was possible to document. This led to going line by line through every available published account and existing public document. The lists are still far from being complete. Sometimes, rescuers arrived at the scene of the disaster in their capacity as police, firemen, etc. only to discover that they would be trying to save their own family along with the strangers. Many people who would never have considered becoming involved were drawn to the unfolding tragedy and rose to the occasion in ways they could never have anticipated. Of course there were the usual scoundrels and bandits. There were legal ramifications, and significant changes in legislation. If some source of information has been overlooked, it was due to the sheer volume of data available and the diversity of its locations to be visited and recorded.

For those who care to investigate further, an extensive bibliog.-raphy has been included, along with a resource list of organizations with collections of materials related to the *General Slocum* disaster. In the appendix, lists of vessels and crews who participated in the rescue, police, fire, and private citizens who were given commendations for their unflinching efforts to save as many as they could, ministers, medical personnel, etc. who aided in the recovery, have been compiled. Finally, in 1904, New York was a city of newspapers. Several of the English language and the German language papers reported almost minute-to-minute for the first several days. Long lists of victims, the missing, and persons seeking information, and those who contributed to the Mayor's relief fund were published. At times it seemed as if every faction of the city's populace from Mayor to peddler, grandparent to school child was either involved or wanted to be. Thus, sorting out who did what when was the greatest challenge of this project. At the end of the appendix are three tables of data of particular importance to genealogical researchers.

The Validated Names List includes those names and addresses of persons aboard the *General Slocum* on June 15th who have been confirmed by cross-checking as described here. The original list was compiled from various newspaper accounts, but was then cross-checked with the *Department of Charities Report for 1904* in which the Chief Medical Examiner prepared a compilation of all names and addresses associated with the disaster for whom death certificates were issued. It is arranged by household and includes the names and condition of persons associated with the dead; who were missing, injured, or not injured. While not complete, it is the most completely researched of the contemporary lists. This was then checked again against the victims' families' suit for a hearing before the Court of Claims (Bill 4154). Because this document contained sworn affidavits from the surviving family members, we know that such details as name spellings and addresses in 1904, jobs/ salaries, and medical conditions were true for the individuals mentioned. It also gives a brief glimpse of where the family was and how they were doing in 1910 when the suit petition was heard by a committee of Congress. The two shortcomings of this list are that it only includes about 200 families and that some older parents or grandparents, when filing, were a little confused by the wording of the standardized affidavit form used and reported current ages instead of ages in 1904 for family members, but this was easily determined on a case by case basis. Again, this is far from complete, but does provide valuable clues to further research. The Validated List was also cross-checked against the *New York City Directory for 1904*, which was published in July of that year but, had been prepared in the latter half of 1903.

The Validated Names List is followed by a list of the Non-Validated Names. These names, again from newspapers and other period texts could not be validated by either the Charities Report or the Affidavits attached to *4154 A Bill before the Committee on Claims of the Congress of the United States, April 1910*. This is not meant to imply that those individuals who could not be validated did not actually play a part in the tragedy. Some the author is quite convinced were there, but their roles have yet to be confirmed by a valid reference document or the word of a descendant. Where, in a few cases, the author's interviews with descendants did validate particular individuals; they were moved to the Validated Names List.

During that dark day and those that followed, New York shone brighter than ever before as all her people truly came together to comfort and support each other. It is our hope that this work will stand as a collection of snapshots of a fateful day and those that followed to the logical conclusion of events. Neither book nor marble statue can stand as a memorial to the residents of "Kleindeutschland" and all of New York City, who suffered a disaster of such great impact that it eclipsed all that had happened in the previous 350 years, and would stand unrivaled for ninety-seven more.

The author wishes to thank the many people who have contributed their knowledge, their time, and items from their collections of *General Slocum* memorabilia. First is Tom Wicks, who was also captured by the tragedy and the lack of certainty concerning the number of victims. Determined to compile a complete and reliable list, Tom spent thousands of hours on the computer creating a database from every record, article, or comment he could find and then offered his database to the author. This was an invaluable tool when cross-referencing compilations of families and their interrelationships

Pat Eckhardt and Ed O'Donnell both kept the flow of information racing to my mailbox and onto my email as if they thought that to let up on me might mean the project's demise. At times this seemed the case if only by my drowning in information! Next is a gentleman I met on the Internet while researching my own "Slocum family," Rob Kolsch. His web site guestbook is a clearinghouse for descendants and researchers trying to reconstruct their ancestors' connections to the *General Slocum* tragedy. His gracious offer to allow me to use his guest book as a point of contact with Slocum descendants saved me thousands of hours of detective work and added more than he can know to the final product.

When trying to sort out the conflicting and often incomplete data associated with the determination of "what happened when" and "why", I must also thank James and Susan Lamberton, whose deep understanding of ships, marine safety, and their knowledge of sailing through Hell Gate made my passage through the many references to those troubled waters considerably easier. While working on how the hospitals, and especially Bellevue Hospital and the temporary morgue on the 26th St. pier, played a part in the tragedy, I met Lorinda Klein, Bellevue's Historian and Sandra Opdyke, both of whom have researched the New York hospitals at the turn of the 20th century extensively and were both generous and gracious with their time and information. Finally, *Thank You* to all the descendants who sent clippings, scoured their own collections of materials for tidbits of interest, and bombarded their elder relatives for more; to my daughter Jennifer Hoover and my cousins, Lyda Spear, Carol Bollinger, and Joann Schmidt, for their encouragement, and abiding interest in our family's history which led me to begin this project.

---

[1] There is no accurate head count available to prove or disprove this number; however, since this vessel was licensed to carry 2,500 and often carried more passengers than allowed, it is reasonable to assume that she was carrying a load of between 1,300 and 2,500 passengers. The low number of 1,300 represents the total of the City's "official" count of 1,031 dead +   +/- 300 survivors.

INTRODUCTION

Die Seibzehnte Wasserfahrt (The Seventeenth Boat Trip)

As a child of fifty years ago, I remember scraps of German – a mealtime grace, a bedtime prayer. I remember as a child, sitting at Sunday dinners and listening to the elders rehash gossip of the neighborhood from when they were young. But no one spoke of the *General Slocum*. My mother described how beautiful their Lutheran church was at Christmas and how her ever-thrifty mother and aunt made ends meet when they were young and single. My Dad reflected on how hard it had been for him, as third generation in this country to learn the German language. The Lutheran church his family attended required all confirmands to study in German. Learn or face the family disgrace of not being confirmed as a member!

It was not until I was a parent that I heard bits of the story. Going through old photographs one Sunday afternoon with my folks, we found a boy in knickers with a man in a straw skimmer. "Yeah, that's me and your Grandfather. We were going on a Sunday school cruise", Dad said. "All the churches did it back then", Mom added, "Like the *Slocum*." The room fell oddly silent. I looked from one to the other and saw that expression usually reserved for ancient relatives when they are remembering their youth, that look of seeing themselves in a time and place no one younger will ever know. Dad added, "What a tragedy, but that was before our time!" and quickly changed the subject. It was not mentioned again. Years later I asked Mom and she gave me the barest of details. She spoke in the third person, as if physically keeping the tragedy away from her family. Later still when I began researching a family history for my own enjoyment and for my grandchildren, the *General Slocum* steamed back into my life. But this time, like the ship, ablaze with terrified children, it would not be ignored or denied.

The story of the *General Slocum* has been retold many times. First person interviews done within hours of the tragedy were included in the newspapers of the day. The Chief Examiner of the City of New York published a report which listed the fatalities, the injured, and the missing as a book–sized report within his annual report to the City of New York for 1904. Several authors have recompiled the data over the years describing in infinite detail the horror of the moment, but none have collected the descendants' stories. This book places the tragedy in the context of the family. One hundred years later, the fire and the water, the grief and the guilt of the survivors have passed from memory, but the tales survive and still challenge us with words first written for the commemoration of the General Slocum Survivors' Monument....."Let us not have died in vain".

While researching my family history, a cousin sent me a letter. I had asked her to identify the people in a picture taken in 1952. In the picture, I am only visible as one little white shoe amongst a dozen towering adults. My cousin sits in front of me, another grins from the corner of the photo. When the picture was returned, it had been annotated with the names and relationships of all the members of the family present that day. One note stood out from the rest. Cousin John, his father, mother, and three sisters had been aboard the *Slocum*[1]. Only he and his father survived. There was further information on "Uncle John's" second family, but I remained with the first family wanting to learn their story. The 97th Memorial Service for the victims of the *General Slocum* Disaster gave three of us cousins the impetus to get together and work on the family history. As we three cousins sat down to compare notes, I mentioned that I had part of a letter from a mutual ancestor laying out the earlier generations of our family tree. How I wished I had the entire thing. What a treasure it must be. To my surprise, one of the cousins had the complete letter with her! That letter, written in 1948, does not mention the *General Slocum* tragedy per se, but in alluding to it, concludes with a statement, which beautifully defines that generation and their legacy to us.

> "We have all inherited from those great grandparents of ours who made such stirring beginnings in a new country, a stamina and backbone to endure and carry on no matter what inscrutable mysteries and happenings come our way."

This project contains the uniquely personal recollections of the survivors and their families, the good Samaritans, the heroes, the caregivers, the passers by whose families were touched by the largest maritime disaster and most deadly fire in New York City's history[1], prior to September 11, 2001. Because the *General Slocum* was a family outing, it destroyed both the lives of the 1,000 plus victims *and* the interrelated webs of countless other families. In the case of my own family, there were Muths, Hessels, and Schnitzlers aboard, comprising a party of fourteen grandparents, parents and children; all one family, but from four separate households. In order to honor each family's experience, each family's story has been compiled separately. Each story is unique, just as each household was unique. By contributing their stories, these descendants have chosen to honor their families in a uniquely personal way.

Post Script to the Introduction

Since September 11, 2001, every person contacted either for their story or for research assistance, has felt compelled to reiterate the eerie coincidences of that tragic day and the events of June 15, 1904. Although these events are separated by ninety-seven years, each is a cataclysmic event, which defines New Yorkers, Americans, and indeed the citizens of the world at large, in a peculiar light. Both events began in the flash of a fireball and ended in the largest single loss of life at the time (excepting only battles of war[2]). Both brought together every resource of the city, the charity of the state and the nation, and the condolences of the world. One must wonder what our descendants will think of us ninety-seven years from today.

Notes on Names Listed Throughout the Text

So many people played rolls in the *General Slocum* disaster that to name each one every time they appear in various accounts, would be a monumental task, however, those included in a family story or who appear on the crew list also appear in the Validated/ Non-Validated names lists. Various individuals may also be recorded multiple times if they were connected with any of the vessels that came to the rescue, the responding uniformed services, public officials, or medical personnel who attended at the scene of the disaster. Persons honored as heroes may also appear multiple times.

Each contributor's page stands alone as one descendant's memorial to their *Slocum* Ancestor. Besides the story of the ancestor's family, the reader will find a descendants' family tree. In some cases, this may begin with the original immigrant of the family, in others, it will only reach back one or two generations beyond the ancestor's generation. In the succeeding generations, family lines have been dropped to illustrate the position of the contributor in the family structure to the *Slocum* Ancestor.

---

[1] Both in contemporary materials and in interviews with descendants, the actual name of the vessel, the General Slocum, and the more colloquial version, the Slocum, were used interchangeably and the author has chosen to follow suit.

[2] September 11, 2001 is considered a peacetime event in the context of the time when it happened.

The following abbreviations have been used throughout the work as follows:

(Name) A name in parenthesis is either an alternate spelling or an alternate name reported for this person in the press coverage
(?)   Age as reported at the time
(D)   Died
(s)   Survivor of the disaster
(m)   Remains missing
(n)   Not aboard – this is usually used for relatives or friends who made identifications of victims
Id'd.   This person identified the body or bodies of relatives listed below his or her name
T   Called to testify at the Coroner's Inquest or the Captain's trial
indicates a number assigned by the Coroner's office; 3-digits are the body number given at the time of retrieval, the 4-digit number is the official Death Certificate number issued when the body was released for burial. Unless otherwise noted, this number is for Bronx County, N.Y. because that was the scene of the disaster, a few who later died in hospital were given Manhattan or Brooklyn death certificates. If the death certificate was from a borough other than the Bronx, it is noted beside the number.

Thus a child who was onboard the *General Slocum*, died, and was identified by someone who was not onboard would be listed with the identifying individual as follows;

James Smith, father (n) -324 E. 7th St. Id'd.
    James (4) (D) #333   #1234

If this death had occurred later in Manhattan, the #1234 would be followed by "Man.".

---

[1] Over the years, as various authors have compared this event to others in New York city's history, the *General Slocum* disaster continued to elude an adequate peacetime comparison until the terrorist strikes on the World Trade Center complex in 2001.

Family Relationships

When I first began working on my family's history, I was aware of the fact that my grandfather had married twice, but only knew the name of his second wife, Emma Daug, my grandmother. Trying to track down the church records for this marriage, I contacted several Lutheran Churches in New York City. One referred me to Zion-St. Mark's uptown in Manhattan. Although this church's address was miles from where I thought my grandparents had lived, I sent off a letter. The response included all of the surnames except Schreiber, Fischer, Christ and Smith, which would be linked to me by the disaster. Several years later, a newly reconnected cousin sent me a letter and family picture which included "died on the *Slocum*" in reference to several deceased members. This led me, first to the *New York Times* accounts of the tragedy, which completely consumed the paper for almost a week after the disaster. Among the many articles, I found several interviews with my grandfather. In those stories he mentioned most of the surnames listed above. Then I found another account about a police officer who came to the rescue only to find his own family involved. His third person account was reported with the note that he had just identified the bodies of his wife and children. As I began trying to document the deaths through death certificates and cemetery records another of the names, Christ, reappeared as a "cousin" in the news accounts. The cemetery records revealed a paper trail for a family plot which belonged to the Christ family; being transferred to the Hessel family and then including a Schnitzler burial. This link has not yet been researched, and should lead to another branch on the family tree. Again, the Smith "cousin" also has not been fully researched, nor has the relationship between Christ and Smith. But what this truncated tree illustrates is that the roots of the *General Slocum* disaster run much deeper than the perfunctory victims / survivors lists. As the reader reexamines his own family tree, the clues to unknown branches and shoots abound aboard the steamship of fate for so many families. During the research of this book, several descendants were able to connect their lines and have discovered new cousins or ones long absent from the family fold. Even more unusual and rewarding have been the connections made between descendants of survivors and the descendants of the heroes who rescued them. All too often readers of history think that what happened so long ago holds little direct connection to the present, or if it does, it is an abstract thing of only passing interest to the current generation. In this instance, that is not the case. There are thousands of descendants, many looking for that one lead which will open the book of their family history. It is the author's hope that all who seek will find and that this book may be a conduit to the answers.

Postcards to relatives showed the immigrants' first sights of the city....
The Statue of Liberty and Ellis Island.

NOTES ON KLEINDEUTSCHLAND, THE GERMAN COMMUNITY OF NEW YORK CITY, 1904

> "...Sturdy German race.... industrious, self-respecting and frugal...among the best citizens we have."
> -George McClellan, Mayor of New York City

The German ethnic community of New York City came into existence almost as soon as there was a colony on the soil of the New World. They arrived with the Dutch settlers of New Amsterdam, and for centuries, Germans from Prussia, Hesse, Russia, Poland, and other provinces and principalities continued to come. Like their Dutch, English, and other European neighbors, they brought strong moral values to match their strong backs. Unlike other groups, the Germans generally were not from the most disgruntled, politically incorrect, or economically disadvantaged classes. They arrived already educated, either by practical experience or by schooling. During the American Revolution many arrived as professional soldiers leased to the British by their governments. Some slipped away in the aftermath of battle or at the end of their tenure as prisoners of war, and several thousand chose to re-enlist as American troops[1]. As these individuals and families settled into the "frontier life" they realized that here was opportunity unparalleled in Europe. Their letters began to glow with their success and supplied encouragement for friends and relatives to join them. At home in a new land and at the dawning of a new century, they moved into the life of their communities, either rural or urban, with skills and confidence which made them desirable assets to the communities in which they settled. Although the Germans' positive traits were noticed and appreciated, their continual trickle into the American population went unnoticed

In the early years of the nineteenth century, conditions in Europe both encouraged and discouraged the *Auswanderer*[2]. Over the centuries, the family farms of the German states had served the population well. The new century dawned with a middle class of tradesmen, merchants, and farmers anticipating their ability to support their children for generations to come as they themselves had been raised. But forces were already coming to bear on this situation and would change the Germans' hopeful outlook. The population of the principalities had been steadily growing. This meant that as a farmer's sons reached adulthood, the farm had to be sub-divided to make room for new families. And the more the land was divided, the less it could grow and the poorer the farmers became. At the same time, the reduction in land caused prices to rise for the crops that remained. Much of Europe was engaged in one struggle or another with bordering nations, so taxes were on the rise. There were waves of unrest and then, as in Ireland

and several other countries, the potatoes, which had been encouraged as an alternative to rising grain prices, began to blight. Although not as heavy an impact as in Ireland, the potato harvests' failures made the population, in general, and the middle classes in particular wonder how they would survive. This was not the only factor, but definitely a pervasive one for the now rising tide of German *Auswanderer* making their way to many European ports for transportation to America, Canada, South and Central America, even Australia. In other areas, it was the destitute poor who were shipped abroad by their governments. In the German states, the poor were excluded by rising transportation costs. Even some middle class Germans found themselves standing on the dock as the ships sailed. Feeling it prudent to purchase their passage before leaving home, some *Auswanderer* spent more of the remainder of their resources than anticipated, to reach a point of embarkation; then found themselves without pre-paid passage due to rising prices between their purchase and their arrival in the port.

Between 1830's and 1850's, middle class Germans looked across the sea and saw a chance to maintain or improve on what had always been an established living. There were now thriving communities in every frontier hub and eastern city. The United States was an easy place to settle without the overbearing restrictions of the old European systems. By the mid-nineteenth century, Germans made up the largest and fastest growing ethnic communities across the country of which New York City was the largest and would remain so until the start of the new century. New York was growing exponentially. The Erie Canal had opened the Midwest as a market basket for the city. The railroads hissed their way into the heart of lower Manhattan. The bustle and commerce gave the impression of plenty, but the cityscape was a Dickens nightmare. Children roamed at large, many sleeping in doorways, slums had become a reality, and illness roamed unchecked. By 1848, 160,000 immigrants, many of them German, were pouring into New York annually. But they did not remain long in the unsavory atmosphere downtown.

German immigrants had first settled in lower Manhattan, but by the latter half of the nineteenth century, the swelling enclave had moved north of Division Street and east of Bowery. As the community grew, the northern extreme moved into the area of Tompkins Square. By 1900 it was straining at its unofficial boundary of Fourteenth Street. Immigrants who arrived with technical school educations soon became engineers, architects, and other well-heeled professions "uptown". After leaving his position as Police Commissioner (1904-1906), William McAdoo wrote <u>Guarding a Great City</u>, in which he describes the German neighborhood.

"…you can go down to one of the police precincts in this district and find more people in it than in a whole list of American cities.....Just think of it! You could put Boston and all its culture down here and it would not have as many people as this one police district, and the same for Baltimore, Cincinnati, [and] San Francisco..."

As the Germans grew in numbers and in economic mobility, many began to migrate again, this time into the Yorkville area of the upper Westside and into the Bronx between Simpson St. and Westchester Ave. Another segment was moving into Brooklyn Heights. Others moved west across the Hudson River into Hoboken and Jersey City, New Jersey.

The Germans also brought longstanding traditions from their homeland, most resembled their neighbors, but in two regards they tended to stand out. Unlike other populations, they did not spend the majority of their Sunday in worship. Church service was of paramount importance, but when it was over, they either retired to a family dinner, or the *Bier Garten*. Some writers have described this institution as a tavern or bar but, it was much more. Activities there usually centered on a meal as well as a beer but, during the several hours set aside for a *Table d'ote*[3], friends and relatives moved among the tables greeting each other and exchanging the news of the day. This was also a less serious venue in which to reaffirm business relationships. During the Nineteenth Century, the rise of the mutual aid societies further reinforced the community's ability to retain its natal culture, and acted as a safety net to the less secure members of the community. For instance, if a child was ill, and the family could not pay a doctor, the society would find one among its members and pay him to attend. If a man lost his job, without immediate prospects of new employment, the organization supported his family until he found work. Even after death, a man knew that his family would at least be able to give him a descent burial, thanks to the Mutual aid societies. Between their tenacious inclination to succeed as individuals, and the insurance gained from penny dues in a mutual aid societies, the *Deutsch* gave themselves a foothold to climb out of the lower eastside with its hoards of humanity crammed into lightless and airless slums. The German Community took care of its own and was proud of it.

By the end of the nineteenth century, New York's *Kleindeutschland* or *Weiss Garten*[4] could boast the largest concentration of Germans in the country. In the 1900 Federal Census for the State of New York, 480,026 New Yorkers were listed as German; out of this number, 322,343 or one third of the total, were living in New York City. The neighborhood was in actuality a small town in dimensions as well as facilities. St. Mark's Evangelical Lutheran Church was the pulsing heart of the neighborhood. Other Lutheran churches dotted the eastside of Manhattan, as did other denominations, but the congregation on 6th Street that began in 1847 from the overflow of the congregation of St. Matthew's Lutheran Church on Walker Street, was the centerpiece of the community. In the 1870's the congregation had swelled to between 1,200 and 1,300 souls and still attracted an impressive number of congregants to its services and Sunday school classes. The gregarious Germans hardly closed the doors of the church at night, with many congregational clubs holding meetings in the basement. The church saw itself in the classic German Lutheran role of the Good Shepherd, responsible for the wellbeing of its congregation. A meal, a room, a few cents to tide the new-comer over, perhaps a job could be found here, and in times of trouble, personal or communal, this was where the community gathered. Indeed, the newspapers on the day of the *General Slocum* disaster ran stories about how the first person to arrive in *Kleindeutschland* immediately after the *General Slocum* burned was a woman who ran from the streetcar crying. She mounted the steps and pounded on the church door, too upset to realize that everyone she assumed would be at the church was aboard the flaming wreck she had come to report. And it was to the steps of the church that the bereaved, the confused, and those who wished to help came for weeks afterward. Mayor McClellan, in eulogizing the community's tragedy in the <u>New York Times,</u> stated "they are taking away my best citizens". He then recited the Germans' enviable record in service to the community, their ability work on all levels of the city's business structure, even the tidiness of their neighborhood. He then noted that no place in the city was more appropriate to house the city's relief fund for the victims, than St. Mark's. The Mayor's Select Committee worked side by side with the congregation, among the basement clubrooms and kitchen, identifying the missing, posting the lists of the dead, and distributing the relief.

In 1904, the Germans were prospering and becoming "middle class" by contemporary standards. Some were moving uptown to become "more American" but, in the old neighborhood, a new generation of German real estate brokers, and contractors were changing older tenements into tidy apartments fit for the growing families of their merchant neighbors. With the family they had raised securely entrenched around them, these "modern" Germans could look with satisfaction at what a life of hard work had brought them. Their children were generally expected to survive their infancy and young childhood and most would go to school until the end of the sixth grade, or more. The family was not hungry and bank accounts were not uncommon. These middle class families sometimes even allowed themselves a bit of jewelry –something befitting their status in the community, but never gaudy or ostentatious. Ironically, these little luxuries would serve their wearers well in those first dark hours of the chaos after the disaster.

Then, between the sailing of the *General Slocum*, and her grounding as a flaming hulk[5], everything changed. Some families were lost

Tompkins Square Park had recently been outfitted with an outdoor gymnasium by the city for Little Germany's children[6]

entirely. Others were left to simultaneously bury the dead and tend the injured in hospitals across the city. Fathers widowed and childless couldn't stand the quiet of the neighborhood where children had played and neighbors conversed. Their burdens overwhelmed them made heavier by their determination to take care of their responsibilities by themselves. To accept charity would be a shame on the family. Many were forced into bankruptcy and moved in with relatives elsewhere in the city. Some moved to escape the silence. The Bronx put distance between the survivors and their old familiar neighborhood; Middle Village, Queens allowed them to quietly remember at the graveside. The true number of victims may never be known - some tried to kill themselves when they found that neither wife nor child was coming home again. A few succeeded in the weeks and months that followed. In the flash of a spark into flame, families ceased to exist, a community dissolved, and a church was uprooted. One unnamed gentleman quoted in the New York Times, having lost his wife and children, voiced the thoughts of the generation; "I will bury my dear ones that are dead from their old home…then I will take a last look at the church, bid good bye to Good Pastor Haas, and turn my back on this part of the town forever." But if this expression of desolation seems to fly in the face of the Germans' more usual characterizations it does not detract from another virtue of the German character, courage. Writing after the disaster, Munsey's Magazine[7] described the people lost that day as follows;

> "These women were not afraid. Women afraid do not rush with passionate eagerness into a red cavern of fire because they see a boy's cap or the flutter of a yellow curl."...."
> Such is not fear; it is love. And there is no pain in the world like the agony of the love that is losing its dear ones."..."but it was not a death-grapple for personal safety. It was a struggle of husbands to save wives, of mothers to save children, of children to save parents. This family devotion crowned the tragic scene on the *Slocum* with a halo of heroism that should ever be remembered."

As in every disaster, before or since, New Yorkers of all ethnicities and every status, rallied to aid the victims of the *General Slocum*. Nearly every organization, social, political, or commercial lined up to help in some way. Donations poured in from around the world. Although well meaning, The Women's Health Protective Association, convened a meeting with representatives of the Mayor's select committee, which showed how misunderstood the Germans were in some quarters of the city. These well meaning matrons were ready to take in the orphans, send food baskets to the hungry and set up a settlement house to get the neighborhood going again. Accompanying the committee that day was Dr. Semken, a member of the St. Mark's congregation. After thanking the ladies for their good intentions, he explained that these were middle class Germans, who if fate had not intervened, would have been able to handle whatever came their way. It was only the magnitude of the disaster that had overwhelmed their own resources, and regardless of their current situation, they would not accept outside charity. He suggested instead, that a deaconess of the Lutheran Church be found and hired for one year to aid the families as needed. This he explained would be better accepted and more appropriate to their needs. As ever, the German community would take care of its own in its own way. In the end, of those individuals / families eligible for charity, less than 75% accepted any aid. Those who did drew only the minimum needed to cover burial or hospital expenses. By August 1904, the Mayor's Committee disbanded. Unable to give away all available funds, they set up a fund for the future needs of the survivors made dependant by the disaster. Eventually this became a scholarship fund for the victims' children.

> "The loss of nearly 1,000 such people was a loss of a great Moral force…" -Justice Morgan J. O'Brian, Appellate Division,
> Supreme Court of New York

The New Williamsburg Bridge over which the funeral corteges passed

THE IDEAL EXCURSION of

TABERNACLE M. E. SUNDAY SCHOOL,

To Cold Spring Grove, on the Sound,

ON WEDNESDAY, JUNE 19TH, 1895.

The Finest Excursion Steamer Afloat, GENERAL SLOCUM, will leave Java Street Dock promptly at 9 o'clock.

ADULTS 50c.   **CHILDREN 25c.**

Remember the date and don't fail to be there.

Special Attractions—Particulars Later.

A typical ad for an excursion aboard *General Slocum*.

---

[1] Schwalm Family Association

[2] One who emigrates

[3] A meal of many courses that is served with a period of time between each course. Thus, such a meal could last for 4-5 hours.

[4] White Garden referred to the white picket fences which surrounded the front yards of many tenement buildings.

[5] For a detailed account of the sailing of the *General Slocum* and her end as a burned and sunken hulk, see the Appendix.

[6] The American Monthly Review of Reviews, *The Playground City* by G. W. Harris, [Month unknown] 1905, Pg. 580.

[7] Munsey's Magazine, *The Story of the Slocum Disaster*, December 1904, Pg. 44.

A Gay Excursion to Hell

The story of the disastrous burning and sinking of the General *Slocum* Steamboat has been the gist of fiction as often as fact. In the confusion of the disaster every reporter in New York City vied for the best account. Many made their way to the scene and had to be restricted from hampering the rescue and recovery work. Just as today, the police had to set up barricades to keep reporters, relatives, and the curious out of the way as doctors, nurses, and rescue workers fought to save the victims.

Deadlines had to be met; stories had to be called in to the pressmen whose flying pencils scrawled columns for the typesetters. The chaos and shock, the unfamiliar spelling of German names, the sheer magnitude of the situation all contributed to the beginnings of a first magnitude urban legend[1]. Each year beginning in 1887, St. Mark's German Lutheran Church hosted an outing for the Sunday school members, their relatives, and friends. In the months preceding the event, the planning committee met and discussed what the seventeenth Sunday school excursion would be like. Would they again hire the *General Slocum* steamboat of the Knickerbocker Steamboat Company? Would they go to Locust Grove, Long Island for a picnic? Would there be a band? What treats for the children en route? What small concession could they negotiate with the steamboat company, so many details to handle!

By the time the 1904 contract was signed, the congregation already knew that there would be ice cream served on deck on the way up river, for the children; clam chowder would be cooked aboard and served with lunch at the grove. The bar would serve several brews, in the church's own glassware. The band had agreed to a small reduction in their fee. The leader, a friend of the committee chairperson, had agreed to play all the way up and back; their families would also attend. Members of the congregation had fanned out through the neighborhood streets soliciting contributions and selling advertising space in the event's program. The steamboat company was even donating rolls of tickets, which could be sold in advance as a fund raiser. Some of these, purchased by local merchants, were then given to the neighborhood's non-congregants. The local schools would not finish their semester until well after the 15th, but any child who asked for an exemption to attend the outing was given permission. As the day approached, the excitement in the Eastside community became palpable. Mothers dickered with grocers for the best delicacies with which to fill their huge picnic hampers. For days the many aromas of German and American cooking hung thick in the summer air. Children were cautioned not to become overexcited, to be good, or they might not be allowed to go. Fathers proudly smiled at bouncing little ones happily and loudly anticipating the wonderful day to come. For most of the men, it would be just another Wednesday at work, but knowing that their success had paved the way for their families' enjoyment was their special satisfaction.

In 1904, the German community was moving out into the general American population. Enclaves had spilled out to areas as remote as lower Connecticut where German culture mixed with American customs and language. The excursion was a grand excuse to bring these seldom seen friends and relations home.

On the night of the 14th of June 1904, every bed was overflowing with whispering children, couches with aunts and cousins. With barely room to move, mothers ironed "good" dresses and hair ribbons, counted plates, and packed their hampers. Some would not have time to sleep that night. By dawn, Papa was making ready for his workday, carefully stepping around his slumbering relatives. Mamma was preparing breakfast, her last duty before an entire day away from work! By Seven O'clock, all were up, dressed and eager to leave. Goodbyes were said on the doorsteps and the fathers headed into the maze of the city and work, while mothers, grandmothers, children and friends found their way down to the foot of East 3rd Street, and the recreation pier, where the *General Slocum* had just tied up. Those who stayed home noted later that the children's laughter could be heard for blocks.

The steamboat companies used trade cards, like this one, plus newspaper ads to advertise their ships. The Grand Republic was the *General Slocum*'s elder sister

The first to board was the band, who took up their places on the middle deck and began to play. As the gates to the gangway swung open, the church's minister, Rev. Haas, took up his usual position at the top of the ramp to welcome all aboard. On the opposite side of the deck, Police officers Kelk and Van Tassel took up positions at the rail lest a rambunctious youngster fall overboard in the excitement. Baby carriages, lunch hampers, handbags and hats jostled and bumped aboard as the women headed up to the top deck and to the back for the best sightseeing seats. Children scampered around under foot locating their friends. All felt secure in the knowledge that they were safe and so, free to relax and enjoy the day.

A crewman later testified that 982 tickets were collected by him on the gangway but, exactly how many people came aboard is not known. Children under twelve did not require a ticket, and some tickets had been purchased as a donation to the church and given away, or just not used. Some of the single adults came alone, and so were not missed by friends until many days later. Some people were only known to those who invited them, and so had no one to identify them afterward. By a little after nine, the ship was ready to leave the pier. They would be in Long Island Sound by mid-morning and at the Grove for lunch. But as she made ready to cast off lines, one report states that, a Mrs. Straub suddenly grew pale with a strange and frightful look in her eyes. "We must go, we will all die" she was reported to have told her companions, and when they laughed at her concerns, she abandoned her picnic basket, grabbed her children and scurried off the boat, then continued to harangue the passengers from the dock that they would soon die if they stayed aboard. Only one late arriving passenger, a man with two small children, who encountered her on the gangway followed her advice and beat a hasty retreat. When questioned later, she said that looking down the deck at the children playing; she suddenly saw small blond skulls among the waves. Fact or fable, no conclusive proof of her tale has been found, the reader must decide for himself.

By nine thirty the *Slocum* was heading up the East River with the tide pushing from behind. As she reached a point between 86th and 92nd street, she was already doomed. The previous evening, the gentlemen from St. Marks had boarded the *Slocum* with several barrels of glassware for the bar, packed in salt hay. The glasses were removed and the hay returned to the barrels for the repacking after the excursion. The barrels were then stowed in a forward cabin below decks, which was used for storing supplies like kerosene and metal polish. This was also where the crewmen cleaned, trimmed, and filled the lamps, and occasionally stole a smoke. This was also against the federal regulations governing this type of vessel. Although there would be several theories on how and where the fire started, this seems to be the most likely place. Other theories included cooking oil being used to fry clams on the main deck for the passengers, which might have tipped over and then been accidentally ignited, or some other grease fire might have been sparked in the galley. In on-scene interviews and later at the inquest, the crew maintained that the forward cabin was the first place fire was discovered. It is also quite possible that the cabin's porthole had been left open, and would have given the smoldering fire oxygen. Forensic evidence examined after the disaster pointed to a carelessly tossed cigarette butt which landed in one of the barrels. By the time the first smoke was spotted by observers on shore and on other vessels, the ship was probably already at the point of no return. The overhanging design of the lower deck, which gave the passengers a more expanded promenade area, and the stiff breeze over the bow as she plowed north, may have pushed the first puffs down along the waterline where it was not noticed by the passengers or crew, and was certainly not visible to the captain and pilot three decks up and well back from the bow in the pilothouse. When smoke finally began snaking up the companionway from below decks, a young boy[2] who tried to report the fire was dismissed as a prankster by the captain. He finally convinced a crewman to take a look, but it was already too late and opening the cabin door only fueled the fire into a raging back draft which exploded up the stairs like a bad Hollywood special effect.

    The time was about ten minutes after ten, and the *General Slocum* was entering Hell Gate, one of the most difficult passages on the east coast. Passengers, at first stunned and silent[3] became a stampede up to the second deck to find their children; then up to the Hurricane Deck hoping to outdistance the blaze. People began crowding the rail and preparing to jump. The band kept playing a la Titanic, hoping to calm the crowd. But when hope was gone, they ran up to the Hurricane Deck and began pulling down life preservers for the passengers. As the press of the crowd grew, many of the band members were pushed over the rail and into the water. Those excursionists, still in control of their senses, also tried to pull down life preservers from the overhead cages. The linen coverings, long exposed to wind, sun, and salt, shredded in their hands. Those that survived the drop of three stories into the water, split on impact or in a worse turn of fate, became anchors, as their filling of pulverized (instead of blocks) cork became waterlogged and dragged their struggling victims down. Later, people felt that the captain, who had sailed the New York Rivers and harbor for forty years, should have immediately stopped his engines or turned broadside to the current and sought a place to run aground. Most of the people who put forth these ideas were not watermen and did not understand, as a pilot does, the treacherous water the *General Slocum* sailed that day. This steamboat, which traveled faster than nearly every other ship in the harbor, was going at something between ten and fifteen miles an hour by her engines. Add to this the approximately twelve miles per hour of current pushing her through Hell Gate Inlet[4], and there was no stopping by virtue of killing the engines. Her forward momentum would have had her nearly into the Long Island Sound before her headway was abated. The *Slocum* was 250 feet long and sixty wide plus the width of her paddle-wheels. The channel of Hell Gate varied from about 275 to almost 400 feet in width and was lined, sides and bottom, in rock so hard, that it resisted all but the most overwhelming attempts to blast it out. The *Slocum* would have had just seventy five feet of room to turn around at the widest point in Hell Gate channel. So large a vessel was also subject to having her belly torn out without reaching the safety of the shore. Some of the authorities who loudly called for better thinking from the safe distance of shoreline interviews, questioned the logic that sent the *Slocum* flying passed the Sunken Meadows off 128th St., but here again, before safety might be reached, the ship could have foundered on rock ledges while still in swift running deep water. The proof of this had happened in 1878 when the *Seawanaka*, another large paddle-wheeler, had grounded and burned there with great loss of life. Still others felt he should have put in at one of the several city piers along the waterfront, in the Port Morris area. The captain testified later, that indeed he had veered toward the most likely one only to be waved off by a tug because of gas tanks nearby and wooden structures on the wharf, which might have caused an even greater conflagration. The captain's solution was to steam full tilt for North Brother Island. His objective was to beach on the western side of the island near the main buildings of Riverside Hospital.

    As she approached, the staff saw the disaster bearing down on them and sprang into action. Calls went out to the police and fire houses on both sides of the river to send help. Then Doctors, nurses, administrators, cooks, and patients headed down to the beach. Fire hoses were strung along the hospital's pier and human chains formed from the beach out to chest deep water, ready to hand back victims as they drifted by. Many vessels which had fallen in behind the *Slocum*, picking up jumpers, now came in as close as their drafts would allow and continued in the rescue. As the *Slocum* grounded, and without warning, the Hurricane deck's stern section, whose stanchions had nearly burned through, collapsed. Perhaps as many as 300 passengers were either thrown into the water or fell straight down into the boiling cauldron of fire in the ship's midsection. The hundreds of passengers who had either been caught in the bow ahead of the fire, or who took refuge there in the early minutes of the inferno, now jumped overboard and waded ashore from chest deep water. Others along the starboard side managed to cross a teetering plank hastily thrown between the *Slocum* and one of the rescuing boats. Triage stations sprang up along the beach and neat rows of bodies grew from the carefully tended lawn; dazed survivors shivered in groups or wandered the waterline looking for missing relatives and friends.

By eleven thirty there was nothing left to do but collect the bodies. The recovery went on through the night and then continued for several more days. Cannons were brought in and exploded over the wreck to help dislodge victims trapped in the wreckage. Hard hat divers returned time and again with armloads of children. Sometimes they came up tightly clenched in each other's arms, or as a mother still holding a baby to her breast or a child by its clothing. In the final analysis, the city decided that 1,031 people had died, hundreds more were injured, and sixty one were buried as unknowns. These were either too burned to be recognized or had boarded without family or friends to seek them out after the disaster. How many bodies were carried out to sea with the tides could not be estimated, there was no precedent.

The *General Slocum* had passed her annual safety check only a few weeks earlier; life vests and preservers, lifeboats and rafts, fire hoses and pipes, valves, etc. were all listed in good condition but the regulations of 1904 did not mandate their inspection as ground for issuing the yearly safety certificate needed to sail. Her decks and compartments were properly fitted to her use. She was licensed to carry up to 2,500 passengers but, sometimes carried twice that number[5]. The search for answers would take the next several years. Laws would change and the public would realize that they had to be aware and alert to protect themselves. Swimming lessons, for the first time became a childhood necessity. There would be an inquest, several trials, prison, and a presidential pardon before the story of the *General Slocum* came to a close in 1912.

As difficult as the story of the *Slocum* is to read, it is also difficult to realize that we view those events from a very different world. 1904 was the age of business having its own way regardless of the cost to the public; government was not the regulatory authority we think of today; graft and corruption throughout the bureaucracy was common. If there is good to be found among the ashes of the *General Slocum* it is this; the United States began to view safety as a prerequisite of business where the public is concerned. Other ships had burned, some in the same section of river, with terrible numbers of deaths. But this time the numbers were so tremendous; the devastation so complete, that the effect was immediate and electric.

President Teddy Roosevelt demanded an investigation by the Department of Commerce and Labor, reporting directly to him. The Steamboat Inspection Service, the agency responsible for the ship's in-spections, investigated in hopes of discovering a way to avert blame to no avail. The state and the city also launched investigations in an honest attempt to identify the person or persons responsible. And some among the German-American community, long considered good natured of per-sonality and up-standing in character, galvanized themselves for a fight. Although the outcome of so much scrutiny was not what we would expect in similar circumstances today, it was those very differences that led to the type of legislation that we would expect today; an American consumer movement was born.

In her glory days, the *General Slocum* was the biggest and fastest of the paddle wheelers in New York's harbor. On any summer day, she might have set sail with 2,000 or more people.

---

[1] A century later, when surviving accounts are compared a credible compilation of newspapers, survivors' memoirs, and public records can be used to reconstruct the events of the *Slocum*'s last voyage.

[2] Accounts vary as to whether both reports were made by the same child or several different ones at about the same time.

[3] Survivor, Albert F. Frese, reported this reaction in several interviews conducted many years after the disaster. (See Albert F. Frese in Our Family Stories)

[4] See tide charts and Navigational charts for New York Harbor and the New York Times weather forecast for June 15, 1904.

[5] During the Coroner's Inquest, it was noted that at times she carried up to 4,700. Furthermore, this was not an isolated situation; most of the harbor steamers did the same thing

Our Families' Stories

On June 14, 1904, the families whose stories appear in this section considered themselves "Americans" not Germans, not German-Americans; just Americans. They lived in a large city neighborhood, much like other neighborhoods in other cities. Their family origin, their first language, their occupations were much the same as their neighbors and their fellow citizens in other regions of the United States. They were proud of their successes and their families; most were confident in their faith. The same could be said for those who came to the rescue, or aided in the care of both the living and the dead following the tragedy. They were not just "Irish cops" or "society matrons: they were members of the largest identifiable community in the country; they were "New Yorkers".

In researching these stories, it was particularly interesting to realize that the "poor Germans" of the disaster were also Irish, Italian, Polish, Russian, etc. as were their rescuers and all those who impacted their lives as a result of the tragedy. News accounts may have focused on the young Irish-American nurse-helper who became a heroine but, missed the story of the British, Canadian, and other hyphenated Americans who worked tirelessly without thought for their own safety. In the end, the following stories show that they were also just Americans going about their normal routines.

While many of these stories highlight the actions of victims, survivors, and rescuers, there are also the stories of those who "missed the boat". Sometimes this was the result of other obligations; sometimes it seems to have been the intervention of fate. While each story is different, the results are the same. In fact, whether victim, hero, first to help, or too late to board, all of theses stories have similar outcomes. We, who trace our lineage back to this singular disaster, have received a legacy in the form of a family story. These stories not only tell us what happened to our families on one day in history, they give us a blueprint for success born out of tragedy.

Andrew Andrews, Pilot of the *Franklin Edson*

Andrew Andrews   Bertha Louise Smallwood

My father, Andrew Andrews, was born and raised in New York. He attended school until either the sixth or eighth grade and then began helping to support his family. He worked for the McAlpin Tobacco Company driving horse drawn wagons of tobacco to and from New Jersey on the ferry. Then he worked for a funeral home, driving a four-horse team and hearse to Lutheran Cemetery. He talked about eating after the funerals at Niederstein's, a German restaurant outside the cemetery. I am not sure which job he had first but, the boss had a brother or relative who was a harbor pilot. Dad took an interest in this occupation and eventually earned his license.

On the day of the disaster, my father was aboard the *Franklin Edson*. They had come northbound on the East River and were at the North Brother Island dock having just discharged passengers. Their steam was up, and so they could pull out into the river and let the *General Slocum* dock. Dad was the pilot and First Officer of the *Franklin Edson*, serving under Captain Henry Rick. They first lay alongside the aft end of the *Slocum* and Captain Rick went overboard pulling several victims from the water as Dad maneuvered the *Edson*. Captain Rick came back on board and Dad jumped onto the *Slocum*. He could not swim and so decided to help from the burning deck of the steamer, where he passed women and children onto the *Edson* until she had to back off and Dad returned to the *Edson's* deck. By the time the *Edson* retreated, Dad had sustained burns to his back.

Dad used to mention the bodies on the shore at North Brother Island. He was very critical of the captain of the *Slocum* and said he should have grounded earlier than he did and that his decisions had only increased the spread of the fire. He was critical of the Steamboat Inspectors and talked about the life preserver issue.

Later that day and the next, he returned to North Brother Island to continue in the recovery effort. There he met my mother, Bertha Louise

Smallwood, who was a nurse working on the island. She and her sister had arrived in this country earlier that year from Conway, Wales. They had not gone through Ellis Island because their uncle, Thomas Smallwood, who lived in Brooklyn, had sponsored them. Dad used to tell me how one of the men who worked in the morgue would offer Mother orange juice in order to impress her. He may have already been dating her. My father finally countered this by telling her that her suitor kept the juice cold by keeping the container on a tray under the head of a corpse in the icebox. Apparently the story took hold and they married in 1906. This story was retold often as I was growing up. Mom would insist that I drink orange juice laced with that awful tasting cod liver oil and Dad would tell the story.

Years later, when my parents' friends came over to play pinochle on Friday nights, there would be conversation about the disaster. These friends had come to this country in the 1920's from Germany and Dad would relate some of the details of the fire but, it certainly was not braggadocio.

Mom, on the other hand, was very proud of his heroism. She would brag about his bravery but, if he was present, he tried to minimize his involvement. He certainly had a right to brag, after the disaster, he was awarded the U.S. Volunteer Life Saving Corps Medal of Honor for bravery on 15 June 1904, and I believe another medal from Kaiser Wilhelm. He also kept, as a souvenir, a bunch of pennies that had melted together in a gumball machine on the *General Slocum*.

I recall, especially when W.W.II started, my father recalled the events at the beginning o W.W.I. the U-boats were a problem and the U. S. Navy was using sub chasers, a vessel about 80 to 120 feet long with high power engines, to protect the coast. The government also confiscated many of the large luxury yachts that were owned by the wealthy on Long Island to be used in defense. Dad tried to join the Navy in a program using local pilots and captains to operate this fleet. He was bitter because they turned him down for having received the medal from the German government. He was a very patriotic person and served in the Civil Defense Corps as an Auxiliary Policeman during W.W.II.

When I was about ten or so, we traveled to New York, Dad, Mom, my sister Edna, and I. We stopped at the Slocum Monument in Lutheran Cemetery. We all got out to see, except Dad. He said that he had been there the day of the funerals and did not need to see it. However, he did insist that we stop at the German restaurant! He never mentioned, and I did not know until 1996, when I researched my family's history, that there is an Andrews plot only yards from the monument. How sad.

Contributed by: Gene Andrews, Ocala, Florida

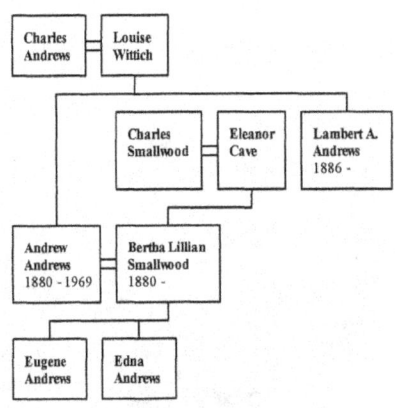

Andrew Andrews & Captain Rick aboard the *Franklin Edson* many years after the disaster

Andrew Andrews Medal of Honor Certificate from the U.S. Volunteer Life Saving corps for his heroic rescues at North Brother Island on June 15, 1904

Henry Borsum
Address Unknown

Henry & Henrietta

My memory of what happened to my family on the day of the tragedy comes from my mother Emma E. Borsum. Her father, Henry Borsum died in the disaster. The rest of the family, her mother Henrietta (Schwann) Borsum, and the children Emma Elizabeth, and Paul (who died in infancy) didn't go on the trip that fateful day because someone was ill. We can only speculate why Henry boarded the *Slocum* without his family. He may have been on the *Slocum* with members of a congregation other than St. Marks, since I was baptized at Emmanuel Lutheran Church at 88th Street and Lexington Avenue.

After the tragedy, Henrietta worked as a housekeeper for two gentlemen, a Mr. Levine and his son Arthur until her death in 1913. Emma became an orphan at eleven years of age and Mr. Levine became her legal guardian. Since he was not in a position to raise a child, she was placed in the care of Carl and Jenny Rutishouser. This couple was childless and had been friends of the Borsum family. Emma remained with them until she married my father, Andrew Schafer, Sr. My mother Emma, who was born October 28, 1902, was not yet two years old on June 15, 1904, and she never had much to say about the Slocum disaster except that she remembered being told that my grandmother Henrietta tried to find her husband's body.

Contributed by: Andrew Schafer, Newton, New Jersey

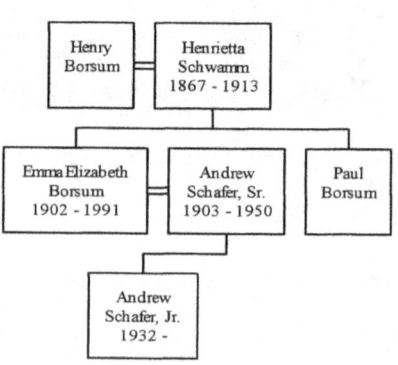

Albertina Burkardt
Tompkins St., Manhattan

Albertina Burkardt

My five brothers and I, Walter Burkart, lost our greatest treasures our great grandmother, grandmother, and aunt in the *General Slocum* disaster. Our great grandmother, Anna Burkhardt, was sixty-six years old, my beautiful grandmother, Albertina Burkhardt, was thirty-nine, and our Aunt May was four. Albertina left four sons, my Dad, Edward, Arthur Charles, and George. Edward and Arthur were on the *Slocum* that day and Arthur lost his right eye.

After the *Slocum* disaster, my grandfather, Peter Burkhardt, a civil engineer, became very sick from the loss of his mother, wife, and daughter. He was admitted to Metropolitan Hospital on Roosevelt Island, which was under the 59th Street Bridge where he died at the age of forty-something. You know how God does things; my grandmother's sister, Katherine Herrmann, couldn't have children of her own so she raised Albertina's remaining children. We called her Grand Aunt Kate. I think we were a blessing to her.

My parents came from Prussia, and settled in the lower East Side of New York That was the place all the German immigrants settled. When I was a little boy, my Dad never liked to talk about the *Slocum* disaster but, would bring us to the Lutheran Cemetery in Middle village, Queens to see the final resting place of his family. The plot stands only a block away from the Slocum Monument. All of us children felt cheated by not knowing or having grandparents. The only thing we did have was a cherished picture. From the time we were children we have searched for lost family and I have found cousins that I have not been in touch with for fifty years.

Contributed by: Walter Burkhart, Whitestone, New York

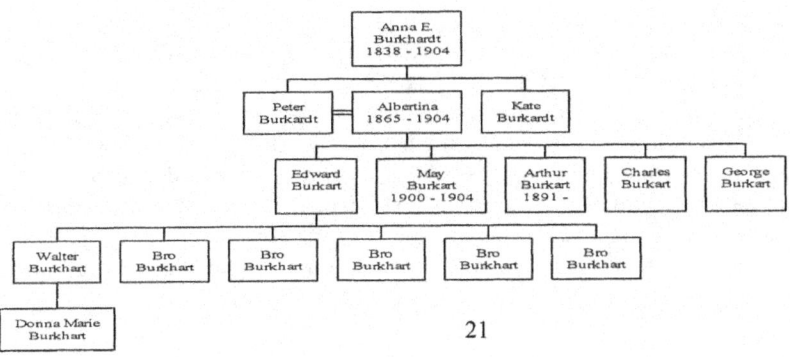

Catherine (Uhlmeyer) Gallagher Connelly
424 East 15th Street, Manhattan

Catherine Uhlmeyer was born in 1893 to a family living in Little Germany. But the joy they felt over her arrival was soon shrouded in grief when her Papa died the following year. Her mother, Veronica, soon married again; this time to an Irish gentleman, John Gallagher, who worked in a local clothing store. Soon two more babies arrived, Walter in 1895 followed by Regina Agnes in 1903. Life and the family were doing well and Catherine was growing up. As 1904 began, she made her first communion in a beautiful new white dress. Then, as the school year drew to its close, she looked forward to the pleasures of summer. Out of the blue, her mother was given tickets for the St. Marks German Lutheran Church's annual excursion and picnic. In other years she had watched the boat depart on this fabulous trip to a picnic grove with many of her friends but, she was Catholic and so not a part of the trip. Now she was determined to go!

Veronica was given only three tickets and planned to take the two younger children. Catherine would remain behind with her grandmother. First she schemed, then she took herself to Mrs. Woodhouse's grocery store where Veronica had been given the tickets. On the threshold she screwed up her face and forced a few crocodile tears to dampen her lashes before she stalked in. It must have been quite a performance because she left with her mission accomplished! There would be no keeping her from her friends, the free treats, and her very first boat ride.

On the morning of June 15th, over the protests of her stepfather that the boat was dangerous, the Gallaghers left for the pier. Along the way Catherine met her friends and they giggled and chatted about the free milk and soda onboard and the games they would play at the picnic grove. But they never reached Locust Grove and the picnic. As the ship caught fire and panic swept over the crowd, Veronica tried to hold on to all of her children. Catherine lost hold of her mother's hand and was pushed to a rail where a stranger threw her overboard. One week later, she died her beautiful white dress black, for the funerals of her mother and siblings.

In an interview, many years ago, Mrs. Connelly said that at first everything was surreal; the empty house, the loss of friends, the impossible task of trying to carry on. Then, she said, life felt like a terrible fairy tale made up by her mother. All she remembered was the jumbled impressions of fire and screams and panic. For many years Catherine's life seemed to go nowhere or whichever way the wind blew. She moved from one relative to another and left school at thirteen. A normal life seemed impossible until at age twenty she met and married Thomas Connelly and began a new life with the family they created together. It became a very full life indeed. Eleven children, dozens of grandchildren and great-grandchildren richly filled her years.

At one hundred and nine years of age, Mrs. Catherine Connelly died. Over the decades she had become a local celebrity to those who traced their heritage back to the *General Slocum*. Even in her most advanced years, she enjoyed being interviewed and faithfully attended the *Slocum* memorial services each year until she was no longer able to do so. At the time of her passing there was only one known living survivor of the disaster.

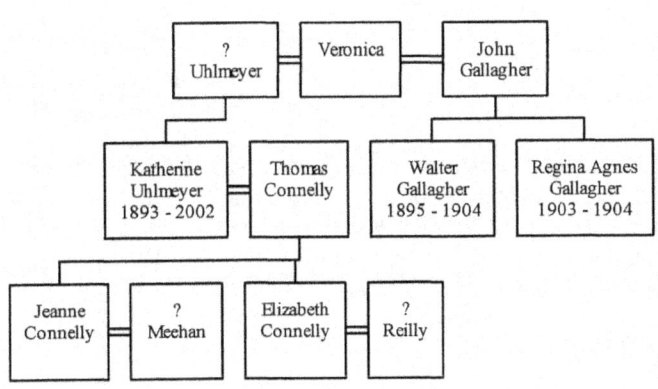

Michael Calderone
Address Unknown

Michael Calderone, my great-grandfather was about eight years old. He was so excited; the whole family would be going on the outing. But the day before the big day, he accidentally prevented the family from going. His mother was working in the basement of their Yorkville home and asked him to throw a broom down to her. He literally threw it, hit her in the head and knocked her out cold! Whether it was because the doctor told her to stay in bed, or because she feared the neighbors would think her husband had beaten her, causing black eyes, we do not know but, the result was that if Mama didn't go, nobody would go! My great-grandfather was definitely in the doghouse! But the next day, he became the family hero when the story broke of the burning of the *General Slocum*.

Contributed by: J. Calderone

Frank Daniels Family
Manhattan, New York

My grandfather's name was Frank Daniels and he was either a ferry or a tugboat captain, probably in Flushing Bay and helped with the *General Slocum* rescue. I'm not sure where he lived at that time, probably Manhattan, near the FDR Drive, it wasn't called that then, but it was near the site of the United Nations.

Contributed by: Nancy Calabrez

Barbara (Rayle) Dillon
172 High St., Brooklyn

Barbara (Rayle) Dillon, was my mother. She was seventeen years old at the time of the tragedy. She waited to board the *General Slocum* because her girl friend hadn't arrived. She knew other people boarding but, although annoyed at her fiend, decided not to go on the trip without her. I only know her friend's name was Marie and I thank God that Marie was late that day.

Contributed by: Edna Dillon Hug

Luigi & Adelena DeLuccia
54 East 7th Street, Manhattan

Rosa & Anthony DeCesare   Luigi & Adelena DeLuccia

Luigi and Adelena DeLuccia, my paternal great grandparents, had four children; Frank, Agnes, Nicholas, and Rosa, my "Nana" Rose. In later years she remembered running to the side rail of the *General Slocum* as the fire raged. As she got to the rail, it gave way. She fell into the river with two smaller children clinging to her. All three sank to the bottom of the river, where she blanked out. Many hours later, she awoke, in a temporary morgue on the shore of North Brother Island, completely covered with a sheet and presumed dead. A night watchman noticed when she stirred and had her sent to a hospital. Her mother had been taken to another hospital and the family was reunited after their recoveries. "Nana" Rose was the only one of Luigi and Adelena's children to survive the *Slocum* disaster. Frank was found trapped in the paddlewheel, Nicholas was never found. One year later, Adelena passed away.

Rosa married Anthony DeCesare and Luigi bought them a brownstone in the South Bronx as a wedding gift. Rosa had five children of which, Donald, the youngest, was my Dad. Only Vinnie, my aunt, is alive today and living in Connecticut. Vinnie never had children although married twice and later lived with "Nana" Rose until Rose's passing at an advanced age.

An interesting note to this family history is that Rosa's father, Luigi, eventually remarried, sometime after Adelena's passing. His second wife, "Nette" (Antoinette), was my great grandmother's sister, Mary. Although not actually related, my parents considered themselves "distant cousins". Romantic interests began after my Dad served as a U.S. Marine in W.W. II.

Contributed by: Douglas DeCesare, Farmingville, New York

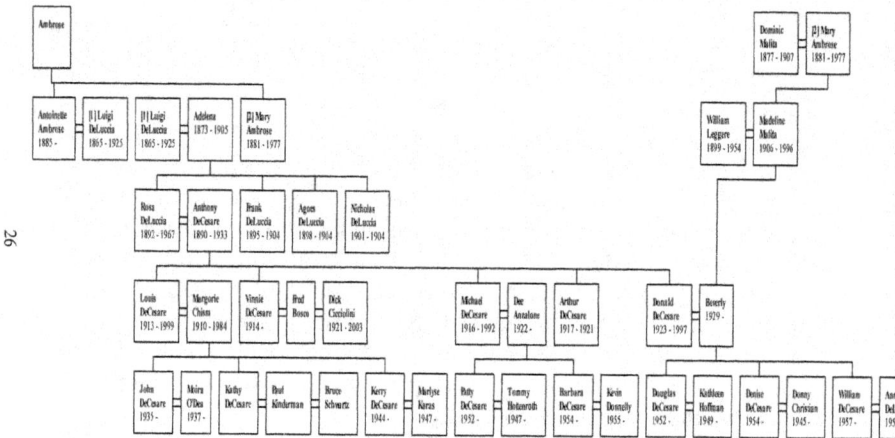

James Duane
329 E. 39th St., Manhattan

My grand uncle, James Duane, was one of the *General Slocum* disaster heroes. He had emigrated to the U. S. from county Mayo, Ireland and was a corrections officer on the *Massasoit*, a ferry owned by the Department of Corrections, which took prisoners from Manhattan to Riker's Island, and was passing the *General Slocum* when the fire struck. According to an old newspaper clipping, which is now lost, James saved over a dozen people, mostly children, by diving into the river and bringing them back to the ship. Some died in his arms. Incidentally, two of the children were the daughters of the owner of "Loft's Candy". Mr. Loft offered my grand uncle $500.00, which he refused. He was also awarded a bronze medal from the City of New York, which he also refused.

We had a news obituary captioned "Hero of the Slocum Disaster Dies", which is also lost. It stated that James Duane succumbed to pneumonia from the chilly waters of the East River but, recovered. The story in the family was that the tragedy affected him deeply. A teetotaler, he later began drinking heavily, which ultimately resulted in his early death. After he died, his sister, my maternal grandmother, Alice Powers [nee Duane], sent her son Walter to city Hall for the medal. The clerk stated that he "couldn't find it". My grandfather, Patrick Powers, sent Walter back with a twenty dollar gold piece. They were then able to locate the medal. Today, the medal is probably in the possession of my first cousin, James Powers [son of Walter], a Suffolk county Policeman, now retired in Florida.

Contributed by: Richard Burns, Ronkonkoma, New York

Note: News accounts reported that Officer Duane was a policeman (he appears on the 1904 Civil List as assigned to the *Massasoit*), and first launched a lifeboat to rescue about seven survivors and another eight victims in the water. After off-loading these, he returned to the side of the *General Slocum* and continued his work as the *Massasoit's* deck crew sprayed him with water.

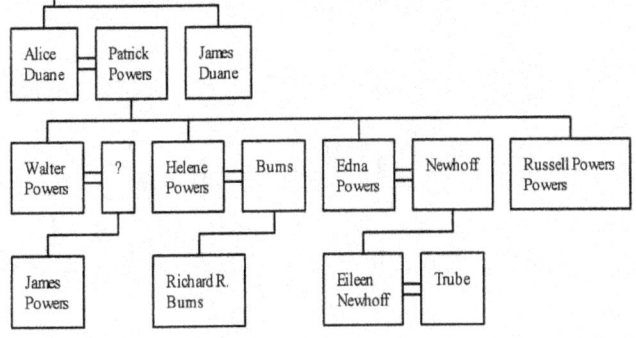

Henry G. Firneisen
40 East 7th St., Manhattan

I am the great granddaughter of Emma Christina Firneisen, a *Slocum* survivor. She was an excellent swimmer and was able to tread water with two of her children under her arms, Marie Theresa, six, my grandmother, and my grand uncle William. Emma and the children were pulled out of the water by an off duty fireman, Patrick J. Lynch. He "borrowed" a dingy and rowed out to the vessel, after convincing a man that he really was a fireman! He is credited with saving forty lives.

My great grandfather, Henry G. Firneisen, a Detective Sergeant in the New York Police Force, was a hero. He had been commissioned under Police Commissioner Theodore Roosevelt and was well known for his ability to "catch thieves", even traveling to Europe to bring one back. He was also responsible for bringing some of the first police dogs to New York City. When he died the headline read "The Terror of Thieves is Dead!" My great grandfather was so grateful to Patrick J. Lynch for saving his family that he presented him with a solid gold engraved watch and pendant. It read "Presented by Henry G. Firneisen to Patrick J. Lynch for heroic conduct in saving the lives of his wife and two children from drowning at the burning of the steamboat *General Slocum*, June 15, 1904."

My great grandparents had another son on the *General Slocum*, their oldest, Henry, age ten, who jumped off the deck and onto a tug which took him to Harlem. From there he made a beeline to his father's precinct where he told him of the disaster. Together they searched Lebanon Hospital until they found the rest of the family. The irony of this story to me is that I owe my existence to the bravery of a fireman named Patrick J. Lynch.

Contributed by: Gail Sirmans Crumbley, Waycross, Georgia
Connie Osborn, Tallahassee, Florida

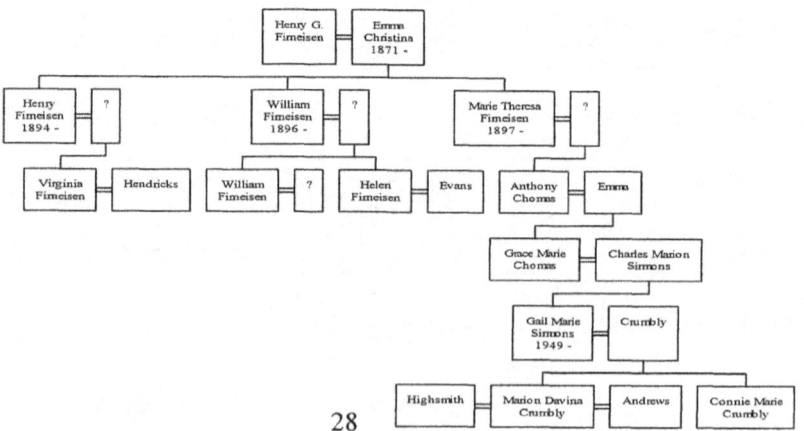

Edwin Fitch
New York City

With the memory of the terrible *Slocum* disaster, in which hundreds of lives were lost and among others his own wife and three daughters, well advanced in years, constantly before him, and recollections of a family circle once complete but in a brief holiday cruelly dissolved, Edwin Fitch, at one time, now almost two years ago, a prosperous decorator and painter in the city of New York, lies upon a bed of sickness in Mercy hospital, suffering from what the medical profession designates "musical heart", when the murmurs of the heart are extremely high pitched.

Before the appalling wreck of the ill-fated *General Slocum*, within sight of the New York coast, nearly two years ago, Fitch was engaged as a scenic painter and decorator in the great metropolis. A native of London, England, he had moved to this country with his wife and in the course of years the family circle had widened and the bonds of happiness had become strengthened, until life seemed sweet to mother and father and their three daughters, grown into womanhood. And then came the fateful day when the pleasure steamer *General Slocum* pulled out from New York into the harbor, and when Edwin Fitch bade farewell to his little family off for a day of pleasure little did he think that he had beheld his wife and daughters for the last time in life.

After the first shock of the disaster had passed, Fitch, unable to bear the pain of his great loss, his home cold and lonely without the presence of those who had made life so pleasant and the home so comfortable, the distinguished decorator resolved to leave the great metropolis and seek the great west country in an effort to forget the great sorrow which had blighted his life. He has traveled many miles since that time and seen many sights and many cities in this country of his adoption. For nearly six months Fitch labored in Cook County, near Chicago, in pursuing his trade, but just about the time success seemed to crown his efforts, his heart became affected, and the husband was obliged to take to the hospital for a long rest.

Recently Fitch came to Dayton, and scarcely had he started again with his work when his heart trouble necessitated removal to the city hospital in that city. Five days ago the unfortunate man came to Hamilton and yesterday afternoon he was removed to Mercy Hospital for a lay up indefinitely. Without relatives or friends in this country, the man's condition is pitiful. He is 58 years of age, of English parentage, and although he has many relatives in the great metropolis of London, not a person in this country can he claim as even a friend. It is not believed that his condition is serious, and with the proper rest Fitch may be able to leave Mercy Hospital within a week.

–The Republican-News, Hamilton, Ohio, 15 March 1907.

Michael J. Fitzgerald
Astoria, Queens, New York

Grandfather Michael J. Fitzgerald was born February 7, 1874 to Patrick and Mary (Halligan) Fitzgerald. A third generation Irish family in New York City, the Fitzgeralds were living at 345 East 45th Street, Manhattan, when Michael was born. The family soon grew to include five children as Patrick continued in the family butcher business.

Son, Michael entered the New York City Police Force in 1895. Later, his younger brothers, John and Thomas followed in his footsteps. Although his career included many promotions and honors, the one the family always pointed to with the most pride, was that received for his heroism on the day of the *General Slocum* disaster, for which he received both a certificate and a Medal of Honor. Although the exact details of Michael's bravery have not survived until this writing, we do know that he was at the scene and assisted in the rescue and recovery efforts, and may have been personally responsible for the saving of lives. His citation from the U.S. Volunteer Life Saving Corps notes only his "heroic conduct in assisting in rescuing and resuscitating the sufferers of the *Slocum* disaster, having been fully attested to this office by satisfactory evidence our Board takes pleasure in awarding to you this Certificate of Honor as a memorial of your bravery and efficiency." In 1905, he was promoted to Sergeant and Lieutenant in 1911. Michael had also held the position of Acting Captain at the Police College. At the age of only 56, he died of a massive heart attack, while working in the District Attorney's office under Deputy Chief Inspector John J. Hennessey of the Bronx. Michael, and his wife, Mary (Sands) Fitzgerald, raised two children. Their son, John Francis, my father, was also one of two children.

Contributed by: Genevieve (Fitzgerald) DeGroat, Howard Beach, Queens, New York

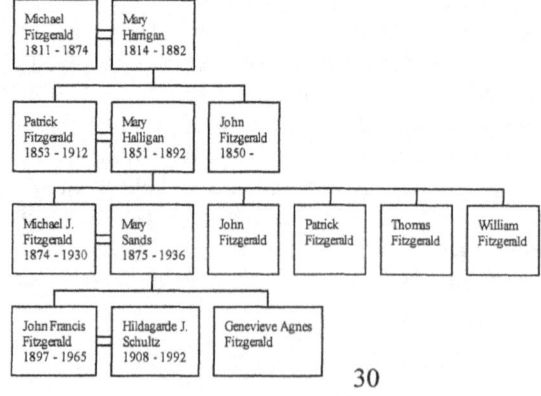

Edward Flanagan
245 W. 28th St., Manhattan

Like all good Irish fairy stories, this one has more turns than a mountain road. My great granduncle, Edward Flanagan could have been a character straight out of one of those tales. His niece, Catherine "Kate" Flanagan Zeisler thought him only one step below saint and as for the *General Slocum* tragedy, he was a hero and for no good reason much maligned in the press! But to begin to understand how Edward Flanagan came to be on the ship that day is to follow one story inside another.

Stefen Larken and Mary "Jesse" Shields met, married, and celebrated the birth of daughters Margaret and Lucy (also known as "Jessie") Larken. Heartbreak followed when the girls' father either died or deserted his family. Soon after Mary died leaving her babies orphaned. Mary's brother, known in the family as "Boss Slattery", stepped in and raised the girls along with his son, John, in Manhattan. When they were old enough, the girls were sent to Cuba for an education. Obviously, the "Boss" was well-heeled. In fact he even had his own footman!

The girls returned to the city and began their adult lives at about the same time as a young man by the name of Edward "Ned" Flanagan took over as the "Boss'" footman. It was love at first sight and "Ned" and Margaret Larkin eloped. Although very upset that his niece had "married below her station", the "Boss" gave the couple two houses in Brooklyn as a wedding gift. Later on, the houses were sold and "Ned" used the profits to buy a boat which became their livelihood. Along the way they had seven children including Edward, Jr. What kind of a boat "Ned" bought is unclear. It may have been a fishing boat or a tug, since his marine activities seem to have been limited to the Hudson and the East Rivers. As Ed, Jr., grew up, he worked alongside his father on the boat. Prior to 1901, Ned had a debilitating accident and had to leave the water, then in 1901 he died. Ed sold the boat and began working on the *General Slocum*.

We have no pictures of Edward Flanagan, but I have seen him described as a "hulk" of a man. Perhaps this was from all those years of working on the water with his father, perhaps heredity. His sister, Kate was almost 5' 10" tall and considered large boned! Kate was strong enough as a girl to swim the Hudson!

Grand Aunt Frances Zeisler Ralston described her uncle Edward Flanagan as a champion who saved a lot of people on the *Slocum*, "Everyone thought him a hero!" time and again; she said that Ed was haunted with memories of the *Slocum* disaster. Given that Frances would have only been a child when Edward died, she must have been repeating what she overheard others say. What we do know is that he took to drink and died of pneumonia and alcoholism in 1918. He was only thirty-nine years old!

Submitted by: Nancy Pitters, Boynton Beach, Florida

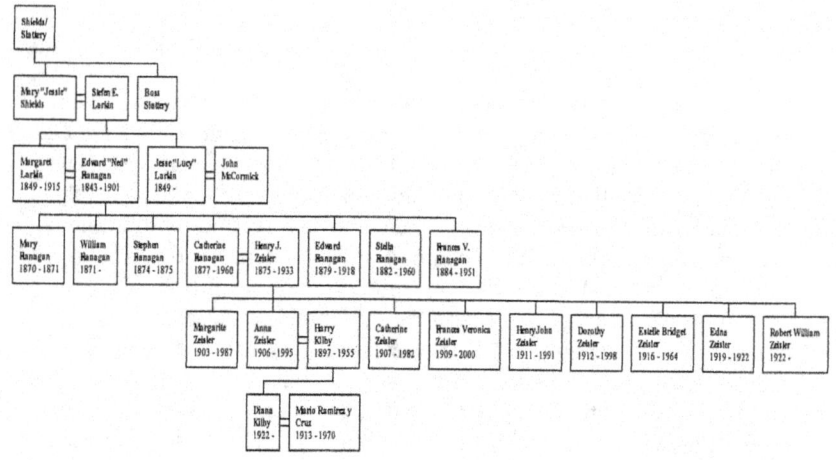

Albert F. Frese
15th Street, Manhattan

Albert F. Frese was only a teenager in June 1904. He had boarded the *General Slocum* with Elsie Eller and a group of friends including Charles Kuenstner, Charles Cordes, Anthony Schwartz, Katie Klem, Clara Helmke, and Grace Bruening for a day of fun and if the boys were lucky, well, you know. Albert's mother and sister, Anna, were also aboard. Albert later reported that he saw flames leap from the forward deck; to his amazement, no one cried out at first. People seemed to freeze. Then the boat vibrated all over and shook as the captain began pushing the engines for all they were worth. At the same moment, a man ran by "... don't go down there, it's a furnace down there!" he yelled over his shoulder. By this time panic had erupted. On the passing shore of Blackwell's Island, people were running down to the beach with planks for stretchers and buckets for water, anything that might serve as makeshift rescue gear. As they ran, Albert remembered they were yelling and waving their arms trying to signal the ship to put in. Separated from his family and friends, Albert went over the rail near the bow and survived. His good friend Charles Kuenstner and his mother were also rescued. The rest of his friends and his sister died.

Note: Albert retold this story several times over the course of his life, and it remains one of the most lucid descriptions of what happened and when it happened. Per Albert's recollection, the fire erupted to the first deck just as the captain was accelerating in order to control the vessel better in Hell Gate. This was the shuddering felt by the passengers. The passage along the western shore of Blackwell's [Roosevelt's] Island takes only a few minutes at a speed equivalent to the *General Slocum*'s 10 knots when the tide is running up the East River, as it was that morning. Albert, like a handful of other passengers who were caught in the point of the bow, ahead of the companionway that the fire ascended, were relatively safe as the wind blew back over the bow. They could await the grounding before having to jump for their lives.

Louise Gailing
Nutley, New Jersey

Our mother, Louise Gailing, who later to become Louise Gailing Joiner, was in her early teens at the time of the *General Slocum* disaster. On the day of the St. Marks excursion, she was accompanying the Erklin family from Hoboken, former St. Marks' members, as a mother's companion, [mother's helper]. Mrs. Erklin had several young children ranging in age from six years down to Stephen, a newborn.

Our mother, Louise, was minding the children when the fire broke out. She tried to grab a life preserver but, saw that it was defective and filled with sawdust. Somehow she managed to hug the children to her and jump overboard. The most amazing part of the story is that she did not know how to swim. Somehow, she managed to keep the two youngest children afloat until someone, described in the press as "an officer" from one of the rescue boats came to her aid. According to that report, he helped to support Louise and the children until a rowboat could pick them up. Her own memory of the event was that she found something, a log perhaps, that helped keep her and the children above water until help arrived.

As we understand it, the two babies, Gertrude (one year) and Stephen (only seven weeks old) survived the disaster. There are records of a reporter interviewing Mom (Louise) shortly after her own rescue; "I had no thought of what might happen to me. I had never swum a stroke in my life, and I didn't know the slightest thing about how I should begin. I only knew one thing, and that was that I must save the babies. So I took one in each arm and jumped overboard and kicked out with my feet and held them up as best I could." It is believed that baby Stephen was the youngest child to survive the disaster.

Another reporter arrived at the Gailing house in Nutley, well before Louise made her way home. The reporter surprised our grandmother with news that her daughter was a heroine. Grandmother did not read the newspapers, and did not know what her daughter had done!

Louise was honored in the New York press and by her hometown as a heroine. She was given a citation and a monetary award. She used the award to purchase a sewing machine, and helped support her family as a seamstress. Eventually, Louise had a family of her own and, when urged, told the story to us, to in-laws and grandchildren over the years. She was never boastful; instead, she told the story in a way that conveyed her fear, but, more importantly, made her listener proud of her courage and resolve. All her children loved her greatly, as did the many grandchildren who were lucky enough to have known her before her passing in the mid-1960's. Today, her great-grandchildren and great-great-grandchildren are told the story of her bravery, and are proud to be descendants of Louise Gailing Joiner.

Contributed by: Edith Joiner Shelly, Iselin, New Jersey
James Gailing Joiner, Hampton, New Hampshire
Lee Ann Brady, Skillman, New Jersey

NOTE: Apparently the Erklins managed to elude the press, and little is known about them. The New York Times reported that they were former members of St. Marks and that the family consisted of Otto, age 30; Anna 32, who was injured but survived; Theodore age six who died; Gertrude, age one year who survived uninjured; and Stephen, seven weeks, who survived with injuries. The family address was given as 1028 Hudson St., Hoboken

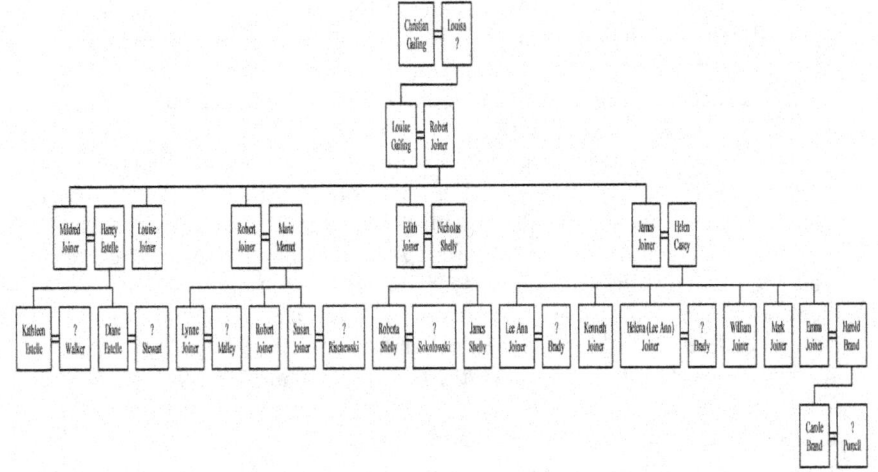

Emily Halley
235 East 82nd St., Manhattan

My family was supposed to be on the *General Slocum* but, due to my great aunt Emily Halley's premonition, they did not go. I cannot imagine how they felt when the news reached them. Surely many friends died. I think about the *General Slocum* I realize that there might have been three generations of my family wiped out had they decided to ignore Emily's warning. Charles James and his second wife, my great great-grandmother Florence, and her daughter, Lillian, might all have died, not to mention other members of the family. I find this a very scary and sobering thought.

My grand aunt Ginny and I are related to Emily's father, Charles James Halley, of Cheltenham, England. He married Mary Ann / Sophia Blakemore, from Bordesley Parish, Birmingham, England of which she was very proud and included this in all her formal papers etc. They had five children, whom they raised at 347 Albert Road, Birmingham. One of these children was my great great-grandmother, Florence Eveline Halley. It was Florence's sister, Emmeline Mabel, "Emily", who saved our family from the *General Slocum* disaster.

The family was split up for a time when Charles and Mary Ann moved to Queens, New York. Charles was a painter and decorator and had a workshop near the Saint Peters German Lutheran church, which is the only tie we seem to have to the church as we are neither German nor Lutheran. Only young Charles Eugene seems to have come to this country at the same time as his parents. The girls came over later. There were still Halley relatives in England at the time and perhaps they stayed with them. We have some family pieces from the Birmingham area and know that one relative was a bandleader and a composer; we have a lovely piece of music he composed, which I had a local musician record for me, from an old piece of sheet music.

Over the years, the girls grew up and married. Then in 1898, they lost their mother. Emily married in Saint Peter's German Lutheran church on 16 august 1899 when she was twenty-two. Her husband, Henry Black, was from Mitchen, England. Henry may have been a bartender by trade. Eventually, they had a family of four. I do not know much about this part of the family but, a family legend states that they had odd pets and a motorcycle that they kept inside their rental apartments. No one really believed this until I produced photos that had always intrigued me; some young Halley cousins, among them my grandmother, as a very young woman, in a backyard with a motorcycle and a canoe. Ginny and I believe that this was Henry's famous motorbike.

Several of the family, including my great grandfather, Armand Pierre Saffert and several of the Halley boys, made their living working for the Louis Comfort Tiffany Factory in Corona, Queens. The factory building is still there today. The family remained in a small enclave of Corona and Elmwood, Queens until Charles Eugene moved to New Jersey; several others moved on after Charles James died. The family still stayed close to New York but, moved to the suburbs. My great great-grandmother Florence remarried and moved to Pennsylvania but, when she was widowed again, she moved back living first with her father and then her daughter. These two ladies lived into their 90's. I was fortunate to have spent most of my young years living in the same house with them.

A few years ago I stopped in at a local antique shop. As I entered, I spotted a lovely hand-painted plate on a sideboard. The painting was of a bunch of grapes a fairly typical subject but, done in an interesting style. Of course the first thing anyone does with this stuff is turn it over and I am no exception! The signature on the back was M. A. Halley. I brought it home and I wonder every time I look at it if it was painted by Mary Ann Halley, my third great grandmother.

Contributed by: Judith Loebel

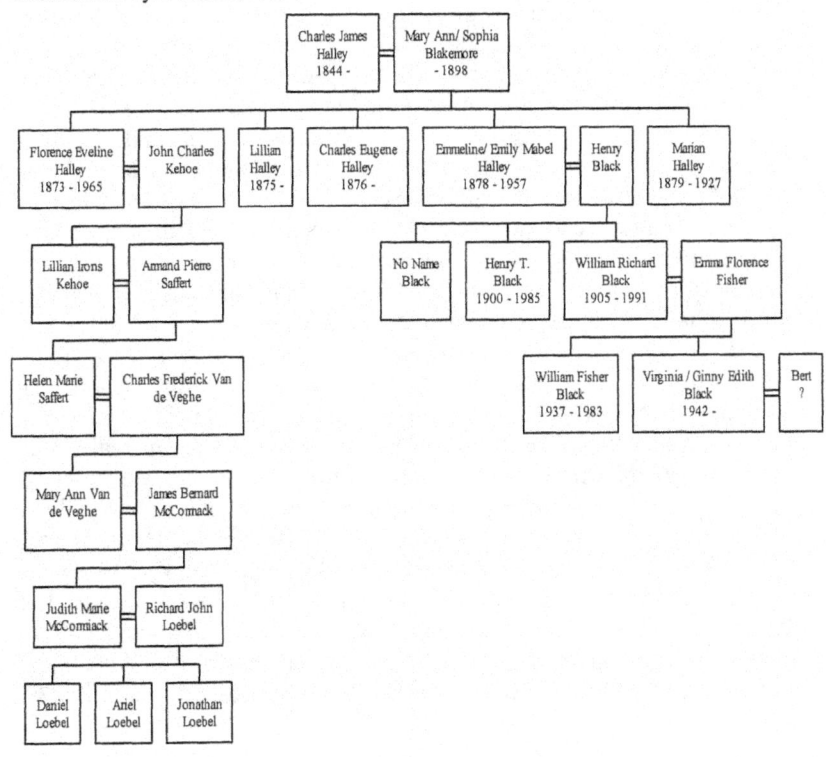

David Harris (nee Rosenholz)
242 5th Avenue, Manhattan

My grandmother, Deborah nee Block Rosenholz, (1875-1959), wrote a brief memoir of her family's involvement in the *General Slocum* disaster. While not a direct descendant of the disaster, I am a first cousin, twice removed of Silvia Harris, age ten, who did not survive this tragedy. Silvia's father, David, and his brother Jacob Rosenholz were born in the area of Russia / Poland now known as Kalvaria, Lithuania, while Lena nee Bliss, Silvia's mother was born in Ohio. Lena's parents were Prussian / German. Here is an excerpt from my grandmother's memoir.

"About this time a charitable organization, St. Mark's Lutheran Church Sunday School, arranged a boat ride to finance their various activeities and Mr. David Harris, my husband's uncle, allowed his adopted daughter, Silvia, to go on this trip with some neighbors. David and Lena Harris not only loved this child, they adored her! I have never seen such devotion given to any child, even by its natural parents. In fact, it was always commented that this devotion was beyond understanding. To them she was like the gold a miser gloats over. If anyone accidentally brushed against her in the street, Mr. Harris would readily have killed the offending party but, his small stature prevented this from actually happening. She was so spoiled that had she not been only ten years old, no one could have tolerated her. And the adoptive parents themselves had engendered all of this! David's business was prospering and nothing that Lena wanted was ever refused her. It was his habit every day to leave the flower factory in the afternoon, to attend the racetracks. He did not believe in banks and always carried a small fortune on his person. Of course his business must have been very good, since I do not believe a man can consistently gamble and still have money. When they lost this child, it took only two years for him to go broke. He lost all interest in his business and everything around him. As trouble never strikes singly, Mrs. Harris gave birth to a son, who was blind. They became so poor that they had to move and lost the business and it was up to us to support them, for they had become really old people before their time. The added burden of a blind child was no small trial. The boy lived with his parents but, they never recovered from the loss of their adopted Silvia or their standing in the community. This was a terrible disaster and caused great loss of life. Many children were being blown to bits by an explosion of the boat and many others burned to death. This was called the *General Slocum* disaster. As one looks back on this tragedy, how can anyone forget? It caused so much sorrow and bereavement at the time. But, of course, it is not forgotten, just set to one side so life can go on."

Contributed by: Mark Rosenholz, Albany, New York

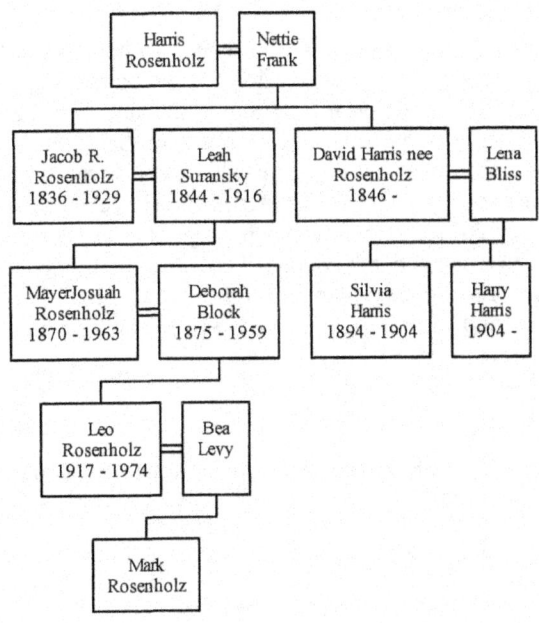

Archibald Alexander Hill
Address Unknown

I have the gold watch that was presented to Archibald for his part in the relief effort. I inherited it and a brittle copy of the report, from his son Archibald Anderson Hill, two exceptional men. Another family distinction is that Mildred and Patty Smith Hill, both sisters of Archibald Alexander Hill wrote "Happy birthday to You."

Contributed by: Muriel Wright

Meta Jaeger
Address Unknown

My grandmother, Meta Jaeger, is a hero in our family for helping out at the scene of a steamboat fire. She was a nursing student in 1904 and graduated from a two year program at Lebanon Hospital in December of that year. Although I'm fairly sure it was probably the *General Slocum* disaster rescues in which she was involved, my documentation does not confirm her presence at North Brother Island (Riverside Hospital). Growing up, in the 1960's, I remember seeing a photographer's picture of a ward on which she worked but, someone else in my family now has the picture. As a child, she had been a member of the Evangelical Lutheran Church of St. Matthew, and so she might also have been aboard on the Sunday school picnic herself.

Contributed by: Dorothy Landfare

Note: Lebanon Hospital was one of the primary medical providers for the disaster and was also the hospital where Grace Mary Spratt worked as Superintendent of Nurses. When word arrived of the disaster, she packed up supplies and a squad of nurses and left for the island, where they worked the entire day. Quite possibly, that squad included nursing students since they would have been less necessary on the wards but, the experience would have been invaluable in their careers.

Mary (Hicke) Kershaw
6th Street, Manhattan

My grandmother, Mary (Hicke) Kershaw, lived across from St. Mark's church. In 1904, she was eleven years old and looking forward to the picnic excursion. On the morning of the trip, she was rude to her mother who, as a punishment, refused to let her go. Because of this, she was spared. Her family continued to live in the neighborhood when it became predominantly Jewish and the church became a synagogue.

A few years ago my mother, Mary (Kershaw)Barclay, was driving with a friend, and told my grandmother's story. The friend was stunned; her own grandmother had spilled her breakfast down the front of her only nice dress and couldn't go on the trip either!

Contributed by: Carol Barclay

John W. Klenck
113 St. Mark's Place, Manhattan

Mrs. Bauer & John Klenck

Our father, John William Klenck, was born and spent his childhood at 441 E. 6th St. On the day of the *Slocum* disaster, he had boarded the boat with his mother, Bertha (Eichler) Klenck, age forty; his brother Charles, eight; half-brother William F., age twenty; and a cousin Minnie Klenck, twenty. This was a treat for William, John, and Minnie, who were already working. Their father (Minnie's uncle) had died in 1898, and the family had moved from John's Childhood home to 113 St. Mark's Place.

At some point, about an hour after beginning the cruise up the East River, fire became apparent aboard the *General Slocum*. Bertha tried to put life preservers on the children but, the straps kept breaking, so she herself went overboard without one. The family had joined hands as they stood at the railing and planned to jump together but, the weight of the passengers pushing against the rail caused it to break loose. John lost his grip on his brother's hand as he fell overboard. John's face was burned but, he survived by clinging to the wreckage of the ship while a friend sat on his hands so that he would not slip off the flotsam.

Eventually, an off duty fireman, in a small boat, pulled him to safety. He passed out and as he came to, the fireman was telling him to help other passengers who were clinging to the boat, or John would be thrown back into the water! He helped for what seemed like an eternity and again passed out from exhaustion. This time he awoke on the deck of a tugboat on the way to Riker's Island, where he was cared for in the hospital*. An uncle found John there and told him that his mother and brother were missing. When our father returned home, the house was silent; he had become an orphan at the age of twelve. After the disaster, he moved uptown with his aunt, Mrs. Clara Bauer, and stayed with her until he grew up. For the rest of his life, John Klenck believed in the necessity of maritime safety precautions and that if remembering the disaster kept this need on people's minds, then something good had come from such tragedy.

Submitted by: Marguerite Klenck Lovejoy, El Paso, Texas

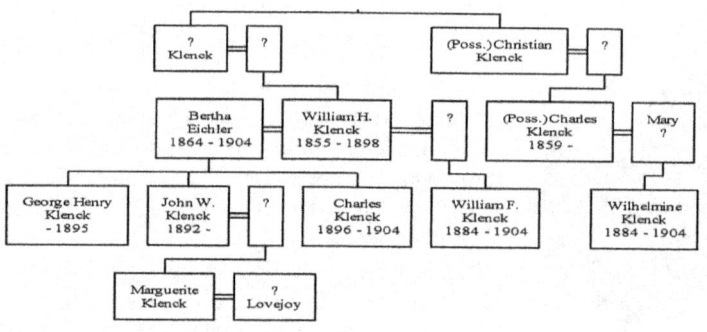

* In 1904, there was no permanent facility on Riker's Island. It is more Likely that John Klenck was transported to one of the other city hospitals.

Peter Knell
83 Thomas St., Manhattan

    What some might consider to be a tragedy involving my grandfather as a youth, turned out to be a blessing in disguise as far as our family is concerned. Had George Knell not been struck by a trolley car, he would most definitely have been a passenger on the *General Slocum* excursion boat on the day it burned.
    Peter Knell, born 8 Feb. 1857 in Weinheim bei / Alzey, Rhine-Hesse, Germany and Alwine Horn, born 24 June 1855 in Stojentin, Kreis Stolp, Hinterpommern, Prussia, were married on 14 September 1878 at St. Matthew's Church on the corner of Broome and Elizabeth Streets in New York City. This was just on the edge of Chinatown and the Jewish and Italian sections of Manhattan. The happy couple took up residence in what we now call Tribeca. Peter had a bakery on the first floor and the family, which grew to include six children, lived above it. Alwine ended up living in that apartment for forty four years. And through it all, the family remained faithful to St. Matthew's. All the children were baptized there, three by the Reverend John Sieker, and three by his son, the Reverend Otto Sieker. George, my grandfather was the youngest.

When George was about seven and one half years old, he was run over by a trolley car just a block from home. His left foot as well as two fingers were cut off. This was probably in June of 1902. By June of 1904, he was learning to cope with a prosthetic foot but still in recovery from the accident. So, to the chagrin of his siblings, it was decided that the family would have to forgo the excursion on the *General Slocum*. The children were not only disappointed that they weren't able to join their friends, they were somewhat resentful. This would be the first year ever that the family did not go on the excursion, and all because of their little brother! Had they gone, I might not be here today. Many of their friends and fellow parishioners were among the fatalities. The Knell family is buried in a plot at Lutheran Cemetery near the monument to the memory of the many children who died that day.

Contributed by: Joan Colvin, Homer, New York

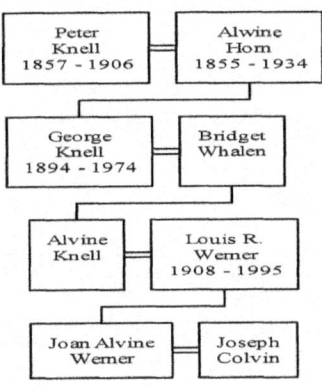

Elizabeth Kuhn
64 Driggs Ave., Brooklyn

Elizabeth Kuhn, age seven years, was aboard the *General Slocum*. She was with an aunt but, we have not been able to determine the aunt's name, however, we believe she was also a Kuhn. Elizabeth's brother, my grandfather, Carl, and the children's parents Carl L. Kuhn and Anna Kopta were watching from shore when the *General Slocum* caught fire and went down. It must have left my grandfather with horrible memories of the water. He never went in the water or on a boat for the rest of his life. The family continued to live in and own the same house at 64 Driggs Avenue until my parents moved the family to Pennsylvania in 1969. Grandfather died in September 2000 at the age of ninety-nine years eight months.

Submitted by: Debra Poling

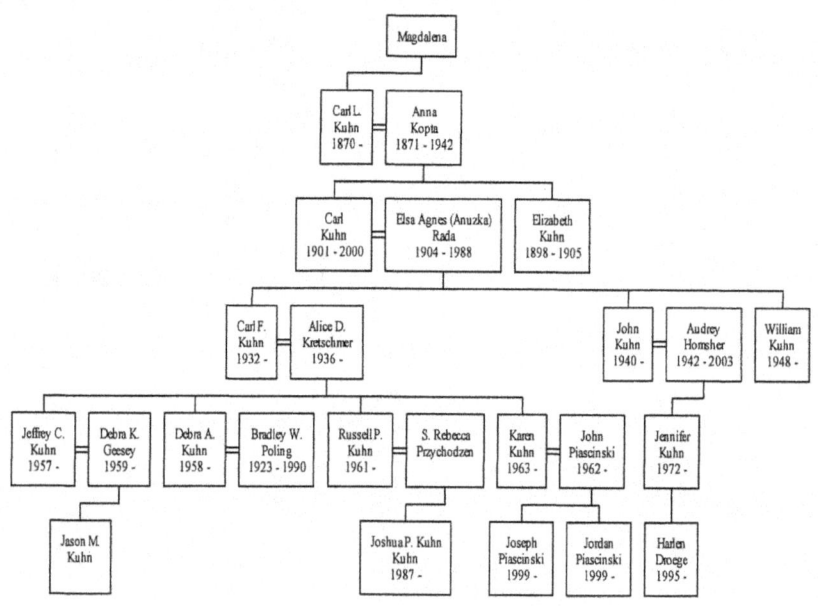

Anna Charlotte Lange
73 1st Ave., Manhattan

Anna Lange, twenty-two years old, was invited to go on the *Slocum* on June 15, 1904 for the picnic that day with her friends. But she could not go, as she had to work in the family owned candy store and ice cream saloon; the F. Lange Confectionery, on First Avenue. The Lange family lived above the candy store and it must have been wonderful growing up with candy and ice cream available whenever you wanted it. Our family still has some of the vanilla bottles from the store that are still full (I tasted it once and it was fine), and a wooden one-cent candy box.

I have a letter written by my aunt, in which she states that the Lange family friends, the Cortes, were on board. (Because of the similarities of the names, I believe she just misspelled it, I believe she meant the Cordes family based on a comparison of the spelling to lists of victims and research by the Cordes family descendants.) The Cordes lived at 417 East 16th Street and had a retail bakery on Steuben Street in Brooklyn; later they moved uptown. Because of a copy of the by-laws of the United Confectioners Association, 561 Greenwich Street, New York City, I know that both my great grandfather Lange (Anna's father) and John F. Cordes were Directors/ Owners, so it makes sense that they might have been friends.

In the same letter, my aunt mentions the Adickes family and that they might have been either friends or neighbors of the Langes. The United confectioner's Association was a rather large group, and they held annual dinners to which wives and families were invited. I have photos taken at these dinners but, we can only identify the Langes.

The Lange family was no stranger to tragedy; Anna's older sister, Dora, had died at the age eight when Anna was only six years old. If Dora had lived, she would have been twenty-four at the time of the *Slocum* incident. Anna also had a brother named Freddi. No one knows why Freddi was not on the boat that day; perhaps he too had to work in the store, since later in life he took over the family business.

The Lange family is buried in Lutheran Cemetery in Middle Village, New York, and as a child I recall visiting there, hearing the story of the *Slocum*, and seeing the tall monument placed there as a memorial.

Our family was always grateful that my grandmother had to work that day or we might not have been born.

Anna married my grandfather, August Krauss in November of 1904. I suspect that the wedding was timed as a result of the tragedy. They lived on Poplar Street in the Bronx and had five children. I am the daughter of one of Anna's sons.

I hope that future generations will read this and come to appreciate those that survived and carried the grief, as my grandmother did, and feel great sorrow for those that were lost and the families that were destroyed because of the terrible tragedy of the *Slocum*.

Contributed by: Charlene Johnson, Stratford, Connecticut

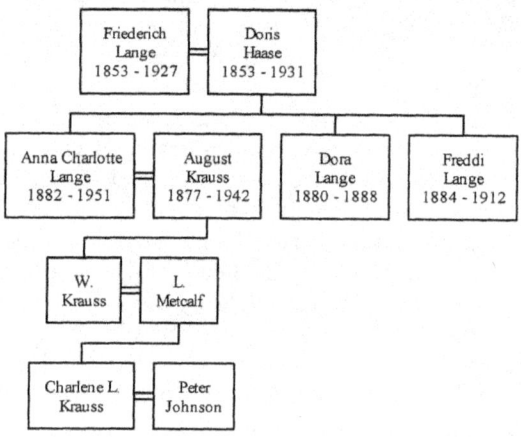

Adella (Liebenow) Wotherspoon
113 East 124th Street, Manhattan

Adella (Liebenow) Wotherspoon would be known for other things throughout her life but, until she reached that certain period of life when one feels the pull of time and the need to tell one's story, few people had any idea of the tragedy and grief which surrounded her younger years. But once she decided to connect with that dark beginning, the light she would bring to the events of her in-fancy and the celebration of her family's triumph over them would inspire all who knew her.

Adella was one of the youngest survivors of the *General Slocum* disaster; one of the few from her family party of eleven to survive, and the longest living survivor of the disaster. Although she was only six months old at the time, her mother made sure that Adella would never forget the burning of the *General Slocum* and their miraculous escape from imminent death. When the fire started, Adella and her mother, Anna, were in the bow section of the boat. As the fire roared to life, Anna held her baby of six months away from the flames with one arm and held onto the ship's rail with the other. Like many others, mother and child, had climbed over the rail and either hung or stood on tip-toes along the edge of the gunwale, hoping to be rescued. By the time Anna decided to let go, her left arm, shoulder, neck and head were badly burned. She covered Adella's face and, with her own clothing on fire, jumped into the East River.

As Anna was preparing to jump, Adella's father, Paul Liebenow, was searching for his wife and children amid the panic and smoke. He had gone to another part of the boat just before the fire and could not get back to where they were. Paul and an uncle stayed on the boat until their burning clothes forced them to jump. Mother and child were picked out of the water by the crew of the *Franklin Edson*. The details of Paul's escape are unknown but, he later spent many hours searching for his family.

As he went from hospital to hospital, in search of his loved ones, doctors at each stop tried to detain him to treat the serious burns around his head but, he refused. Adella's parents were eventually reunited in Lebanon Hospital where both were brought for medical attention. But Adella would have one last battle before her reunion. Another mother crazed by her near-death experience loudly and tearfully insisted that Adella was her baby. The mistake was soon realized and Adella was returned to her family, sooty and scared but, otherwise well.

Once the family was reunited, Paul resumed his search for Adella's older sisters, ages three and six. Anna, the three year old, was later found but, Paul fruitlessly searched in hopes of finding Helen. She never was identified and the family finally concluded that she must have been buried among the sixty-one unidentified victims in Lutheran Cemetery, Middle Village, Queens. Two of Adella's aunts and two cousins also died that day.

During his initial quest, Mr. Liebenow happened into the 138th Street Police Station. As he viewed the bodies laid out on the floor, Coroner O'Gorman approached to interview him concerning his family. The poor man was so distraught that he did not realize that the coroner only wanted to help and he lashed out, attacking the coroner physically, fearing that he would be detained. The two men did not meet again until the ceremony for the unveiling of the General Slocum Monument, in Lutheran Cemetery, one year later, where they greeted each other cordially and shook hands.

Adella was given the honor of pulling a silken cord that released a flag covering the face of the monument. Until his death, several years later, Paul Liebenow carefully collected every article and picture he could find on the *General Slocum* disaster and arranged them in scrapbooks which still survive at the New York Historical Society.

Adella passed away at the age of one hundred, the last surviving victim of the disaster. In her later years she made a point of attending every memorial service and accommodated every writer who cared to contact her. "Well, of course, I don't remember that day, I was only a baby!" she often told the curious and then retold the story she had learned from her mother.

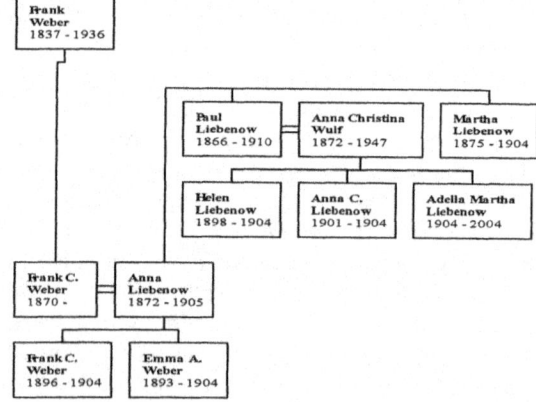

Gustav Adolf Lutz
148 2nd Ave., Manhattan

Julius, Gustav, and John Lutz c.1900

Gustav Lutz, also known to friends as Gus, and to family as "Beep", was seventeen years old when the *Slocum* burned. He was born in Switzerland but, grew up between Manhattan, New York and Demarest, New Jersey, where the family had a boarding house and farm. He also had two older brothers John and Julius "Jack", and a younger sister Martha The. family story goes that Gus had gone on the church picnic on the *General Slocum* with his girl friend. They were on the deck and he noticed the fire or smoke or something and so told the young lady to wait for him while he went to find out what was happening. It was the last time that he saw her. "Beep" wound up in the river where one of the rescuers fished him out and threw him into a boat with a lot of dead bodies. Something in the way they threw him in or the position he was lying in expelled the water from his lungs and he survived. In fact he survived to a ripe old age! Three years after the *Slocum* disaster, he met and eventually married my grandmother, Elizabeth Anna Frederick.

Contributed by: Ruth Firth, Petersham, New South Wales, Australia

Note: According to the Coroner's reports and newspaper interviews, Gustav's girl friend was Matilda Merseles of 6th Avenue, Brooklyn. A John D. Lutz identified her; he was probably of the same Lutz family but, no confirming data has been located to date. Furthermore, it was probably this same John D. Lutz who gave Gustav's address to the coroner. Gus and Matilda had joined a group of friends aboard the *General Slocum*. Another member of the group was Ellen Breden of Brooklyn. She later testified that she and Matilda were together when Ellen decided that they should jump for a tug which had come alongside. Matilda refused for fear of being crushed between the *Slocum* and the tug. Gustav also tried to keep Ellen from jumping but, she broke his grip and went over the side possibly taking Gustav over the rail with her.

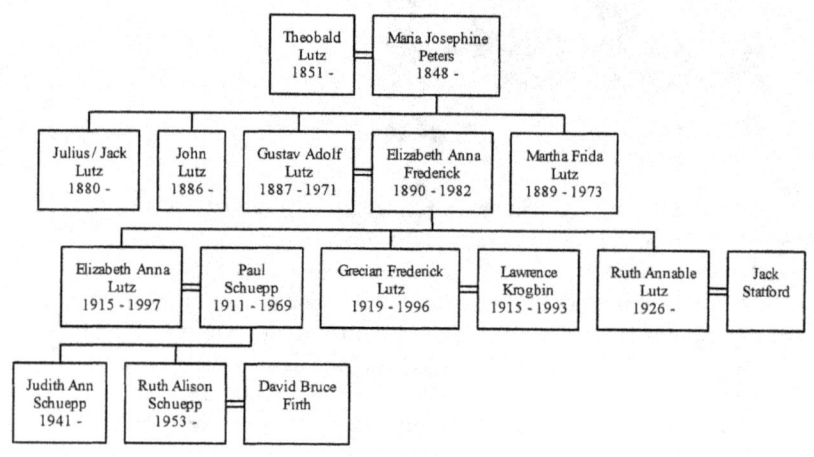

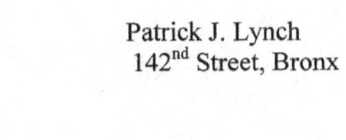

Patrick J. Lynch
142nd Street, Bronx

My great grandfather, Patrick J. Lynch was a New York fireman attached to Engine Co. 60 as an engineer. At the time of the *General Slocum*, he was married to Jane Callanan and they had six children; Estelle, Grace, Alice, Jane, Mary, and Eddie. Estelle was my Grandmother. She later married George Moonan and had my father, Raymond and his brother George.

Ever since Patrick had joined the fire department, he had been a hero to all who knew him. His first assignment was at Engine Co. 80 on Cannon Street where he made his first life saving rescue in 1879, while still a private in the department. By 1904 he had been decorated several times for his bravery and tenacious rescues. More than once he was carried out of a burning building more dead than alive!

In June of 1904, Patrick was an Engineer of Fire Company 60. June 15th was his day off, and as he later told the papers, he had intended to swim at the foot of 141st St. but, family tradition says that he was actually going to work as a lifeguard that day. Which ever is the truth, he was an avid swimmer and looking forward to this recreation. But as he approached the East River, he saw the *General Slocum* plowing up the river fully engulfed in fire and smoke from stem to stern. He jumped aboard the schooner *John T. Russell*, which was at the pier, and attempted to untie a rowboat to get to the impending disaster. A deckhand threatened him, if he did not let go the rope but, the fireman convinced the man of his mission and rowed out into the stream in pursuit of the *Slocum*.

At the scene of the grounding, he made several trips under the paddle-wheel to rescue men, women, and children hanging onto their last refuge from the fire. Each time his boat was filled, he handed over his cargo to the tug *Wade* and returned. In all he was credited with saving forty lives. Both the American League of Honor and the United States Volunteer Life Saving Society gave him medals. One gentleman, a police detective, was so grateful he gave great grandfather a gold watch, chain, and locket engraved with his thanks.

Contributed by: John Moonan, Clearwater, Florida
Christine Moonan Wichard, La Jolla, California

Estelle Lynch

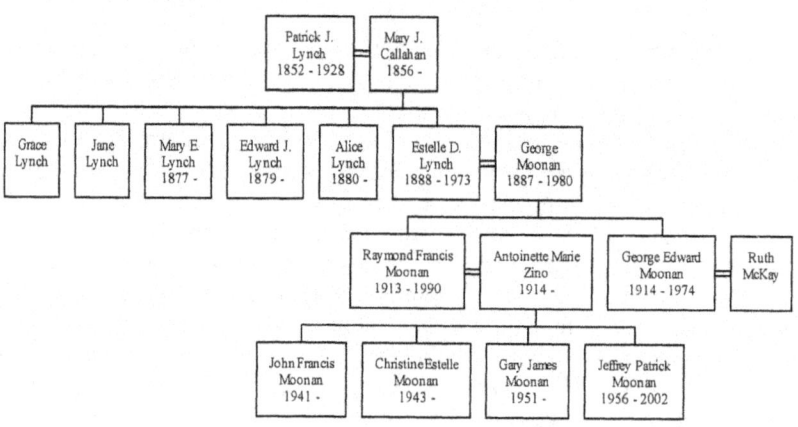

Henry Neats Mallabar
North Brother Island, New York

My great uncle Henry Neats Mallabar, a descendant of William Mallabar (1755-1838), was born in 1849 at Stockton-on-Tees, County Durham, United Kingdom. His family had lived in this county for at least four hundred years. He immigrated to the United States sometime in his youth. After his marriage in 1869, he returned to England with his wife, where they stayed until after his two children had been born. The little family recrossed the Atlantic and settled in New York City where he eventually became the Chief Clerk at the Riverside Hospital on North Brother Island and an American citizen. Being a strong swimmer, when he saw the disaster unfolding, he felt compelled to help and was successful in rescuing eight of the passengers. While bringing his eighth rescue ashore, a two hundred pound woman, he had a stroke and was partially paralyzed until his death in 1907. For most of his last four years, he retained his position at the hospital and worked as much as his condition allowed. The volunteer Life Saving Corps awarded him one of the few gold medals awarded for rescues associated with the *General Slocum* disaster.

Henry was one of eleven or twelve children, at least three of whom immigrated to North America. Henry's brother, my grandfather, John, settled in Winnipeg, Canada and married. He had four children and died of TB in Mexico in 1899. His widow, Sara (Scott) Mallabar, went on to found Mallabar Costumer, which is still operating after 100 years, and was the largest operation of its kind in Canada for many years. Henry had one boy and two girls who continued living in the New York City area for some time after his death but, disappeared from the records sometime between 1915-1920.

Contributed by: Ross Malabar, Surrey, British Columbia, Canada

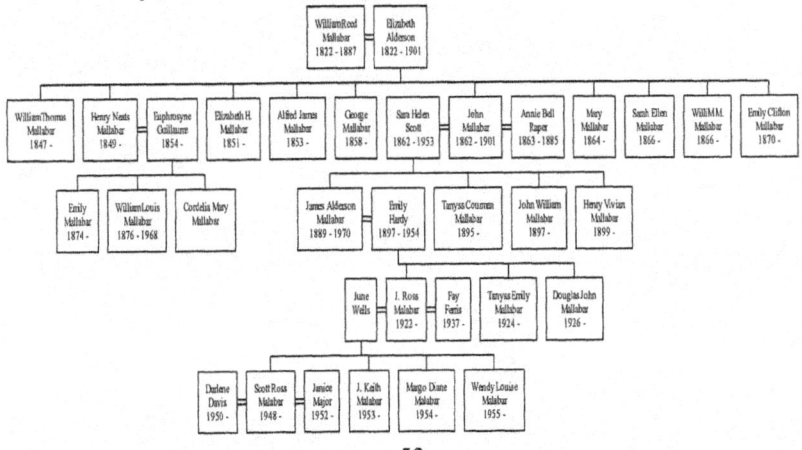

John McAllister
92nd St., Brooklyn

John McAllister was a thirty-seven year old immigrant from County Antrim, Ireland. He was a crewman aboard on of the tugs following the *General Slocum*. He operated a towing line and in spite of the intense heat, rescued 75 people. He was awarded a gold medal from the Survivors' Association and another from the Life Saving Service.

Contributed by: Elizabeth Thelen

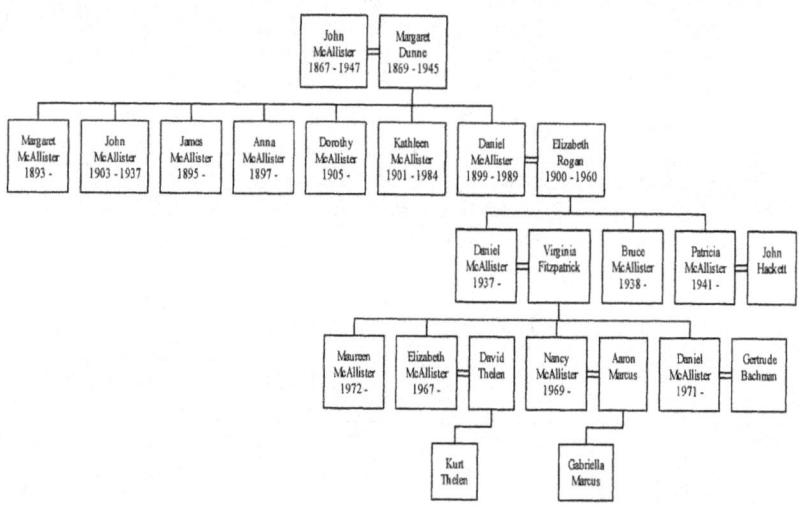

Michael McGrann
2161 8th Avenue, Manhattan

What follows are reminiscences of life in the McGrann family "post-*Slocum*" handed down through the family from my grandmother, Frances, to my aunt, Anne (McConnell) Levy.

Michael McGrann, my great-grandfather, was born in Ireland and immigrated to America. He married Anne Deane in New Haven, Connecticut. They settled down in New York City, started a family and lived a very comfortable life, which included five children who grew to adulthood. The first, Joseph, married Kate Mauer. As children we always referred to her as "the Aunt Kate", she lived past 100 years! His three other children; Agnes, Florence, and John, never married.

After the death of Michael McGrann on the *Slocum*, his wife, Anne, started a little luncheon-sandwich shop business but, never worked in it herself. She used her income and insurance to keep things going. Her sons Joseph and John were the managers and they had counter help and young men went out to the local parks and summer amusement areas to sell ice cream, candy, peanuts, etc. to people who took walks to enjoy the nice weather. The business ended before World War I. During this time Grand Aunt Kate took care of the children, I believe, although she was a lifelong spinster and from what I recall, had a very short temper with kids like me! Regardless of that, the family worked together to make a living, each in their own way. Kate was stricken with a heart attack on the steps of her church on Good Friday, 1972, and died.

Agnes McGrann was one of the first career girls. She worked in the main office of Hearst Publications; she never married, and was the prime support of the family. She died a few months before my grandmother, Frances, gave birth to her first born, Anne, in 1928. My grandmother Frances was always very close to her sister Agnes, and never wanted to talk about her death. Since she was sick for a few months prior to her passing, the family always thought it was cancer.

Florence McGrann never worked steadily at one job (as I've stated before, I always thought she did some sort of child-related job). She was the single sister who always stayed home with her mother, Anne (Deane) McGrann.

Frances, my grandmother, was a trained business machine operator. She had taken a business course and operated one of the first "Elliot fisher" bookkeeping machines. She worked for the American Sugar Company in the Wall Street area. She worked there until her marriage to John McConnell in 1924. They had three children; Anne, John, my father, and Lorraine. She remembered the story of the *General Slocum* disaster, and passed it onto us. She died in 1984.

Contributed by: Patricia Lawrence, Groton, Massachusetts

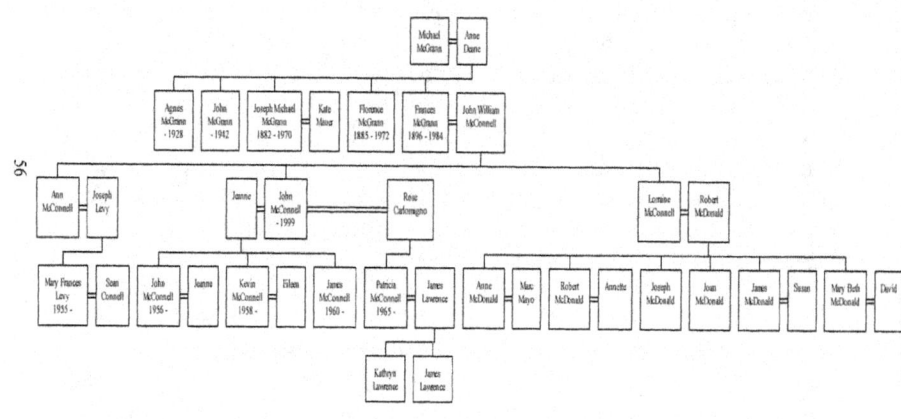

Franz William Moller
45 E. 2nd St., Manhattan

My grandfather, Franz William Moller, lost his wife, Catherine, and two of his children on the *Slocum*. William was a grocer and wholesale tobacco broker. The family was on a church outing that day with the First Lutheran [Church] of New York. William had some good stogies with him and took some of his pals to the bow to smoke away from the crowd. When the disaster began, he was unable to get to his wife and children due to the panic onboard. Being at one end of the boat, he was able to swim away when she rolled. William had two remaining children. One was killed by a trolley car in New York. The remaining son was gassed in WWI and did not last long afterward. Grandfather married my grandmother, Annie, in 1906. Like him, she was a German immigrant.

Contributed by: Hank Moller

NOTE: Those passengers who were forward of the main cabin, in the bow, were shielded from the intensity of the blaze by being in front of the point where the fire burst up on to the main deck and by the stiff breeze blowing over the bow as the boat moved forward. Although the *Slocum* did eventually list to one side as she settled to the bottom off North Brother Island, she did not roll over.

Charles and Louisa Motzer
405 or 406 6th Street, Manhattan

This article was written by Louisa Motzer:

Being a member of St. Marks Church, we had an annual excursion and, of course, everyone was really anxious to go. Most of the children got new dresses and shoes and the boys got new blouses. Everyone was up bright and early, we had breakfast and Mother said, "I really don't care about going today," but Father said, "Well, you just can't disappoint the children." He had ordered the coach to take us down to the pier. Well, there certainly was a lot of excitement. Soon we started to sail. Mother and my sister, Lena, were sitting with friends right by the steps coming down from the top deck. My Sunday School class was getting ready to sing, *Eine Garten Geat* [In the Garden where a Thousand flowers Bloom], when we heard the clanging of bells and people hollering, "Fire! Fire!"

I tried to get down to my mother but, people were coming up to the top deck. My friend, Frances, and I held tight to one another. How we both got to the end of the railing, I'll never know but, in back of us was a very stout woman. She gave us one shove and over we went, still clinging to one another. A tugboat came along; the man pulled me up by my hair (which was very long). They took us to North Brother Island; they gave us each a bathrobe. While waiting there, the butcher we dealt with spied us and offered to take us home. When I got home, I remember that I asked for my father but, was told that he was at the hospital looking for us.

Well, I lost my dear mother and also my sister, who would have been five years old the next day. Our family doctor found my mother but, she could not be saved. We never found my sister. This tragic day was June 15, 1904. Mother was buried in Lutheran Cemetery.

Contributed by: Nancy Cunningham Emmons, Washington, New Jersey

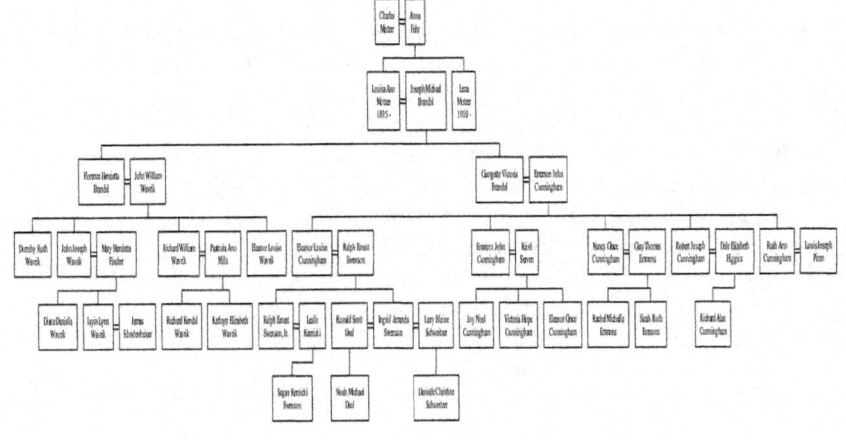

Edward Mueller
368 Bowery, Manhattan

Edward Mueller emigrated from Rossbach (now Hranice, Czech Republic), a town in the Sudetenland region of Bohemia. This is on the far western tip of the province, where Bohemia, Bavaria, and Saxony come together. He was a woodwind instrument maker who traveled steerage class to New York in 1884 after completing a four year apprenticeship. The firm he joined, Penzel, Mueller & Company, Inc. eventually located at 6 Cooper Square in Manhattan.

Grandfather Edward realized that this would not be enough to follow his dreams and so he enrolled in free public night school for English and bookkeeping. By 1904, he was a partner in the business with Lois Penzel. This little company would rise to great prominence in its industry and count among its customers many of the great musicians of the twentieth century.

The story is told in our family, that the day of the *Slocum* disaster, the wife of his business partner died of a heart attack when she learned of the deaths of Edward's wife and children. His first wife's maiden name was Weber but, I do not know her first name.

Edward tried to return to work after the disaster but, was not emotionally ready. Instead he returned to Germany to visit his mother and sisters. When he did return, he eventually married Anna Rau, sixteen years his junior and moved uptown to Yorkville and then later still to Astoria. That second marriage produced five children but, the family had not seen the end of tragedy. In 1918, his eldest child, Edna died in the great flu epidemic at the age of twelve. Edward died in 1956 at the age of ninety.

Contributed by: William H. Manz, Rockville Centre, New York

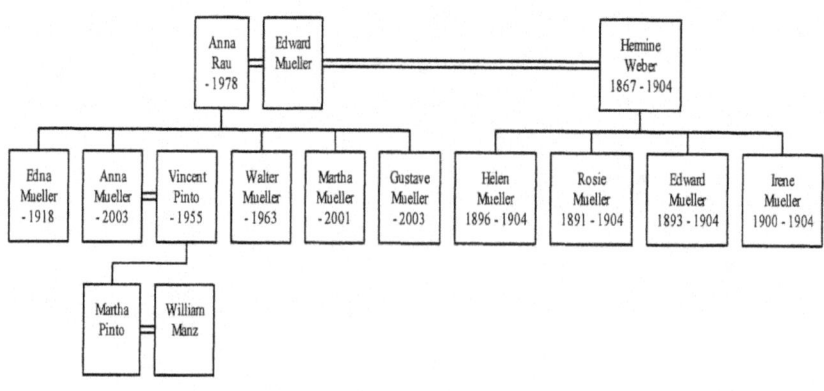

Agnes Mundle
11 East 7th Street, Manhattan

Agnes Mundle, age sixty-four, was aboard the *Slocum* with her daughter-in-law Lillian, age thirty-eight, and two grandchildren with whom she lived, Lilly age eleven and Arthur, age eight. My great-grandfather Harold Cloeren, was Agnes' grandson and he was also age eleven in 1904. I remember him telling me stories as a child of the disaster. He looked forward all year long to the outing and was ill that day and was very upset that his mother (Caroline Mundle Cloeren) stayed home with him. His grandmother Agnes was pulled from the river by her long hair. His cousin Arthur was not as fortunate and perished on that fateful day. I remember my great-grandfather showing me original articles that he had saved. Besides Arthur, he lost several other cousins and aunts. Agnes Mundle later testified at the Slocum inspectors' trial. My great-grandfather passed away in 1987 at age ninety-four, when I was just fourteen years old.

Lillian & Agnes Mundle

Contributed by: Lynda Wagner, New York, NY
Carolyn Ole, Bergenfield, NJ
Tracy Dunn, Port Saint Lucie, Fla.

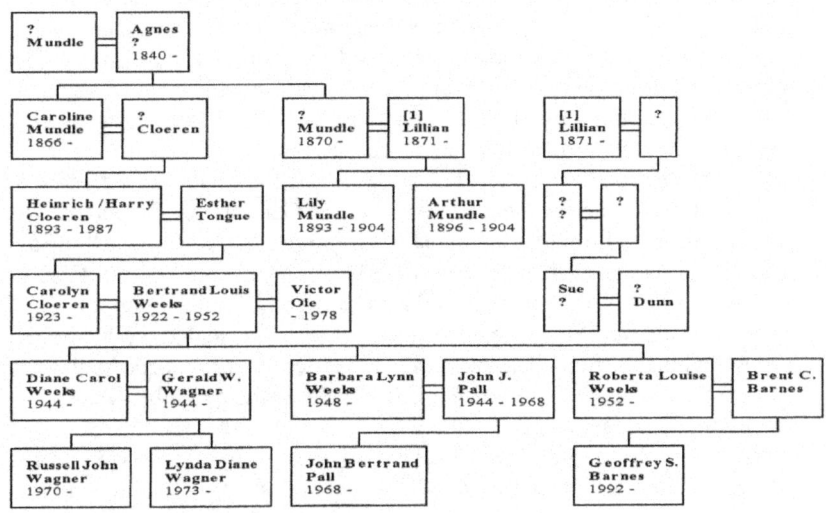

Conrad J. & Connie Muth

Conrad J. Muth, 1254 Lexington Ave., Manhattan
John Muth, 785 146th Street, Bronx
Edward Schnitzler, 10 Gouverneur Pl., Manhattan

After emigrating from Germany, probably in the early 1860's, Johan Muth and Anna Koch traveled to Chicago, where they each had family and were married. Conrad J., my grandfather, was born in 1865. The family moved east to New York City, where John Muth was born just three years before the Great Chicago fire. The family grew until, by 1900 there were six children, four surviving to adulthood. In 1904 Conrad and John were both married with children of their own.

Conrad's wife, Henrietta Margraf had given birth to a son, Conrad H. "Connie", in 1893, and then died from the complications of a miscarriage in 1898. My grandfather and his son moved uptown, from Houston Street, to Lexington Avenue and moved in with Conrad's mother and unmarried sister, Anna Catherine Elizabeth. On June 15th, Conrad wished his mother and son a good time, said goodbye and headed off to work as a garment cutter. John and his family were going on the *Generals Slocum*. Anna, was supposed to go, but decided to spend the day with her future fiancé. No one suspected that what began as a party of fourteen would return as a party of three.

Somewhere between their apartments and the dock, the Muths met other relatives including Hessels, Schnitzlers, Christs, and Smiths. We do not know exactly how Anna Koch Muth died, but it was during the tragic hour of the disaster. Her body was found on Riker's Island. Neither do we know exactly how Connie escaped but, he survived his burns. Conrad told a reporter, while searching for relatives, that his brother John had survived by jumping overboard but, had hit his leg on a rail and broke it, so Conrad had spent two days searching for the family.

John had been married at St. Marks in 1895. He was a dressmaker and his bride, Catherine Hessel, was from the German neighborhood. The couple had three little girls and a baby boy, John, Jr. They had followed Conrad uptown but, chose to live in the Bronx. Catherine's sister, Christine, had married Ed Schnitzler and they also had a young daughter. The cousins would have lots of company on the boat ride and at the picnic! By noon, the horror of the *General Slocum* disaster had spread across the city and anxious fathers were hurrying home to see if their families were among the lucky survivors.

Officer Ed Schnitzler, of the NYPD, was on duty; he walked a beat along the East River above 108th Street and saw the flaming ship as it passed. He hastily commandeered a small boat and rowed out into the river to help. It was only as he drew alongside the paddle-box, that the smoke cleared and he saw the ship's name. What a horrible decision he had to make; to search for his family, or to rescue those who came his way. His decision was the latter. It was noted in an article on the 16th that he had identified the remains of his family at the Alexander Avenue Police Station.

While Officer Schnitzler was helping people into his skiff, John Muth was frantically trying to find his own way off the burning boat with his family. In the confusion and smoke, he had lost touch with his wife and children, now as the smoke momentarily shifted direction, he saw his baby boy. A family story says that to celebrate the day, the proud Papa had made a little red jacket for his son and its bright color now caught his eye. Grabbing the child,, he hurried to the rail and jumped. But in his haste he misjudged and hit the rail below. He lost his grasp on the baby. Both landed in the water and were rescued by separate boats. Both John and "Little John" were found in the hospital, both returned home that night. All the rest were dead.

In August of 1905, Conrad remarried. His bride, Anna Caroline Daug, came from Hoboken, New Jersey. In 1914 their only child, Thelma Marie, my mother was born. In 1925 at the age of sixty he died of a heart attack. Family photos taken during his marriage to Henrietta show a vigorous happy man in his thirties; when his daughter was born a photo was taken; he looks trim, fit, and happy, but by the 1920s he looks far older than his years in spite of a beaming smile and stately bearing. It is hard to imagine how he carried on in the face of losing first, his father in 1892; a wife in 1898, and his mother, sisters-in-law, and nieces on the *General Slocum.*

John also remarried and had a second family. His sister Anna married her fiancé, Henry Schreiber, within the year. But just as her rendezvous with him had been a secret, so is her wedding date. Her wedding ring was inscribed December 1904 but, her wedding photo, taken in the groom's family parlor, is dated March 1905. According to German custom, the groom gave his bride a ring at the engagement inscribed with the engagement date, which was moved from one hand to the other during the wedding, thus became the wedding ring.

"Little John" Muth & his sisters, 1904

Contributed by: Karen Brendel Lamberton, Suffern, New York
Carol Schreiber Bollinger, Washington, D.C.
Joann Fisher Schmidt, Clinton Corners, New York

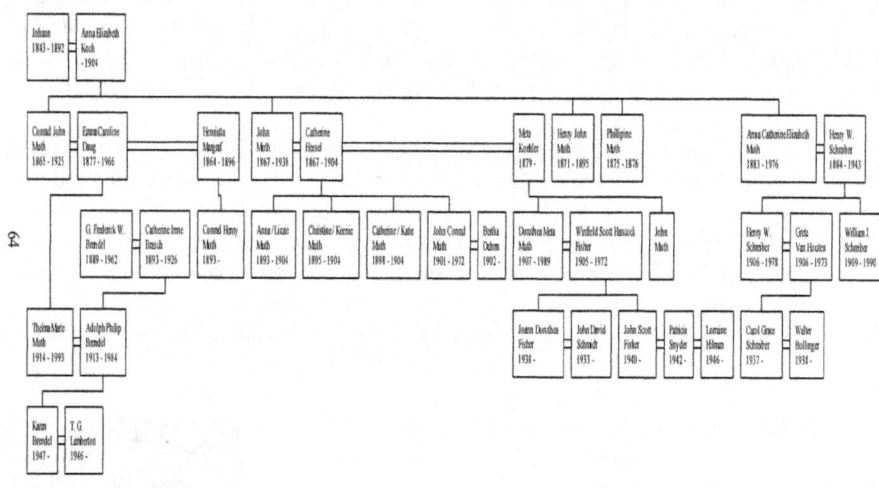

Officer John T. O'Connor
Manhattan, New York

Grandfather, John T. O'Connor was born in Ireland in 1850. I am not exactly sure of when he came to America but, he was here prior to October 19, 1888 when he married my grandmother, Mary Agnes Martin in St. Augustine's Roman Catholic Church in Brooklyn, New York. He was appointed to the NYPD, shield number 5633, in January of 1891 and was on duty on the day of the *General Slocum* sinking. The family story says that he saved several victims and helped to recover others. He received an award for these rescue efforts.

Contributed by: Tom Mulvihill, Belle Harbor, New York

Eileen "Nellie" O'Donnell
Riverside Hospital, North Brother Is., New York

Nellie O'Donnell, my grandmother, died long before I was born, so most of what I know about her comes from a front page story in The New York Times, June 16, 1904. That was the day the news broke of the *General Slocum* disaster. My uncle located the newspaper some years ago and sent copies of it to family members.

Nellie had emigrated from County Cork, Ireland and was about twenty years old and was working at Riverside Hospital on North Brother Island at the time of the disaster. Apparently, as the *General Slocum* neared the island, an attempt was made to run ashore but, instead, the burning steamer ran into a bank that dropped at a steep angle into deep water. Hospital employees ran to the water's edge to attempt a rescue. According to the Times:

> "Nellie O'Donnell, assistant matron, was first in the water. She had often remarked to the other attendants that she wished she knew how to swim, as she believed everybody ought to know how to keep afloat in case of accident but, she had never had the necessary courage. When she saw the people jumping into the water and many drowning before her eyes, she forgot all about the necessary courage and plunged in to save whom she could. To her subsequent amazement, though she thought nothing of it at the time, when she got into the deep water she found that she could swim.
>
> "Miss O'Donnell is strong and when she realized that she was really swimming she grabbed a small boy by his collar and towed him ashore. Then she went back again and again to the rescue until exhausted. In all, she brought ten persons to safety. Afterward she would take no credit, saying that it was all a miracle. As to her new found art of swimming, it would be of no value to her, as, after what she had seen and heard, she would never again dare to venture into water over her depth."

Augusta Viktoria, the German Empress and wife of Kaiser Wilhelm presented Nellie with a diploma during a ceremony, which honored all the women of Riverside Hospital for their prompt response to the disaster unfolding around them. She was also awarded a medal for bravery and was sent a collection of beer steins by the German government, which was always proudly displayed on the mantle piece in her home. Nellie subsequently married my grandfather, Henry Campbell, a police detective, moved to Philadelphia and raised four children; Harry, Eileen, Rosalind, and Norman. True to her word, she never set foot in the water again. Nellie died in an auto accident on New Year's Eve, 1929.

Contributed by: Nancy Hendrickson Riley, Reseda, California

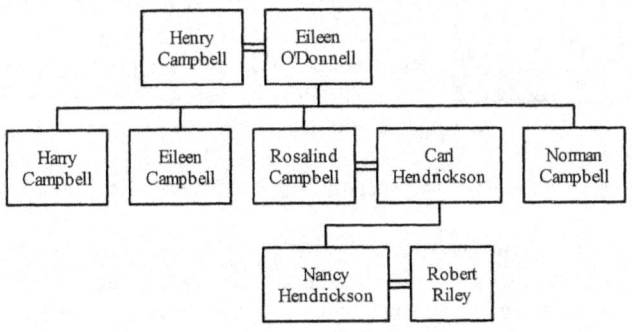

William John O'Gorman
755 East 139th St., Bronx

My great grandfather, William John O'Gorman, was a multi-talented man. Family tradition describes him as an outstanding elocutionist with a great sense of humor. He graduated from Columbia University Law School in 1891 and was on the winning crew team. He published a newspaper, painted beautiful oils, and watercolors, wrote poetry and composed music. In 1921 he became Secretary of the Board of Standards and Appeals, a position he held for many years. He was listed in the census as a builder (the family business), and was listed in the Journal of the Irish American Historical Society. But his lifelong love was politics.

His courtship of Addie Elizabeth Bowne was flamboyant! He arrived at her brownstone on horseback and climbed the steps to her door still mounted. He made quite an impression! Gertrude O'Gorman Wickman, my grandmother, was born in 1902. Some of her fondest memories revolved around the glorious summers she spent with her family on City Island in the Bronx. At the time the island was a popular resort for people seeking escape from the heat of Manhattan. Horse-drawn trolleys met vacationers at the station and took them across a bridge to the seaside hotels and summer home rentals. One summer her father painted the ceiling of the present City Island Yacht Club as a magical universe of moon and stars for his little girl. The family still owns a portrait he drew of his wife, Addie, as the lady in the moon. But there were also long separations when great grandfather was away on business, including rumored investments in gold mines out West. These absences eventually contributed to the deterioration of his marriage and family life.

During the period of the General Slocum, his absences were probably attributable to his responsibilities as a coroner of New York. (The family has his official gavel.) A century later, we can see how heavily the events of the week of the disaster, the Coroner's Inquest, and the trial of Captain Van Schaick must have weighed on the family. From news accounts, one can deduce the hundreds of hours he spent at the scene, at the various morgues, and finally preparing for his legal responsibilities. Then as now, the press probably camped out on the family's doorstep but, unlike today, family members would have been less prepared for invasive questions hurled at anyone coming or going from the residence.

We still remember great grandfather's strength and compassion. A photograph of him working at the water's edge with his sleeves rolled up leaves no doubt how deeply this horrific event marked him forever. Further, it reminds us to never let the family stories die, lest we also forget the hundreds who died that day.

Contributed by: Candace Ann Taubner, Pelham, New York

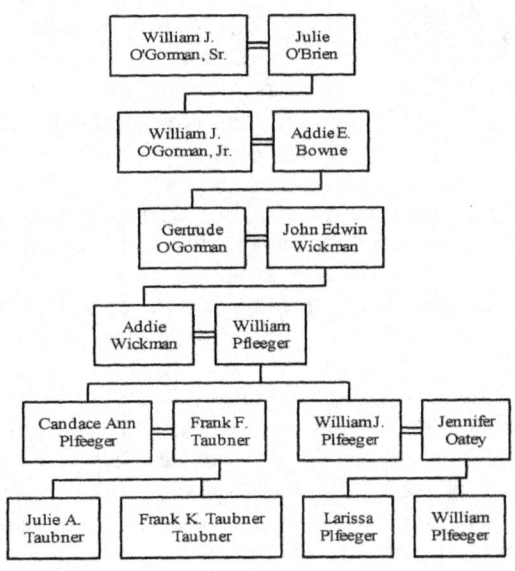

Frank C. & P. Otto Osmers

Dorothea Maria (Bruggemann) Osmers
402 East 83rd Street, Manhattan

Frank C., Emma & Mildred Osmers
449 East 88th Street, Manhattan

Originally the trip was planned for Frank Osmers, his mother, Marie, and Frank's daughter, Mildred. At the last minute, Frank could not go and his younger brother, Otto, my father, went in his place. I do not know why Frank withdrew but, he had a business career, and may not have been able to take the time off. My father was only a teen but, was also working at "office work" for $10.00/ week.

My father could not swim but, was rescued by a tugboat and survived his injuries. His mother, Marie, was very badly burned on the back and side and taken to Lincoln Hospital where her son Frank found her. After an extensive stay she was taken home. But in 1910, in an affidavit before the Court of Claims, Marie is described as still unable to do even housework. Otto was also still having problems and only worked occasionally. Mildred did not survive. Her body was found in the paddlewheel. Emma was so distraught over losing her only daughter that she and Frank moved out of the city almost immediately in an effort to flee the tragic memories. They settled in Englewood, New Jersey, and life went on. But Frank, Sr. continued to follow the subsequent inquiry and changes to maritime law that followed. In 1905, Frank and Emma attended the first memorial service at Lutheran Cemetery.

In 1934, the New York News ran a detailed account of the *Slocum* disaster for the thirtieth anniversary of the event. The News asked my father for his memory of the occasion. My father wrote a three-page account in longhand. He was ready to mail it to the paper but, unfortunately my mother vetoed it. "Did not want any publicity"; Dad's story was thrown out. Instead the paper published only his name and address and his mother's refusal to discuss the tragedy, which my sister, Anne, had sent in. One pleasant result of this article was the fact that a boyhood friend from ball-playing days on East 83rd Street saw my father's name and address and invited all of us to his farm in Sandy Hook, Ct. for several years my sister and I went with Dad for wonderful visits there.

Contributed by: Robert A. Osmers, Flushing, New York
Frank C. Osmers, South Hackensack, New Jersey

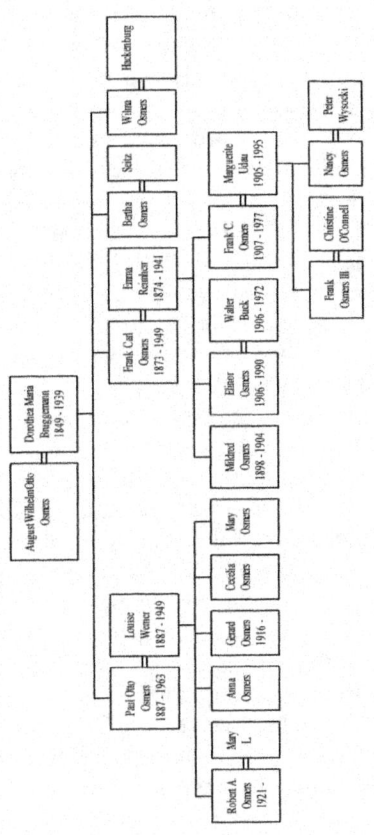

Pauline Peutz
North Brother Island, New York

Pauline Peutz received a gold medal from the Society for the Prevention of Cruelty to Children for her heroic actions at the *General Slocum* disaster where she saved at least six children and five women. She also received a framed copy of the address made by the superintendent of the society. The teenager (18 years of age), who was born in Germany, was working at North Brother Island for only two months as a waitress to the medical staff. She was in the kitchen when the island fire alarm rang. She ran out to the sea wall and over the protestations of others, removed her shoes and skirts and dove in. On the upper deck, a woman stood at the rail holding a little boy. "Throw him to me" she cried, and the woman obeyed. But the boy was carried into the paddle box and under the wheel which was still turning. Pauline held onto the wheel housing and reached into the water. As soon as she pulled him above water, she could see that his jaw was broken and he was bleeding from the mouth. She swam to shore with him and then carried him into the hospital where she set the jaw and entrusted the child to a nurse.

Pauline then returned to the beach and swam repeatedly to the burning steamer, each time returning with another living victim. On her sixth attempt, she was about to reach a little baby when a woman grabbed her and dragged her under. The death grip around her neck nearly killed them both as she swam for shore. She passed out and regained consciousness laid out on the grass. Weak from her exertions and blistered from the heat, she roused herself to seek out the children she had already saved and check on their progress. She then returned to the beach and assisted in saving five women by use of a rope.

At the presentation ceremony, Pauline was accompanied by her mother and was reunited with her first rescue, little Louis Weis in his blue sailor suit with white braid trim. John and Kate Weis had taken in Louis and his four brothers. Her reaction to the accolades was "I didn't expect anything for what I did…at the moment of the accident the little children were standing with their arms outstretched, waiting for help, and I couldn't keep from helping them. I thank you very much." Besides the medal and grateful thanks of the Weis family, Pauline also received several offers of marriage over the next several weeks. –The New York Times

Wilhelmina "Minnie" Rauch
Address unknown

Fred Rauch   "Minnie" Rauch

My connection to the *General Slocum* runs through my maternal grandmother, Wilhelmina Rauch (1888-1969). Minnie, who celebrated her sixteenth birthday on June 9, 1904, had her heart set on going the following Wednesday on that star-crossed Lutheran Sunday School picnic. Her father, Fred, a constable in the NYPD, forbade it. She couldn't recall why but, my mother suspects that he thought the expedition, Sunday school outing or no Sunday School outing, would be inadequately chaperoned. Fred Rauch looks quite formidable in the few photos of him that survive, with his Keystone Kops helmet and eyes like the Ancient Mariner's. In this instance his hardheadedness served a purpose: if ever there were a boat well missed the *Slocum* was it! Minnie couldn't swim and would in all likelihood have been the $1,022^{nd}$ victim.

Her forebears had emigrated from Germany toward the middle of the nineteenth century. Minnie spoke only German till she started school and had vivid recollections of visits to Schützen Park and other German New York locales. In an eerie pre-figuration of the *Slocum* calamity, her maternal grandfather, Fritz Landmann, had perished when the boiler of a river steamer on which he was working as a cook exploded; a heavyset man in the Victorian mold, he was last seen framed in the galley doorway before pitching backwards into the inferno. Perhaps Fritz paid the family's dues, sparing Minnie. She married my grandfather, Wilhelm Hermann Grossmann (1884-1920) in 1909; my mother, Marguerite (b.1917) was their only child. After Hermann's early death, Minnie married William Landmann (1886-1958), Fritz's nephew and her first cousin, once removed; they had a son, William, Jr. (b. 1926).

Marguerite Grossmann and Gordon Cross (1908-1992) were married in 1936. a daughter, Caryl Ann, was born two years later but died in infancy. I arrived in 1940 and my brother, Steve, nine years later. In the course of a nearly forty-year career as an English professor, I've often taught Joyce's *Ulysses*, a book that I first encountered as a sophomore in college and that figures importantly in my own critical work. *Ulysses* is a novel so rich in its particulars that any reader can situate himself in it somewhere. The action takes place on June 16, 1904, and the headlines in the Dublin papers that morning cry out the prior day's disaster on the Hudson. Had it not been for the *dickköpfigkeit*, the mulishness, of Officer Fred Rauch almost a century earlier, I tell my students, they would be listening to someone else interpret *Ulysses*.

Contributed by: Richard Cross, Bethesda, Maryland

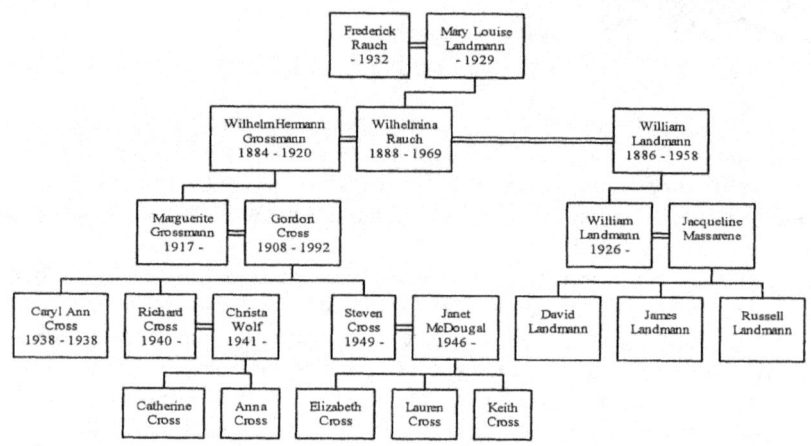

John and Katherine (Wagner) Rheinfrank
343 W. 71st St., Manhattan

John Rheinfrank, Jr.
614 E. 14th St., Manhattan

    John emigrated from Germany in the 1850's and married Katherine Von Waggoner (Wagner), who was from Holland. Katherine's father was already in the United States and ran a lumber and coal business in New York City. John's first job in the U. S. was to import wood for burning in this business. John later bought the business from his father-in-law. The business was called J. Rheinfrank Co., and eventually became Rheinfrank and Sons. There was a main coal yard and two branches. Later on, John founded and directed the Germania Bank on Washington Square. I have in my possession a book in tribute to him from the Board of Directors of the Germania Bank, which was presented to the family after the disaster.

    Prior to coming to America, John was a tailor to the Court of Saint James. My father remembers trunks of patterns in his grandmother's attic that John brought with him when he came to America. Upon his death, John had four daughters and three sons and left $500,000 in his will. I have been told that Katherine's family was also onboard the Slocum on that day but, I don't know if they were saved or not.

Contributed by: Candy Twynam, Columbus, Ohio

Note: There are several stories from news accounts, about this family. One is substantiated in family records. Several of the daughters had been touring in Europe with friends at the time of the disaster. Word reached their ship that a tragic accident had happened on a Steamboat in New York harbor, but it was not until the ship was in quarantine and John, Jr., could come aboard that the girls found out that their own parents had died. One account gave the name of the ship as the soon to be infamous *Lusitania*, another names her *Lucania*. Both were plying the Atlantic at this time. Another story, which is unvalidated, is about two Rheinfrank brothers, perhaps George and Frederick (or Gustave and Frederick), who while searching for fourteen Family members at the temporary morgue came upon an old man who sat head in hands weeping. When asked why he was so terribly distraught, he replied that he had no money to bury his wife; the young men gave him money for a coffin.

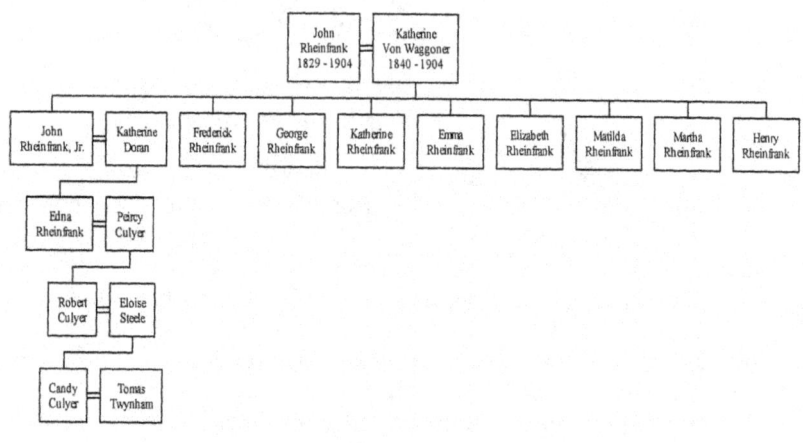

Henry Seifert
215 West 23rd Street, Manhattan

Henry Seifert has no descendant to tell his story. He was an orphan before the General Slocum disaster. Henry had attended Mount Hermon School and was supported by the members of St. Marks Church. In 1904, he was a senior at Springfield College (Springfield Training School) and about to graduate. Traditionally, the school enjoyed a week of festivities leading up to the actual graduation but, Henry was not on campus to enjoy them. Instead, he chose to help look after the children of St. Marks on their annual outing to Locust Grove. Then he would return to the campus in time for "Class Day" and graduation. This was not so unusual a choice as one might think at first glance, since Henry was also a former Sunday school teacher at the church, was a trustee of St. Marks Benefit fund, and had been one of the Luther League's delegates to the City League. After graduation, he had already been assured of a position as Assistant Secretary at the 23rd Street branch of the YMCA.

When word of the disaster reached the campus, the school's traditional class day was replaced by a memorial service for Henry Seifert. President Doggett emphasized in his eulogy how Henry epitomized all that was good in the German tradition among the students. Henry Seifert was interred in Woodlawn Cemetery in the YMCA plot.

John Augustus Scheuing
Jackson Street, Bronx

John Augustus Scheuing, my great-great grandfather was born in September of 1861 in New York City, the son of German Immigrant parents, Christian Scheuing and Catherine Knorr. He joined the NYPD in 1890. On 15 June 1904, he was a patrolman assigned to the Alexander Avenue and 138th Street Station in the Bronx. I believe it is presently designated the 41st Precinct but, in 1904 it was the 35th Precinct.

John, a devout Catholic, was living on Jackson Street, in the Bronx, and was married to his second wife, Rose Twomey, with whom he had a son, Christopher. His three sons Edward, John, and Charles, from his first marriage to the late Matilda Dalymple, also lived with them.

On the morning of 15 June 1904, while walking his beat on 142nd Street, he saw the *General Slocum* burst into flames; he commandeered a soda water wagon to take him to the waterfront. Once there, he cut loose a row boat and proceeded toward the *General Slocum*; he soaked his jacket in the East River and covered his head with it, as protection against the intense heat. According to accounts I have from earlier published works and newspaper stories, he climbed on to the paddle box (despite warnings from seamen on the nearby boats that he was rowing to his death) and pulled five children into the row boat. His bravery in doing this, while pieces of the burning ship were falling all around him, was said to have inspired others in their heroic acts. The heat was so intense that it singed off his handlebar mustache and blistered the backs of his hands. It is unclear how many trips he made in the row boat but, in one account after delivering his passengers safely on shore, he and another policeman, Herbert Farrell, ran to the pier at 138th Street, climbed aboard the New York City fireboat *Zophar Mills* and resumed pulling people out of the water while the firemen were battling the flames onboard the *Slocum*. Later, he assisted in recovering the dead from the river. One newspaper account credited Scheuing and Farrell with recovering over one hundred bodies.

Incredibly my great-great grandfather and Herbert Farrell, after their great heroic efforts, were brought up on charges by their supervisor for "failing to meet their reliefs!" In other words, when their replacements reported for duty, Scheuing and Farrell were busy saving lives and had not returned to the station house. The news of the "charges" and subsequent disciplinary hearing were met with out rage and indignation form the other officers of the precinct, who supposedly threatened to appeal directly to Mayor McClellan about this injustice. I have not found any evidence that the hearing ever took place.

I do know that Scheuing and Farrell were awarded the "Honorable Mention" citation and medal by the NYPD for their "heroic actions at the *Slocum*." They were both also awarded the "Congressional Gold Life Saving Medal" and the "Life Saving Medal from the U. S. Life Saving Service Corp.", the predecessor of today's Coast Guard. Some twenty five years later, during the Great Depression, John A. Scheuing, sold the Congressional Medal, which was solid gold, to feed his family.

John A. Scheuing continued to serve in the New York Police Department until his retirement in September of 1929. He was married and widowed three times and fathered at least seven children who survived to adulthood. One son, Christopher, joined the NYPD and was promoted to Detective for "Valor". He died in a shoot out, trying to stop a holdup at a Lexington Avenue "speakeasy" in February 1931, at the age of twenty eight. He left behind a wife and a seven year old daughter. Christopher also received the department's highest medal. His name is on the Wall of Honor at 1 Police Plaza, New York city and on the Police Memorial in Washington, D.C.

Another son, Charles, my great-grandfather, joined the New York Fire Department and died in October 1924, at age forty four from complications of repeated smoke inhalation. John's youngest son Kenneth also joined the NYPD and served with distinction for over twenty years.

John Scheuing died 2 May 1937 at the "Home for Incurables", now St. Barnabus Hospital, in the Bronx from "cardiac problems" at age seventy-six. His funeral mass was celebrated at Saint Jerome's Roman Catholic Church, which is on a corner opposite the old 35th Precinct, where he was assigned on 15 June 1904. He is buried with his third wife Dora Murphy in Calvary Cemetery, Queens, New York.

Contributed by Eugene F. Kelleher LTC MS, US Army Ret., Lakewood, Colorado

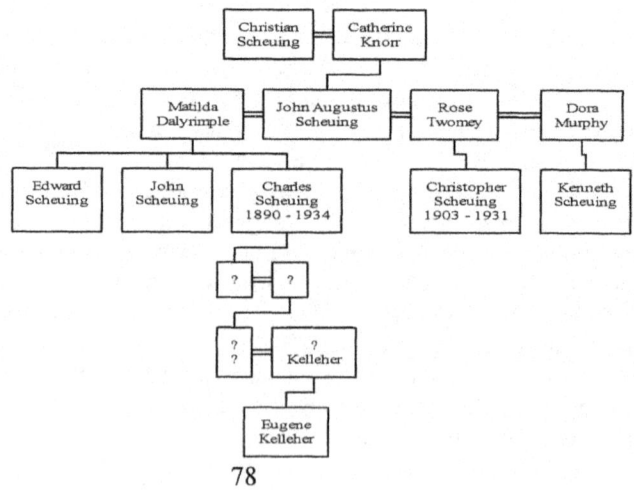

Frances Ethel Stecker
New York City

Frances Ethel Stecker

My grandmother, Frances Ethel Stecker, was born in 1902 in New York and was about three years old when the *General Slocum* sank. The whole family had planned to go on the excursion and were waiting for a streetcar when grandmother's urgent need for the bathroom compelled the entire family to return home, thus missing the boat. Had it not been for this change of schedule, my mother and I would not be here today. The family remained in the German community from which many of the victims came. Great-grandfather, Bernard Stecker, died when Frances was about four. Her mother remarried and her name became Groves.

At the end of World War I Frances was a volunteer nurse in a hospital and that is where she met my grandfather, William Rollins Ammons, from Richmond, Virginia. They married and moved to Richmond around 1920. My mother, Frances Shirley Ammons was born in 1924 and was an only child. She married my father, Ernest Warner Mooney, Jr. from Petersberg, Va., in 1946. He was also an only child. I came along in 1949, followed by my brother, Mark Allen in 1950, and my sister, Maura Lee, in 1958. My grandmother died in 1998. I grew up hearing the story of her near miss. Ironically, I also have an ancestor who survived Pickets charge!

Contributed by: David Rollins Mooney, Pittsburgh, Pennsylvania

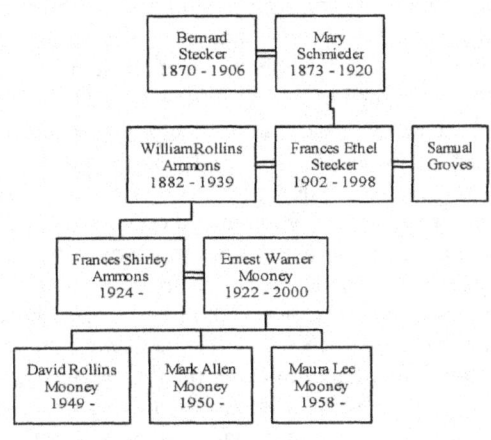

Johann Dietrich Stelling
11 East 119th Street, New York

My grandfather, Johann Dietrich Stelling, was born in Kuenwaldes, Hanover, Germany on February 2, 1887. He immigrated to the United States on the *Friedrich der Grosse* and arrived at Ellis Island on November 20, 1902. He was only fifteen years old. His brother, Heinrich, sponsored him and my grandfather went to live with him at 11 East 119th Street in what is now East Harlem. This is where he was residing when he witnessed the *General Slocum* disaster on June 15th, 1904.

In 1968, I learned about the disaster from my grandfather, who was eighty one years old, while working on a class project about his Ellis Island experience. He told me how confusing it was standing there with people speaking all kinds of languages. He said it was a "tower of Babel". He worried that his brother would not find him and was happy when Heinrich showed up. His New York was a very different place than the New York I grew up in. He told me of trains which ran on steam with smoke coming out of their chimneys. He could get from Harlem to the Battery on a streetcar pulled by horses. He would travel from 125th Street to Central Park and 59th Street. There a new team of horses would take over for the trek to the Battery. He also said that you could go to any tavern and for the price of a nickel beer, dine on a free lunch.

One of my assignments at school was to read "A Night to Remember". The story of the *Titanic* had always fascinated me and I wondered if he remembered reading about it since the survivors were brought ashore at New York. That was when he told me he saw one of the worst ship disasters with his own eyes. I thought he meant the *Lusitania*. That had happened at sea, he said. The one he saw happened in the East River and many German immigrants lost their lives that day. He told me he was just a little older than me (it was 1968 and I was sixteen) when he witnessed this event. There were tears in his eyes as he said that there was a fire aboard and many women and children were jumping into the water, burning to death, or drowning. I could have asked more questions but, he was clearly back in another time and place and, at his age, I did not want to upset him more. I just said that it must have been a terrible sight. Years later, I came across articles about the *General Slocum* and realized that what my grandfather had witnessed must have been a devastating sight indeed.

How did witnessing this tragedy affect my grandfather? I can only speculate that he must have had to deal with some post-traumatic stress. I think also that being so young and so new to this country (here only a year and a half), it must have struck him at the core of his love for his native land. My grandfather was a Lutheran and may well have known some of the people aboard the General Slocum through his church or his hometown. Even if he did not, the sight of so many of his countrywomen and children perishing under such circumstances must have been unimaginable to him. America was a land that was supposed to give you the opportunity to better yourself and that is why these people had come here. Now they were gone. So many hopes dashed, so many husbands without wives; children without mothers or siblings. I can only imagine the mourning in the community! He must have thought that maybe it would have been better if they had never come here.

My grandfather married Meta Wendelken in 1914 and had three sons. My dad, Walter, was the middle child. By the early twenties, grandfather owned his own home and grocery store in the Bronx. For a while, he also owned an apartment building on Ryer Avenue in the Bronx. The family was prosperous enough in the 1920s to go back to Germany for a visit. Grandfather told me of the inflation brought on by high World War I reparations and how impoverished everyone was. He himself would fall on hard times when the Depression hit America but, nothing compared to what he saw in his homeland.

In the early 1960s, my grandfather visited Germany again and there were tears in his eyes when he talked about the Berlin Wall. I thought of him in 1989 when the wall came tumbling down. And I thought of him again on September 11th, 2001, when my hometown was attacked and so many people perished in flames or died jumping to escape them. I was filled with such sadness that this could happen to New York and with such astonishment that the unimaginable had occurred. Until the fire and collapse of the World Trade Center, the burning of the *General Slocum* held the gruesome distinction of producing more casualties than any fire in the history of New York City. And it is still listed as the worst maritime disaster in New York. That day I felt a kinship with my grandfather that bridged the ninety seven years between these two tragedies.

Contributed by: Laura Stelling Simurda, Parlin, New Jersey

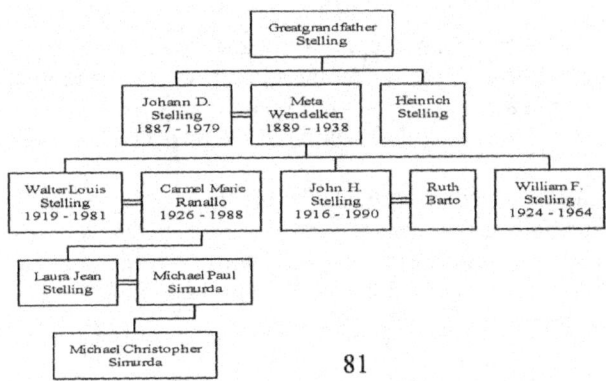

William Bell &
Emma Haas Tetamore

William B. & Sophie (Hansen) Tetamore
1471 Bushwick Ave., Brooklyn, New York.

  Until I began my paternal genealogy some 30 years ago, I did not know about my grandparents' siblings and their families. As I began to seek out photos from the past, I found an uncle with a wonderful album full of the Tetamore family gatherings between 1908 and 1915. These had taken place at a farm in Canaan, New York, which was co-owned by family members.
  In later years my aunt Martha talked about the farm, vaguely mentioning a boat fire in which family members died. She also gave me early Bible pages of family births, marriages, and deaths. On a Birth page is Sophie Hansen, wife of William B. Tetamore and their son, Herbert Tetamore born Dec. 9, 1901. After this entry in a different handwriting – "*deceased by Slocum disaster*".
  William Bell Tetamore was the son of John William Tetamore and Elizabeth Martin Tetamore. He was the youngest of eight children and was born on June 28, 1874, probably in Brooklyn, New York. In about 1900, he married Sophie Hansen. Their son, Herbert William Tetamore was born to them on December 9, 1901 in the family home at 1471 Bushwick Ave., Brooklyn, New York.
  Then I began to ask Aunt Martha in earnest about these family members. On June 15, 1904, Sophie Hansen Tetamore and her son, Herbert, joined other members of her family and boarded the *General Slocum* in New York harbor for a day's outing with St. Marks German Lutheran Church. Their party included Anna R. Hansen Haas, wife of the pastor and sister to Sophie; the pastor, Rev. George C. F. Haas; Anna's and Sophie's mother, Sophie Hansen; and son Herbert W. and the Haas' 12 year old daughter, Gertrude. Emma Haas, sister of the pastor, Jane T. B. Tetamore and Leila Tetamore, William Tetamore's sisters completed the family group.
  Sophie's mother would not jump overboard; Sophie would neither leave her nor throw her son, Herbert, over the side. Jane, Leila, and Emma Haas survived, but Emma was badly burned about the face. Anna Hansen Haas, wife of the pastor, and their daughter Gertrude were also victims of the fire.
  Until I read the accounts of the disaster in the New York Times of June 15, 1904, I had no idea of the magnitude of this fire. My eyes were opened! The survivors of that disaster were the family who spent so many summers and holidays on the farm in the foothills of the Berkshire Mountains in Canaan, New York.

Emma Haas and William Bell Tetamore were married July 11, 1908. When the farm was sold in August of 1920, they were living at 2814 Clarendon, Brooklyn, New York. From there they moved to Boston, Massachusetts. It was thought by older family members that William and Emma went to the hills of Canaan for peace and serenity. However, their tragic experience never escaped them. William died March 28, 1928, and the death was considered a probable suicide. Sadly, Emma died July 11, 1935 in another fire. The blaze consumed the summer home, of her brother, the Rev. John A. W. Haas, in Stroudsburg, Pennsylvania.

Contributed by: Anne Tetamore Rehbach, Rochester, New York

NOTE: At the time of her death, Emma Haas Tetamore had been a resident of Stapleton, Staten Island, N.Y. for about two years. She was at the cottage in Paradise Valley, rented by her brother–in–law, Rev. John A W. Haas, who had also escaped death, on the *General Slocum*. The Reverend Mr. Haas was also the president of Muhlenberg College.

On the morning of her death, the Haas' housekeepers had arisen, set a fire in the kitchen stove and begun their chores. The husband then left the house to attend to other matters. A few minutes later, his wife discovered the fire roaring out of the stove and sounded the alarm. Reverend Haas and his wife escaped in their night clothes believing that Mrs. Tetamore was following them since Mrs. Haas had spoken to her in the living room.

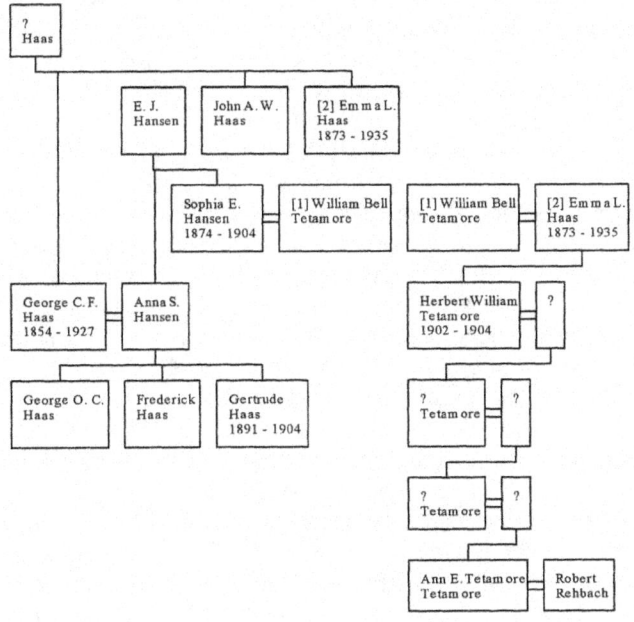

Dora (Laik) Krauss Thoma
90 Avenue A, Manhattan

From a letter written by Carl Thoma:

"...It all started when your great grandfather (my grandfather) Andrew Laik (pronounced Like) met and married Sophia Ernst resulting in eight children; seven girls and one boy. One of these girls, Bertha (your grandmother) met and married Louis Dietrich resulting in six children; three girls and three boys. One of these girls is Anna, your mother. Another of the Laik girls, Dora, my mother (your aunt) married Jacob Kraus and had two children, a boy and a girl. Jacob Krauss and the two children died and Dora married Stephan Thoma and I came along. There is not much information on the Thoma side of the family because when my father was married for the first time in Germany his older brother had married a very rich girl and her people gave them a house. My father then asked his parents to will their house to him since his older brother already had a house. His mother refused to do this saying that traditionally the house must go to the older of the sons. He became angry, told his mother to keep her house, and came to America with his wife where his two children were born. He would not speak of or correspond with his parents after that.

Tragedy struck his life in 1904. There was an excursion steamboat called the General Slocum charted from the Knickerbocker Steamboat Company by the St. Marks German Lutheran Church, on 6th Street, for their annual Sunday School picnic and excursion. They left the foot of East 3rd Street at 9:40 a.m., June 15, 1904, for Locust Grove, on Long Island Sound, where the party was to spend the day.

When about opposite East River Park, at the foot of East 86th Street, smoke was seen coming from the forward cabin but, the steamer kept on for a distance of about three miles before she was beached on North Brother Island, about forty minutes, as near as can be determined, from the time she cast off her mooring with [the] joyous party.

It was [one of] the most disastrous marine accidents that has ever been chronicled in history, for [of] the 1,331 souls, so happy and full of life, who started that beautiful, bright morning in June, only 407 are known to have survived the catastrophe.

That night when he came home from work he found that his wife, age 34, and both children, ages five and eight, had all perished in this terrible calamity. It must have been an awful shock to him.

At about the same time my mother lost both of her parents, her first husband, and her two children. So living in the same neighborhood and knowing each other's misfortunes, they decided to get married and did so in 1905. I was born in October 1905 and in March of 1906 my father died of a ruptured appendix. Mother never remarried and lived with Aunt

Lena (Carolina) and me. That about covers the early history of our family. The rest you know..."

Contributed by: Monalee Washa Bodmer, Ft. Lauderdale, Florida

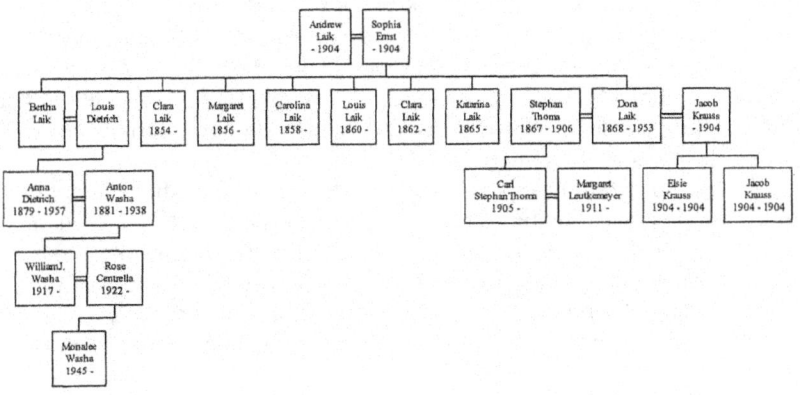

Sophie Ulrich
433 or 443 W. 41st St., Manhattan

Sophie Kreoll Ulrich and her husband, Lorenz, emigrated in the mid-1800s. They first lived in Little Germany but, as Lorenz prospered, they were able to buy their own tenement building and decided to move away from 41st

Sophie Ulrich, seated   Lizzie Ulrich

Street, partly because the thought the people at St. Marks were "too German". (As you know, Rev. Haas spoke only in German.) Sophie thought that since they were in America, they should do things the American way. Katherine, one of Sophie's daughters, (my grandmother) used to tell of her confirmation when Rev. Haas asked difficult questions in German and insistted the children respond in German.

But the family kept up their old friendships and when the notice of the outing came, Sophie signed up right away. Originally she wanted one of Katherine's boys to go with her but, Katherine said that school was too important and he couldn't go. When Sophie's daughter Lizzie (Elizabeth) saw how disappointed her mother was she volunteered to go. Lizzie was a beautiful woman, a hat model, and as it turned out had no work on that day.

When they had to escape the fire on the *General Slocum*, they jumped hand in hand into the East River but, people landed on top of them and they were separated. Sophie did not see her daughter again until she found her, packed in ice, at a makeshift morgue on the Department of Public Charities pier at 26th Street. I did not know Sophie but, my mother visited her often and has passed on detailed recollections to me. Another of Sophie's daughters, Katherine, later married Henry John Harms. Rev. Haas officiated at the ceremony.

Submitted by: William N. Yeomans, Saddle river, New Jersey

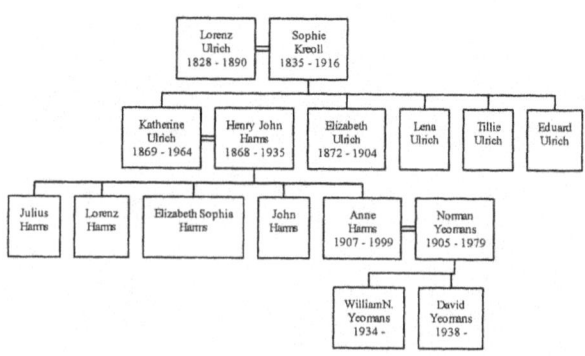

Captain William H. Van Schaick
West Troy & New York City, N.Y.

William Henry Van Schaick was born into one of the Old Dutch families of New York and was born in one of the great old Hudson River Valley mansions. The house had belonged to a revolutionary war officer ancestor and stood on an island sometimes referred to a Van Schaick's Island, near Albany and Troy, New York. His father, Jacob had chosen Mary Jane Berry, the daughter of an equally old and respected Long Island family, for his wife.

William H. was the eldest of seven siblings. He left school at seventeen, and followed his father, a ship's captain, onto the Hudson River. He married Mary L. Laney and by 1870, the family included three children; Elodia M., William H., and Charles, a baby of three months. After Mary's death in 1877, William, his parents, and the children moved to Brooklyn. By 1880, the family included William's new wife, Mary E., and a fourth grandchild Carrie.

By 1904, Captain William H. Van Schaick, Sr., had plied his trade for over forty years. He had received his pilot's license on his twenty first birthday; his Masters Certificate ten years later. Since the 1890s, he sometimes lived aboard the *General Slocum* year round, at other times, he may have kept an apartment on the west side of Manhattan, near the *Slocum*'s North River berth.

"Captain Billy," William's son, was also a captain and commanded the *Cephus,* another excursion steamer, owned by the Iron Steamboat Company. Billy had a family in Brooklyn. The Cora Van Schaick, who helped gather petition signatures for William's presidential pardon, may have been Captain Billy's daughter. The press listed her as William granddaughter.

After the *General Slocum* burned, the press went to work on William's career. He was involved in several groundings on sand bars and collisions with other vessels, but none was considered worthy of reprimand or suspension of his license. His record was not unusual; other captains had had similar run-ins. What were overlooked were incidents like the so-called Anarchists' Excursion, when William was threatened and nearly shot while at the wheel. The first mate and pilot had jumped the man and secured him until they docked, without their help, the *General Slocum* might have had a different deadly demise. Reporters said he panicked and that was the reason for taking the *Slocum* to North Brother Island but, it seems unlikely since he remained at his post until the ship was grounded, most of the passengers had jumped and the pilothouse and his hat were on fire.

The majority of the harbor pilots and captains in New York shared this assessment loudly defending his actions for eight years, until his presidential pardon in 1912. The December 1904 issue of Munsey's Magazine, in an article examining the wreck, the inquest, grand jury indictments, and the opinions of the day concluded that although the captain bore some responsibility, the directors and managers of the Knickerbocker Steamboat Company should bear the greater burden. Further the magazine noted that they "[showed] the criminal indifference of the times", and that society at large, with its eye on the dollar, was ultimately to blame.

In an interview with the Brooklyn Daily Eagle, the captain's sister, Mrs. Lucine Thompson of Brooklyn, noted that the captain was one of a very few pilots who thoroughly knew Rockaway Inlet and Hell Gate. She also related the following to illustrate how unthinkable cowardice or panic was to William. At the age of fifteen, the captain was driving his sister through a suburb of Troy behind a team of horses. The animals took fright bolted. He headed them up a rocky road; then turned them at full speed into a fence, stopping them. When Lucine was seventeen, she went swimming in the Hudson where the current nearly drowned her. Her brother swam out; brought her to shore, draped her over a barrel beating on her back to revive her.

After the trial and his three year incarceration at Sing Sing in Ossining, New York, Captain Van Schaick and his new wife, Grace Mary Spratt, moved to a farm at Amsterdam, N.Y. The couple had known each other since the early 1890's and he had repeatedly proposed. Mrs. Van Schaick said that her objection had been that they both wished to pursue their careers, but when his need was so great, she had to go to him. Indeed, she pushed his wheelchair at the Coroner's Inquest, and sat behind him at the trial. The couple was married at Amsterdam on February 19, 1906. When his appeals ran out, she began the battle to win him a presidential pardon. After two petitions were denied by Teddy Roosevelt, a third was signed by William Howard Taft. She had compiled nearly 300,000 signatures. His parole, one of the first in this country, took place on January 21, 1911 and the pardon on Christmas day 1912. The notoriety was more than the marriage could stand, and only thirteen months after pledging their love, the couple separated.

The captain moved to the Masonic Home at Utica where he lived until his death on December 8, 1927. Out of fear of possible desecration, the family would not reveal the location of his grave until the 1990s, when a great niece placed a stone on his grave in Oakwood Cemetery, Troy. Engraved below the name and dates is the word "vindicated."

Contributed by: John Ackerman, Greenville, Al.
                   Candy Graves, Austin, Tx.
                   Nancy Hahn, Oak Island, NC.

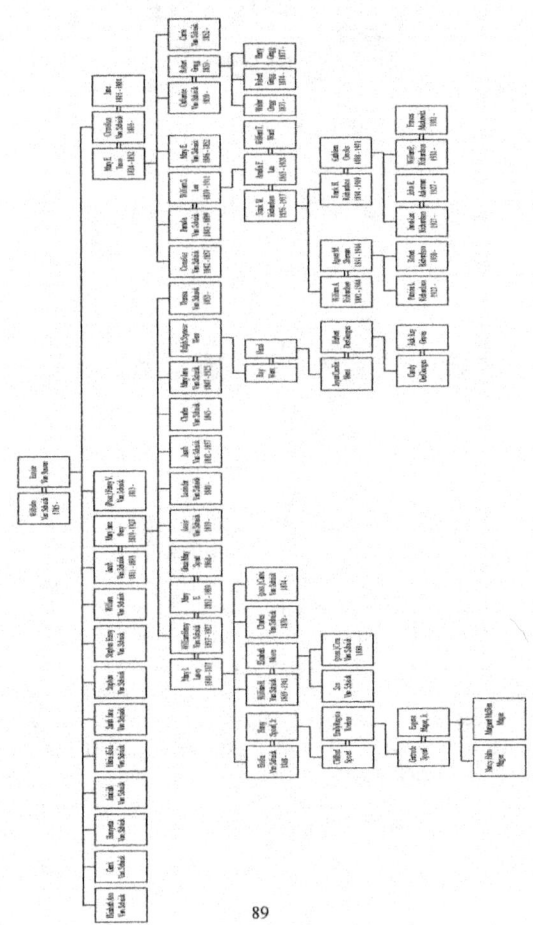

89

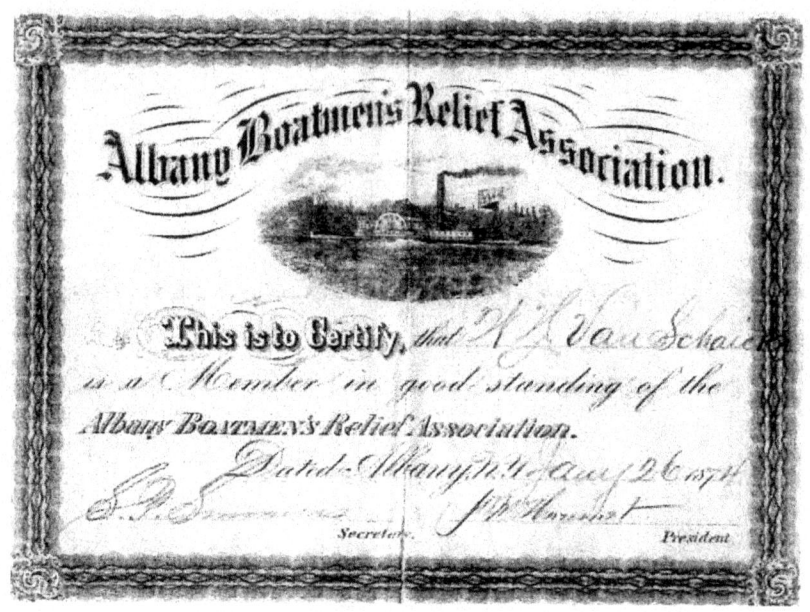

Note: the captain stated in interviews that he had safely carried 1,500,000 passengers during his career. If this is calculated on the 30 years he was a ship's master, the number is not extreme. This calculation does not include years when he was not a master. The year 1904, whose season was only a few weeks old, also is not included. In the 1880's, he piloted the Slocum's sister ship, the Grand Republic, a ship with similar passenger capacity to the Slocum. The calculation is as follows; If for thirty years he made at least one chartered excursion per week from early May to late September, a minimum of twenty trips was made per year. If on each of these trips he carried approximately 2,500 passengers, (trip = round trip excursion charter only), then he must have carried 50,000 passengers per year. At that rate in thirty years, he could easily carry 1,500,000 passengers.

James Celestine Ward
East 136th St., Bronx, New York

Since at least my grandfather's generation, my family lived in the Mott Haven section of the Bronx. The area of New York City could easily be called "another cradle of America" The American flag was proposed at Clason's Point before it was finalized in Philadelphia. The British frigate *Hussar* sank nearby. And the dome of the U. S. capital building was fabricated in the same area as was the "Seated Lincoln", which can be seen in Washington, D.C.

James C. Ward held a maritime license for large craft and was employed by the City of New York for many years in the Maritime Department. He served as the captain of the fireboat *John Purroy Mitchel*, the Riker's Island prison boat, *Massasoit*, and the Clason Point-Whitestone ferry. But, according to the family story, my great grandfather, at the age of twenty-one was the captain of the yacht *Easy Times*, which belonged to a prominent family from New York City.

He was piloting on the East River on June 15, 1904 and noticed that the *General Slocum* was coming up river on fire near Roton Point. He gave three blasts of the yacht's whistle to sound an alert and prepared to move in and help but, the owner of the yacht refused to go close to the paddle-wheeler for fear of ruining the varnish on his hull in the intense heat of the fire. My great grandfather reportedly said to the owner; "Have you ever sailed or know anything about navigation? "That's why I pay you," replied the owner. To which James replied; "That's why you did pay me but, you will never pay me again…You'd best learn very quickly, because you now are the captain of your own yacht." He then took the yacht's dinghy and left the owner and his guests aboard the yacht. He worked, during the next several hours of Hell Gate's high tide, taking many boatloads of people from the port paddle wheel box to South Brother Island. The *Slocum* eventually burned to the waterline near North Brother Island. Bodies drifted everywhere.

James Celestine Ward was later presented plaques and a Medal of Honor in Washington, D.C. but, he modestly declined other compensation fro what he had done. He did, however; accept an honorary captaincy in the Hell Gate U.S. Volunteer Life Saving corps and the U. S. Volunteer Life Saving Corps Medal, which my grand aunt, Virginia Ward, still owns.

Contributed by: Charles Clifford Bothur, Harrison, New York

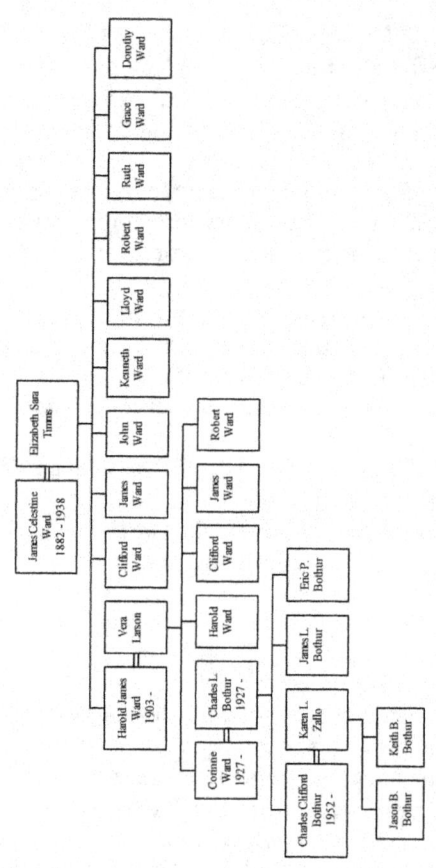

Edward N. Weaver
New York City

My second great grandfather was Albert Weaver. He was the brother of Edward Weaver, the father of Edward Nichols Weaver, the second pilot on the *General Slocum*. At the time of the disaster, Edward N. was only twenty-eight years old. He was one of the few who escaped by jumping into the river, but was seriously injured both by burns and then by striking a rock in the river.

According to the New York newspapers, he lived aboard the *General Slocum* in season and on Twelfth Street in Troy, New York, his old hometown, during the winter. Although his pilot's license was temporarily lifted, during the legal proceedings after the disaster, it was returned and he continued with his marine career advancing to captain, a title he held for the last fifteen years of his career on the New York waters.

Following the disaster, Edward N. lived in the Bronx with his wife Elizabeth; then moved to Jersey City, New Jersey, where they raised a family on Columbia Avenue. When he died in September of 1934, he was survived by a son Edward R. Weaver.

Contributed by: Kim Weaver, Colorado Springs, Colorado

Rush Adelbert Webster
281 Nostrand Ave., Brooklyn

Patrolman Rush Webster grew up in the foothills of the Adirondack Mountains, playing and swimming in the old Erie Canal near his parent's farmhouse. Like many a country boy of his generation he headed for "the bright lights" of New York City to make his fortune. In 1896 he joined the police force, and was on duty opposite North Brother Island as the *General Slocum* ran ashore blazing. Fully dressed in his uniform, he jumped into a rowboat and pulled for the island. He dove into the river and saved one life. Continuing on to the scene, he brought in countless bodies. He became so exhausted that he had to be pulled from the water and hospitalized. His bravery was rewarded with the Silver Medal of Honor from the United States Volunteer Life Saving Corps. He wrote in his diary for March 12, 1909;

> "Permission has been granted for me to accept a medal from Congress in Morris High School's auditorium tonight at 8 P.M. Congressman Goulden and James L. Wells and several others were at the presentation. [I] received the Congressional Medal from Congressman Goulden. The high school orchestra rendered fine music for the occasion."

In her memoirs, his daughter, Adah Frances Webster, wrote how in 1915 he had taken the exam for sergeant and was working one night on the Queensboro Bridge. As a speeding motorist approached, Rush stepped out into the roadway to halt the offending driver but, immediately realized that he would be bypassed. As he stepped back, he was hit by another car and dragged several hundred feet. He was taken to Flower Hospital with multiple internal injuries, suffering greatly. He seemed to be improving and in fact was looking forward to a bedside celebration of his sixteenth wedding anniversary on June 21st. The Websters were cheerful and happy when they parted on the afternoon of June 19th. The morning of the 20th, a policeman came to the house and told Mother that Father had had a cerebral hemorrhage during the night and died. His daughter left us this account of his funeral;

> "He was concerned about mother and had commented how awful it would have been, Had he been killed and left her with four small children to bring up alone. We four children ranged in ages from two to fourteen years.
> "The public school across from our house was closed the day of Father's funeral. Dad lay so still in our front parlor, dressed in his navy blue uniform with all of his medals on his chest and flowers all around as Masonic services were conducted. Rush had been a Master Mason, of the York Lodge, New York City. A wide crepe ribbon with flowers hung on our Front door and an American flag draped the casket as the Police Honor Legion Guards escorted my father to the beautiful Presbyterian Church. Police Commissioner Woods, Deputy Commissioner Lord, 700 members of the Police Department honor Legion, and 300 of his police comrades attended the service.

"A thousand policemen headed by the Police Department Band marched with my beloved father to Prospect Avenue and Boston Road, about a mile and a half from the church. Along the road policemen saluted him as taps was played. He was forty three years old when he was Laid to rest in Woodlawn Cemetery."

Rush Webster, hero of the *Slocum* disaster, continued as he began. He was elected a member of the Honor Legion of the New York Police Department and is still honored today by the city he served on a bronze plaque at Police Headquarters and by fellow officers of the Legion of Honor who, each Memorial Day, decorate his grave with a bronze marker, Honor Legion and American Flags as taps fills the air above his resting place. Thirty-seven years later, at the age of eighty, my mother also passed away. She had never remarried.

Contributed by: Sharon Webster Harvey, Hood River, Oregon

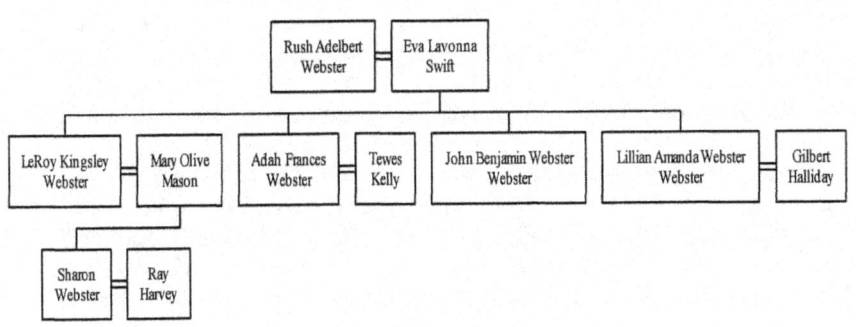

Mathilda Weis, 532 5th St., Manhattan
John & Katie Weis, 167 Avenue A, Manhattan

Great grandmother Mathilda Weis was on board the *General Slocum* with six of her children, Frederick, Jake, Harry, Louis, Sally, Salome, and Emelia, her daughter-in-law, Katie, and her new grandson, John, Jr. Katie's husband, John, was not with them. Katie Weis had emigrated from Ireland six years earlier. She and my grandfather, John, were married at St. Vincent Ferrer Church, 869 Lexington Ave., Manhattan, on February 19, 1903. Katie being of the Catholic faith had reservations about mingling with the Lutheran congregation but, had been convinced it would be a great outing with her in-laws. She was a strong swimmer but, was unable to hold onto her infant while panic stricken people grabbed at her.

After the tragedy, she and John became guardians for the youngest of John's brothers, Harry (12) and Louis (3), who had been orphaned by the disaster. Their father had died a year earlier. Louis was the child whose jaw was broken by the paddle wheel when he was thrown from the deck. He was pulled from the wheel by Pauline Puetz, a nurse from North Brother Island. John and Katie went on to have more children, the youngest being my mother, Anne. They eventually moved to Queens where John supported his family as a chauffer until his death in 1923. My grandmother, Catherine (Katie) Weis lived to be 80 and passed away in 1961. Their other daughter, Catherine, became Sister Catherine Clare Weis, OP, and has spent 70 of her 86 years in the service of the Church. Of Katie's three sons, Albert became the father of Al Weis, second baseman for the New York Mets. During the 1969 World Series, Al hit the winning homerun. Willie became a New York City police officer and Ritchie, who seemed to be a confirmed bachelor, married later in life.

Harry grew to manhood; worked in a brewery and lived in Glendale, Brooklyn, where he married a lady named Eunice. Little Louis also grew up; joined the navy and then later married and had a son of his own. There are so many days I think about the heartache and how difficult it must have been for my grandparents to be the ones to identify their own infant and six members of their immediate family. No grief counseling, just pick up the pieces and move on (although I know many were unable to do this).

Contributed by: Leighanne Mayr, Mesa, Arizona

Note: Samuel Weis, of another household at 167 Ave. A, may be related but, no records have been located to validate the relationship. Another John Weis (age 21) living at 987 151st Street, appears in the coroner's report as Mathilda's nephew but, no other validation has been found.

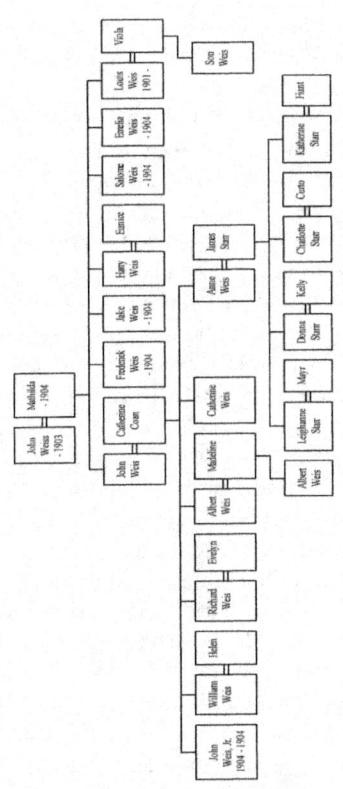

Frederick and Sophie Zipse
335 East 21st Street, Manhattan

My grandfather, William Frederick Zipse "Bill" was born on 15 May 1889, the first child of Frederick and Sophie Zipse who had emigrated from Germany just before the birth of my grandfather.

Sophie Zipse   William F. Zipse

Soon the family grew to six children. Great-grandfather Frederick worked in a bottling company that serviced the Gramercy park area.

Because they were poor, Bill began working while attending grammar school. I was told that he went to the bars in the area and collected bottles to return to the bottling company. While collecting bottles and visiting local bars he noticed an opportunity; the stained wood surfaces would wash away and the bar owners constantly needed to restain the bars. This bright thirteen year old, approached John J. Keller and company, representatives of J. R. Geigy of Switzerland, for permission to sell their dyes to the bar owners as he made his bottle rounds. Of course, he was rebuffed, why would a prestigious company wish to give their best sales territory in New York City to such a child? But Bill was persistent. Eventually he was given permission to sell the product one afternoon a week while making his bottle collections. In time he was given greater responsibility and a territory. After graduating from Grammar School, he used his income to attend the New York High School of commerce and to support his family. Over the years his business experience grew alongside his education as he continued with both his selling and collecting duties.

On June 15th 1904, Bill took a day off to accompany his mother and siblings on the *General Slocum* excursion. He related the story to me many years later. When the fire broke out, he was quickly separated from his mother and siblings. The boat burned for some time and my grandfather, who could not swim, found a crowded corner away from the heat. But the fire spread wildly as the boat continued to sail and eventually everything was being consumed in flames. Then a barge approached the boat and Bill jumped to safety, breaking his leg in the process. The next day, his mother, Sophie, was found alive and still clinging to a log somewhere downstream in the East River. All his other siblings were lost.

My great-grandfather was so devastated that he could not resume work. He lost his job and eventually died of cancer in 1917. The family also suspected a drinking problem, which arose after the disaster but, this was never discussed, just hinted at. Grandmother Sophie died in 1940 and joined Frederick and the children in Lutheran Cemetery. Grandfather Bill remained with Geigy through World War I, rising through their ranks by hard work and imagination until he became a director in 1921. In 1943 he was made president and assisted our military's efforts by helping with the development and early distribution of DDT to our forces overseas, which saved countless soldiers from death by insect borne diseases. My father often related that much of my grandfather's ambition and desire was the direct result of the *General Slocum* disaster. He must have been very hurt by his early years of poverty and family disaster and worked diligently to provide a strong level of security for his family. Grandfather never forgot his ties to the *General Slocum* either and attended every annual memorial service of the Survivors' Association until he passed away in 1961.

Contributed by: Robert Zipse, Burlington, Ontario, Canada
Bill Zipse, Marietta, Georgia

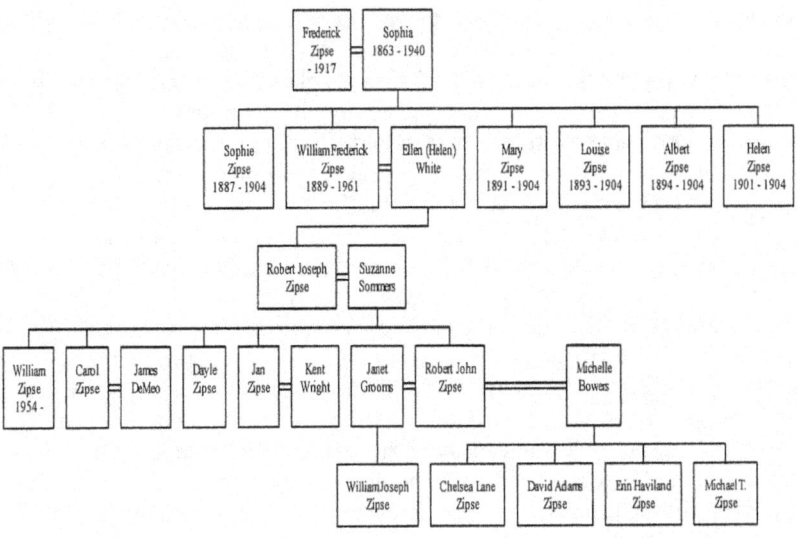

After the disaster, Bill Zipse found these two tokens still in his pocket. He kept them as lucky pieces for the rest of his life.

The Bandelow Boy

Long after the children of the *General Slocum* had been buried, one father continued to believe that his child had survived. Perhaps he had been taken in by a well intentioned, although misguided Samaritan. The trail of clues led him to the far reaches of the city but, to no avail. This may be the most haunting story of all connected with the great tragedy of Hell Gate.

William Bandelow, a clothing designer from Seventh Street, Manhattan, lost his wife Louisa and a daughter of the same name in the disaster. He thought that his three year old son, George had also been lost until a family friend sparked a new search for the toddler. She told the family that she had seen the child in the custody of a well-meaning matron, who may have assumed that the unaccompanied lad, on the pier with other survivors, was a new orphan and taken him into her home.

Mr. Bandelow spent almost a year trying to track down leads all over the city. He enlisted the aid of local officials and the police to help investigate each sighting. Several times during the summer of 1904, a child fitting George's description was seen in the Canarsie section of the city, at the beach, but each time he and the mysterious woman were gone before the father or police could get to the scene.

During the months of agonizing dead-end leads, the father with a designer's eye sent out impeccable descriptions of the outfits the child had worn and those his mother had packed along "just in case". He sent amazingly detailed physical descriptions (down to the condition of the child's teeth), and offered rewards. He traveled the city speaking in public schools before assemblies of teachers and children and handed out posters which included photos of the boy. But all his efforts were to no avail.

In an interview decades after Mr. Bandelow gave up his search, remarried, and had another family, he was asked about his son. The father was still sure that the boy had grown to manhood and probably had a family of his own.

Every house in mourning hung ribbons on the door; black for adults, white for children lost on the *General Slocum*.

very house hung crepe at the door.
Many doors had multiple streamers.
Most also tacked up a black bordered card, carefully lettered:
"We mourn the loss of our friends"

As word filtered into the German community in the late morning of June 15th, relatives began arriving. Men, whose business had them out and about, heard the cry of the newsboys' "Extra!" and stopped to buy a midday edition of the paper. Horrified and sickened with worry they ran for the El trains and headed home. Businesses began to close as proprietors began the gruesome search for loved ones, or in sympathy for their customers' plight. Soon the trains from uptown began disgorging a very different cargo. Saturated and soot stained survivors, some half dressed, some wrapped in blankets and borrowed men's suit jackets, staggered off dazed and in shock. The carriages of upper eastside matrons stopped in front of tenement homes to let off their pitiful riders. The city's charitable heart opened as never before and found ways to help in the first hours after the disaster. Survivors were taken in and clothed from matron's closets; children were handed the fare home; coaches were dispatched downtown full of such misery as would never have been allowed to ride under normal circumstances.

Then the ferrying of bodies began. Tugs, city owned ferries, police launches ran in at the foot of 26th Street with bodies for the morgue at Bellevue, which soon overflowed. A huge temporary morgue was set up on the adjacent pier. First coffins, then wagonloads of ice and the undertakers started coming. Throughout the afternoon and evening of the 15th,

QUEST OF THE DEAD AT BELLEVUE MORGUE.

This is of Bellevue Hospital's Morgue. The scene at the 26th Street morgue, where most of the identifications were made, was worse by tenfold.

the crowd around the temporary morgue grew. By midnight, few had departed for home. On the first night, the police counted 415 bodies at North Brother Island and 50 more elsewhere. The following morning, 41 were found at Hunts Point and in the Sound. Throughout Thursday, the swell of families continued until the side streets were clogged and impassable, but a system evolved to handle the distraught masses.

At the morgue on the pier, the line stabilized at about 1,500 outside and another 1,500 inside, periodically, 200 more were allowed in to join the ritual review of the dead under Captain Gallagher's watchful eye. On the 16th, first, the Sound Fleet making for port in Manhattan, churned up bodies with the vibration of their screws; then cannons were towed over the area where the General Slocum had come to rest and fired off. A ton of dynamite was also exploded trying to dislodge more bodies too waterlogged, stuck in the mud, or held below by debris from the ship to rise on the tide. Several dozen drifted up.

As the identifications began, bodies were removed for embalming and dressing. The brunt of this work fell to the undertakers in the area between Grand Street and 10th Street. In some cases, bodies in boxes stood on the sidewalks awaiting their turn for space and attention inside, as temporary boxes were hauled in by the wagonload from northern Manhattan and Brooklyn.

Friday June 17th, white streamers for young children and black for adults went up on the doors of the houses of mourning. American flags in windows draped in black and purple, or flown at half mast sprouted from the front yards in sympathy. Little was heard in the street, except the rattle of the occasional undertaker's wagon. Fathers waited for hope among the crowds outside St. Marks'. Merchants' windows were filled with mourning bonnets, grave blankets, and other trappings of grief. Throughout the week, roving bands of thieves prowled the district ripe with opportunity. They picked pockets on the street and sometimes broke into the homes of mourners. In several cases, they worked amongst the families as the bodies lay in the parlor. In one case, two teenagers, Joseph Hornstein and Jacob Kolask were caught working a crowd. Kolask was collared with his hand in a woman's skirt pocket outside St. Mark's. Both were taken to the nearest station house and booked. In 1904 there were no "youthful offenders". But these two, and others like them, got off with a trip to the House of Refuge in lieu of prison. The woman, in this case, as in several others, refused to come to court and identify the thief for fear of missing word of her own missing children.

And then the funerals began; 114 were held in front parlors and 37 churches. The lights of Little Germany kept vigil that night as relatives fervently prayed for the return of loved ones still missing and for those who sought them at the morgue. A constant contingent of police officers were assigned to aid the coroner's clerks by taking down the names of those who made identifications and the names of the victims they identified. This information was relayed to the clerks in the Coroner's temporary office on the pier, and Coroner Scholer issued the permits of removal, which allowed the undertakers to remove bodies from the pier. As the pressure mounted for bodies to be identified and claimed quickly, Coroner O'Gorman devised a unique method. Since many of the women wore wedding rings, perhaps their initials and dates might be clues to their identity. He dispatched researchers to the Bureau of Vital Records, their mission – to use the dates in the rings as a pointer to the years to be searched for marriage licenses, and then the initials in the rings to identify the couple and perhaps a relevant address. When a likely couple was identified, a copy of the marriage license was given to Detective Ross, whose men then went to the listed addresses looking for relatives of the deceased.

Saturday 156 more left 88 earthly homes for the ride to the cemetery. For those on the way to Lutheran All Faiths Cemetery, in Middle Village, Queens, it was a ghoulish parade. Police cordoned the streets ahead of each set of hearses and carriages to contain the crowds who silently covered the sidewalks and spilled into the streets. The thrill seekers, who had earlier pushed their way into the temporary morgue, now stole into front parlors for a quick peek. A few of the most brazen even attempted to find a seat in the family carriages as they filled with mourners on their way to the cemetery. As if forming up for a parade, the corteges moved out from the family parlor and down to First Avenue; then onto Fifth Street; then South by East to Delancy and over the new Williamsburg Bridge. It being the Jewish Shabbat, the crowds of spectators were even greater than the day before. The curious gathered on the walks on Delancy Street approach to the bridge. Captain O'Connor of the Delancy Street Station sent an additional 20 patrolmen to keep the onlookers moving. Throughout the funerals, no policeman carried his night stick, and none was needed. By the afternoon, the lines of mourners stretched for miles on both sides of the bridge. Every available hearse and carriage from the city and the surrounding areas was in service, as were many from New Jersey. The pressure was so great, that the number of hearses was limited and the multiple coffins of children were stacked like cord wood; sometimes parents took their final ride with one or more children stacked upon their breast in a final embrace. Carriages were also limited to four per family in order to help control the inevitable traffic jams. The by-ways of the cemeteries, especially Lutheran, were clogged beyond capacity from dawn to dark with many services hastily concluded after the usual closing of the gates at sundown. Before it was all over, 600 households had been affected directly.

On Sunday the 19th, the Cadets from Old St. Marks on 2nd Avenue, fanned out through the neighborhood after Sunday school distributing flowers to each grieving family. Most of their calls were on 1st,

2nd, 3rd, 4th, and 7th streets. As the lines to the morgue, and then to the cemetery moved slowly on, the stream of those driven mad with grief or collapsed from exhaustion began to swell in the hospitals. Although some would soon respond, others were already "hopeless cases", or at best, would be invalids of the disaster for many years to come. Some of these uncounted victims of the disaster never spoke of June 15th again. Some died prematurely.

On Monday June 20, 1904, 200 more funerals found their way to Lutheran Cemetery as the church bells of New York rang in unison; a sound not heard since the Civil War. Across the newly combined boroughs of Greater New York, they rang in peal after peal from 2 to 3 PM as the city grieved in sympathy. Before the temporary morgue was shut down, the police estimated that 20,000 people had streamed through looking for loved ones.

The Memorial Meeting at St. Mark's Church
June 18, 1904

The meeting was held under the auspices of the Lutheran Ministers Association (New York Ministerium). One hundred and fifty ministers from all Protestant denominations, the Roman Catholic Church, and several Jewish synagogues attended.

A representative of St. Marks noted that of the twenty-one officers of the church, only five had been accounted for as of the start of the meeting. Many of the representatives rose to express their congregations' and their own feelings of sympathy for the stricken congregation. Several members of the group were elected as a steering committee to oversee the implementation of whatever resolutions the group might decide to adopt. Various ideas were discussed as to how the group could best support the people of St. Marks and Pastor Haas. Due to Pastor Haas' personal situations of being both in deep mourning himself and at the same time very sick from his injuries, it was decided that the group would take up as much of his pastoral duties as they could, including whatever was necessary to shield him and his family from public scrutiny during this difficult time.

In order to accomplish their mission, half a dozen resolutions were drafted and accepted. All cards and offerings of condolence would be sent to Dr. Heischmann, who would hold them and then forward them to Pastor Haas at the appropriate time. Among the many offers of help from other agencies and private organizations received by the church and the mayor's committee was one from Lutheran All Faiths Cemetery. Their offer of a plot large enough to contain all unidentified victims was accepted as the most appropriate of the several received from various cemeteries around the city. These particular burials should take place as quickly as possible and all at one time. To accommodate this, as bodies reached the cemetery, they would be held in the chapel under the direction of Rev. D.W. Peterson, Pastor of the chapel. No general service for the public would be held until all had been buried. The city's offer to pay for the burial of the poorest and the unidentified was also accepted.

It was decided that on Sunday the nineteenth, memorial services would be held in all Lutheran churches of the New York district. Meeting delegates voted to do the same n contiguous districts and churches of other denominations. Finally, any orphans of the disaster would be placed at Warburg Orphan Asylum, which was a Lutheran agency.

Letters tendering sympathy and assistance from Archbishop Farley, and Presbytery of New York, were read. The representatives of virtually all denominations also gave expressions of sympathy and help. The meeting was concluded following the Lutheran church's custom of casting an assenting vote by rising in song. On this occasion, the hymn [title translated] "Who knows how near our end may be?" was chosen and raised by a chorus of deep and reverent male voices, without accompaniment.

Officiating:		
Rev. DR. John Joseph Heischmann	St. Peters Lutheran Church, Brooklyn	President of the New York State Ministerium; conducted the meeting Member of the Steering Committee
Rev. Dr. Hugo Hoffman	of Brooklyn	Gave the opening prayer
Others in Attendance:		
Rev. Dr. William R. Huntington, Rector	Grace Episcopal Church	Among the first to rise in expression of his sympathy and sorrow.
Rabbi Joseph Silverman	Temple Emanu-El, 5th Ave. at 43rd St. (76th St.)	Rose to extend his congregations sympathies and offer their help "It is our misfortune, not yours alone. Where we can help" we must help, and we will help."
A Representative of	Presbyterian Church 2nd Ave. & 2nd St.	Expressed how deeply affected their congregation was due to the proximity of their church to St. Marks
A Representative of	the Confederation of Churches	Offered the services of their offices to facilitate the communications between the various ministers engaged in the work

Rev. Dr. Joseph Lawton		Noted that due to the extreme number of funerals to be conducted, wherever possible they should be held at the homes of the families, and that those held at the church should be as private as those at home. This he said was to avoid the gathering of large assemblies who might give vent to undue expressions of grief.
Rev. Dr. Dawald		Read a resolution expressing the feelings of grief of the group
Rev. Dr. Hipple of	Methodist Episcopal Church West 104th St	
Rev. Dr. Berkemeir of	Wartburg Lutheran Orphan Home, Mt. Vernon, NY.	
Rev. John Hutchinson	Presbytery of Newark, New Jersey	
Rev. Mr. Perry, Pastor	Collegiate Church, 2nd Avenue	
Rev. M. S. Waters, Secretary	New York & New England Synod	
Rev. Jacob W. Loch, Pastor	Schermerhorn Street Lutheran Church, Brooklyn	Chosen Chairman of the Steering Committee
Rev. E. C. J. Kraeling	Zion Lutheran Church, Brooklyn	Chosen a member of the Steering Committee
Representatives o f	Howard Methodist Episcopal Church	
Representatives of	the Norwegian Church	
Rev. Dr. John B. Remensnyder, President	Lutheran Synod New York & N	This congregation lost 10 members
Rev. Dr. John C. Fagg	Collegiate Reformed Church [Middle Dutch Church], 2nd Ave. near 6th St	This congregation lost 73 members of which 41 were children; 11 from the Sunday School, 13 from the church's Industrial school)
Rev. Alfred Meyers	Marble Collegiate	} all co-coordinated their actions
Rev. E.G.W. Meury	Knox Methodist	} through Dr Fagg and performed
Rev. Edward Niles	Dutch Reformed Church of Bushwick	} many services

Rev. William Schoenfeldt	Immanuel Lutheran Church, Lexington Ave. at 88th St.	
Rev. Julius Geyer	Evangelical Lutheran Church, East Houston St.	
Mgr. Lavelle,	St. Patrick's Cathedral	
Rev. John L. Belford, Pastor	Sts. Peter & Paul, Wythe Ave., Williamsburg, Brooklyn	
Bishop Henry C. Potter	Episcopal	
Rev. Dr. George C. Houghton	"Little Church Around the Corner" [The Church of the Transfiguration] 1 East 29th St.	This congregation lost 3 members
Rev. Merle St. Croix Wright	Lenox Avenue Unitarian Church	
Rev. Henry Ruggles Remsen	Calvary Episcopal Chruch, 4th Ave. at 21st St.	
Rev. Dr. James Oliver Wilson (Wilkins)	Norstrand Avenue Church, Brooklyn	
Rev. Dr. C.D. Case	Hanson Place Baptist Church, Brooklyn	
Rev. Dr. John Lloyd Lee	Westminster Presbyterian	
Rev. Dr. J. Newton Perkins, Asst. Pastor	Grace M. E. Church, 104th St.	
Rev. Dr. Robert S. Mac Arthur	Calvary Baptist	
Rev. Merle St. Croix Wright	Lenox Avenue Unitarian Church	
Rev. A. Steimle, Pastor	Holy Trinity Church, Brooklyn St. Luke's Lutheran Church, Brooklyn	
Rev. W. Ludwig, Pastor		
Rev. J. Ross Stevenson	5th Avenue Presbyterian Church	
Rev. Arthur W. Byrt	Warren Street Methodist Church, Brooklyn	
Rev. W. J. Hutchins	Bradford Presbyterian Church, Brooklyn	
Rev. Macy McGee Waters	Tompkins Avenue Congregational Church, Brooklyn	

Rev. Dr. Morris, Pastor	7th St. Methodist Church	Offered his church for funerals
St. Augustine's Protestant Episcopal church,	Houston St.	Offered the church for funerals
Rev. Mr. Holter	Jersey City Lutheran Church)	Began the collection of donations for the needy for burials & medical care
Rev. Dr. Laidlaw, representative of Rev. Dr. J. P. Holstein	the Federation of Churches of Brooklyn	Assisted with services for the victims after the disaster
Rev. J. Roth Rev. Ernst W. Kaufmann		Ws selected Treasurer of the Committee
Dr. Moldehnke	Beekman Hill church at 50th St	died at the direct result of his exertions helping the General Slocum survivors. Nearly 8,000 attended his funeral service.
Priests who attended at various points where victims were recovered or laid out for identification included, but were not limited to:		
Father Broderick, Chaplain	of the Department of Charities	
Father Donlin	St. Jerome's Church	
Father Boyle	St. Luke's Church	
Father Christian	St. Luke's Church	
Father Donohue	St. James' Church, North Brother Island	Performed 45 last rites during the rescue / recovery

    In the months before the Slocum Disaster, the Lutheran Congregation from 46th Street at Lexington Avenue, was forced to move up to the Beekman Hill area while train track repairs precluded the use of their parish home. Their pastor, the Rev. Dr. Edward Frederick Moldehnke, was among those who came forward to assist Rev. Haas' stricken congregation. The many days of long hours in Little Germany on top of the strain of the move proved too much for the pastor who succumbed unexpectedly while attending to his adopted flock. His standing in the Lutheran community of New York was more than demonstrated by the eight thousand mourners who packed the temporary church home and sidewalks for his funeral. The alter was packed with floral tributes from the outer wall to the communion rail. –Extracted the <u>New York Times.</u>

*Also on the Relief Committee

The Service for Mrs. Anna E. Haas & Miss Gertrude Haas
St. Marks Lutheran Church
323 East 6th Street, New York City
Saturday September 19th, 1904

Gertrude Haas

Mrs. Anna Haas

Due to the sheer volume of funerals to be conducted, few could be held in church. Many families had services in their parlors with as many relatives, friends, and neighbors as could squeeze into the family's home. The extreme number of services to be conducted kept the officiating ministers on the run from early morning nearly around the clock. But the funeral for Mrs. Haas and her daughter, Gertrude was held in St. Marks. No one expected Reverend Haas to conduct or even participate in the service. Not only was the poor man in deepest mourning for his loved ones, he had also been seriously injured in his escape from the *General Slocum*, and could barely move about without assistance. In his absence, four ministers assumed his duties. Rev. Alexander A. Richter, St. Matthew's Evangelical Lutheran church, Hoboken, New Jersey, led the service and delivered a moving sermon; the Reverend Jacob W. Loch of Schermerhorn German Lutheran Church, Brooklyn, gave the readings and the Reverend John J. Heischmann, the President of the New York Ministerium of German Lutheran Ministers, and the Reverend Hugo W. Hoffeman, St. Paul's German Lutheran church, Rodney St., Williamsburg, Brooklyn offered prayers.

A messenger announcing that the body of Mrs. Haas' sister, Mrs. Tetamore, had been identified momentarily interrupted the service. She was brought directly to the parsonage, and when the service was completed, the sisters rode together to Lutheran Cemetery, where they were laid to rest side by side. Twenty-eight ministers followed the cortege. The procession was sent on its way by the German Turnverein Drum Corp, which played muffled drums on the steps of St. Marks.

ST. Marks Sunday School teachers lost in the disaster:
Wm. B. Pullman, Geo. A. Anger, Wm. M. Schlaffer,
Catherine A. Balser, Ida Brandt, Lizzie Ulrich, Mary Abenschein
Henry C. Schnude, Henry A. Kohler, Henry Seifert

Other Members of the Congregation who were aboard

Of the twenty-one members of the Church Council, only five survived. Most had been aboard the *General Slocum* and, of those aboard, only two returned alive. The Sunday school Lost forty-two members of a staff of fifty-eight (forty-eight had been aboard). Amongst the students, three hundred and seventy five out of five hundred members were lost including thirty-nine out of fifty-one in the kindergarten alone; the Women's guild lost twenty-three and the choir ten out of twenty members.

John Holthusen, the Church Sexton and President of the Sunday school for twenty-seven years, had the only family in the congregation not touched by the disaster. John and both his daughters had been aboard and were rescued. Interviewed a few days after the disaster, he was quoted in the New York Times as saying; "There is no need for me to resign, my school is no more."

Board of elders:
    Henry Gerdes
    Charles Anger — Superintendent of the Sunday School, leader of Bible Classes
    William H. Pullman — Treasurer- Was identified by a check in his pocket made out to the Knickerbocker Steamboat Co. for the excursion's charter

Board of Trustees:
    Cord Sackmann — Church Secretary
    George H. Witte
    George H. Brunning
    H. C. Schnude — Superintendent of the English Sunday School
    H. C. Daverheim
    H. Pottebaum
    Peter Fettig — Chairman of the Poor Committee
    Charles H. Schaefer — Secretary, Sunday School worker
    Mary Abendschein — Head of the Excursion Committee, 6 years a Sunday School Teacher, Choir Leader
    Mr. Hiller — Janitor
    Christian Schoett — Church Organist

Sunday School Teachers:

Jeanette Uchlin	Julia Duls	Sophie Zipse
Minnie Burdewick	Emma Muth	Frances Hartung
Lizzie Ulrich	Margaret Lutjens	Emily Schlidline
Kate Gringel	Kate Germann	Elsie Pullman
Freda Cohrs	Sophie Delventhal	Emma Vetter
Minnie Hoffman	Minnie Hayden	Elizabeth Folzke
Katherine Balse	Sophie Rosenstein	

Sympathies Received

Truly, New York's grief was a sob "heard 'round the world." –Munsey's Magazine

THE WESTERN UNION TELEGRAPH COMPANY,
INCORPORATED
23,000 OFFICES IN AMERICA. CABLE SERVICE TO ALL THE WORLD.

ROBERT C. CLOWRY, President and General Manager.

RECEIVED at 201 East 14th Street, New York.

110 Fp Gu 22 Paid 8 Pm

Whitehouse Wahn D C 16

Rev Geo C F Haas

    Pastor St Marks German Lutheran Church, N.Y.

Accept my profound sympathy for yourself and the congregation in the terrible calamity that has befallen you am inexpressibly shocked and grieved

        Theodore Roosevelt.

This Telegram has just been received at the office, 201 East 14th Street, where any reply should be sent.

From the President of the United States;
To the City of New York, and separately to Rev. Haas;
   "To the Rev. George C. F. Haas, St. Marks German Lutheran Church:
   "Accept my profound sympathy for yourself, your church, and your people."
      Theodore Roosevelt

   The following telegram from the German Ambassador at Washington, Baron Speck von Sternberg, enclosing a cablegram from the German Emperor:

"To: Rev. George Haas, Sixth Street, New York:
   "The following cablegram has just been communicated to me by His Majesty the Kaiser: Being most profoundly affected by the news of the indescribably horrible catastrophe which has overtaken the Lutheran congregation, I command you to express to it my inner-most feelings of sorrow."
   "In carrying out the command of my most gracious sovereign, allow me at the same time to offer you my own personal sympathy.
      Sternberg."

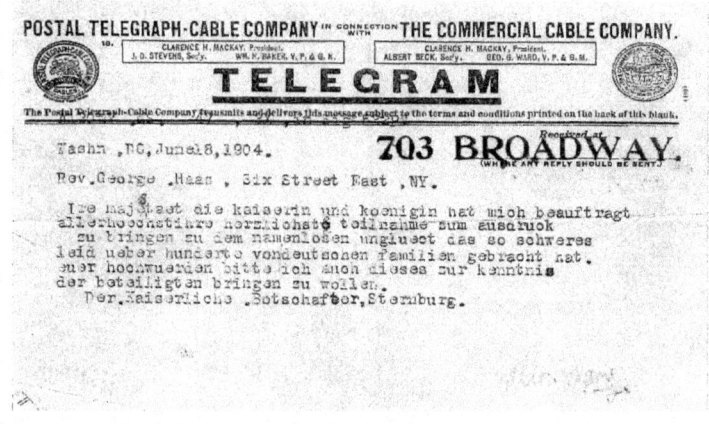

Well known for her interest in what today would be called Women's Issues, the Kaiserin of Germany sent her own condolences, and then later rewarded the hero nurses of North Brother Island.

President Emile Loubet of France in a cable to President Roosevelt:
"Profoundly moved by the awful catastrophe of the 'General Slocum,' I have it at heart to address to your Excellency my sincere condolences, and to send to the families of the victims the expression of my sorrowful sympathy."

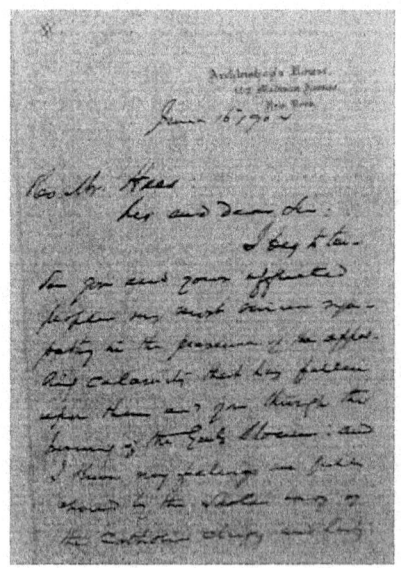

Archbishop of the New York Diocese,
John M. Farley:
"May the giver of all strength comfort you and yours in this dreadful hour of sorrow."

Mayor George B. McClellan, of New York City:

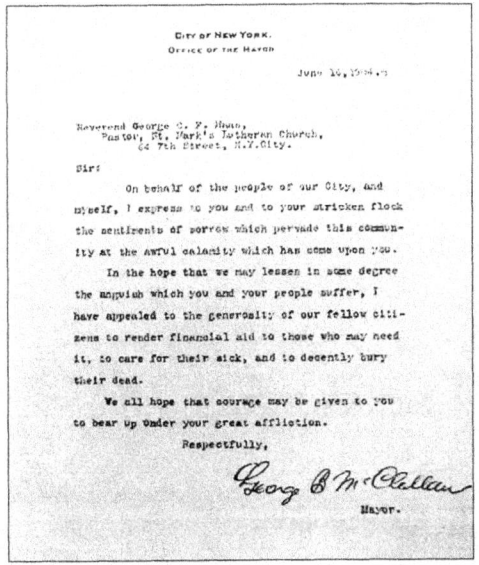

"On behalf of the people of our city and myself I express to you and to your stricken flock the sentiments of Sorrow which pervade the community at the awful calamity which has come upon you.

"In the hope that we may lessen, in some degree, the anguish which you and your people suffer, I have appealed to the generosity of our fellow citizens to render financial aid to those who may need it to care for their sick and to decently bury their dead.

"We all hope that courage may be given to you to bear up under your great affliction."

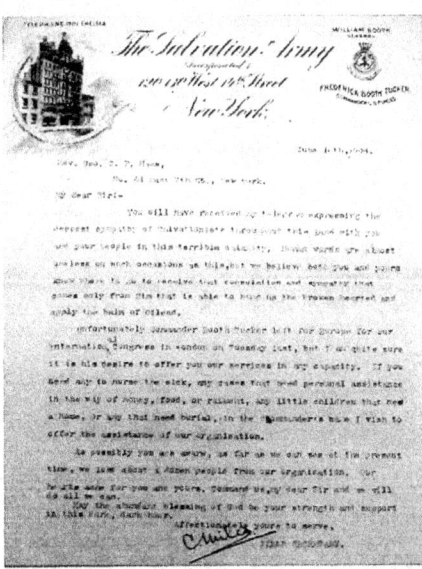

The Salvation Army expressed its deep desire to help the people of Little Germany and St. Marks' congregation however they might.

Other Notable Condolences

Polish National Alliance
B. B. Odell, governor of the State of New York
John Weaver, Mayor of Philadelphia
John W. Holtzman, Mayor of Indianapolis
Chief Magistrate of Glasgow, Scotland, Bailie Sorley, Acting Chief Magistrate
Students, School of Engineering, Mexico
Sir Thomas Lipton
W. B. Derrick, Bishop's Council of the African Methodist Episcopal Church
B. W. Arnett, Bishop's Council of the African Methodist Episcopal Church
Thomas C. O'Sullivan and the Catholic Club of New York
Board of Managers of the Produce Exchange

In the Schools

Superintendent of Schools, Dr. Maxwell ordered that the graduation exercises of the lower East Side schools be abandoned in favor of memorial services. The hardest hit school was Principal Charles C. Roberts',*. PS 25 on 5th Street at 1st Avenue. Of the 2,000 students enrolled in this school, more than 100 of its pupils were excused from class to go to the picnic. When Miss Helen Goldstrum decided to canvass the families of her pupils in order to assess the impact on the class, she found that her class of 20 boys and 6 girls was reduced to 6 alive and accounted for. As the rolls of the victims swelled, so did the lists in the ten schools which comprised the two school districts of Little Germany. Eventually the count stopped at 83 dead, 66 missing, 30 injured, and 68 saved.

The children of the PS 25 had raised enough money for their graduation exercises prior to June 15th, but decided that instead of a party, they would donate the money to the Relief Fund. Each year the graduating class posed for a picture which was then given to the school. When the class of 1904 tried to pay for theirs, the photographer refused to accept their money which was then added to the donation.

The following report of the Girls' Department Service, from the Brooklyn Eagle, excerpted below, was probably held at PS 25 and the number in the description was a typographical error;

Memorial Service at School #24 (Girls Dept.)

Reverend Dr. Burlington, of St. Marks Episcopal Church at 10th Street and 2nd Avenue led the service. There were prayers and then a sermon. Dr. Maxwell also made a short address to the students. Mrs. Helena A.

Hullskamp led the children in the following hymns: "Hope in the Lord", "Nearer My God to Thee", Lead Kindly Light". The following week a similar service was planned for the Boys' Department of the School.

*Charles C. Roberts, 1063 10th Ave., Mt. Vernon, NY.
 Helena A. Hullskamp, 273 138th St., New York.

PASTOR HAAS' SERMON
JUNE 27, 1904

The Rev. George C. F. Haas, pastor of St. Mark's was both badly hurt in the disaster and doubly bereaved by the loss of his wife and daughter. Finally on the 27th of June, 1904, he returned to his pulpit. The church was packed and hushed as he prepared to speak. Each face he gazed upon was a mirror of his own grief. Each mourned for his family as he did for theirs.

As he prepared to speak it seemed to reporters in the crowd that his voice might forsake him, but as he began it grew into the familiar baritone which had for so long filled the sanctuary and all who heard it with the assurance of the Master's loving direction for human life here below. His text was from 1st Corinthians Chapter 13: "And now abideth faith, hope, charity, but the greatest of these is charity."

Rev. Haas began by asking "What is to console us now?" "Shall we continue in our work or give up?" He expressed and confirmed his flock's feelings of mystification that such an extraordinary event should befall so many. He then turned them to face what was to come. "Answering for myself, and I am sure, for the great majority of my people, I can say we will go on."

He then denounced negligence, carelessness, and greed as responsible for the disaster. "I thank God that it has opened the eyes of our whole city and the whole country to what is required to save thousands of others from a like fate. No one on that fatal boat died in vain. The laws of God cannot be violated, even if human laws are."

"My people, I call upon you to put your faith in God and to bear up even though many of your loved ones are gone. Love still lives. Love cannot be killed. We can keep our love and with it the memories of our loved ones who have gone before. In this our darkest hour, with all our burdens and afflictions still fresh upon us, let us look up to God. What is now an awful calamity may in time prove a blessing. Our cross is heavy, but, thank God, it is not too heavy."

Services for the Unknown
June 19, 1904

Rev. Dr. Peterson, of Trinity Evangelical Lutheran Church led the Lutheran Sunday school service at Lutheran Cemetery for the eleven unidentified children. Each child attending carried a small plant, which he or she planted on the grave at the conclusion of the service. Another service for the Unknowns buried at Lutheran Cemetery included over 200 children with arms full of flowers to cover the graves.

For weeks after the funerals ended, Lutheran Cemetery's streets were continually clogged with pedestrians and carriage traffic. Many were mourning families but, almost as many were curiosity seekers, still attending the greatest spectacle in New York. Flowers freshly tended, were either removed for souvenirs or trampled. Neither the cemetery nor the police could completely contain the crowds, until they began to thin of their own accord.

After the initial shock, as families returned to visit family graves, the flower tributes remained as fresh and numerous as at the time of burial. For many months mourners and their friends constantly occupied the lanes of Lutheran Cemetery. The graves where entire families were lost were not forgotten. Many times mourners from nearby graves were seen dividing their families' displays and carefully arranging these bouquets on their neighbors' plots.

One of the hundreds of funerals moves in solemn procession into Lutheran Cemetery

The Most Awful Grave

A trench was dug for the 61 unidentified victims. They are buried in the memorial plot Lutheran Cemetery, Middle Village, queens, New York

Charles Dersch, who would become the voice of the General Slocum Survivors Association for generations to come, had lost a daughter, Helene, in 1890. She was entombed in a family mausoleum, which was constructed for her. When Mrs. Dersch, another daughter, and a niece were lost on the *General Slocum*, they also joined Helene in the Mausoleum. It was the location of this family tomb, which suggested the location for the burial of the unidentified victims and, eventually, the monument constructed over the Most Awful Grave.

The General Slocum Disaster Victims' Monument
Lutheran Cemetery
Middle Village, Queens, New York

On June 19, 1904, the German Societies met in Tompkins Square Hall to plan what was described as a "monster meeting" in memory of those lost. The meeting took place on July 14th and all the German Societies of the city attended. At the same time, a meeting of the German Women of the city convened to raise money for a monument to the victims of the disaster.

One year later, at the dedication ceremony, the United Singing Societies of New York and Brooklyn, five hundred strong, sang of love and hope. Adella Liebenow, the eighteen-month-old survivor of the

disaster, unveiled the monument in front of 10,000 onlookers. Her mother carried her to the place of honor. As they approached, the child dropped her dolly, which was retrieved by an unknown gentleman in the crowd. But her attention was already on her task. With her doll firmly tucked under her arm, she pulled the cord. A flag fluttered momentarily as the child released it and it fluttered softly to the ground and the statues were revealed. "See – pretty." She exclaimed for all to hear. In one second the tableau was complete; the memories of the horrors; the eloquent scars of the survivor imploring the masses for justice; the infant hope for a better future; the Eternal promise in marble relief. The monument, built over what had been described as the Most Awful Grave, the sad mass tomb for the unidentified victims was now a place of peace and hope. Banners in front of the plaza proclaimed, in German, "They were earth's purest children, loving and fair," and "Let us not have died in vain." Neither the families in attendance nor the politicians and editors who decried the horrors of the tragedy could know that these victims had, indeed, not died in vain. The destruction of those innocents was the catalyst for the most sweeping changes in maritime regulations in the country's history.

The General Slocum Disaster Monument in Lutheran Cemetery

In 1912, Kathinka M. Stoss had two side statues erected. One stands at either end of the General Slocum Disaster Monument's plaza. Each is a column surmounted by an angel. One cradles a child in her arms; a reassurance to those who grieve that there is peace in the arms of love. The other blows the clarion call to Judgment. Both challenge us to remember what happened to those innocents buried below and to all who fall victim to the greed and negligence of others.

Other Memorials

Tompkins Square Monument c.1985

On Saturday June 23, 1904, Eastside matrons, Marie Bunsbacher, Marie Scheppner, Katherine Mathes, Dora Rode, Ferninande Knabe, Marie Huhm, and Annie Kiep, who lived near St. Mark's, took an ad in the Brooklyn Daily Eagle. They appealed to all the women of New York to attend a meeting at Tompkins Square Hall on that evening. It was their intention to organize a committee that would begin taking steps to have a memorial tablet or column erected in Tompkins Square where St. Mark's children used to play.

A meeting scheduled for July 14, 1904, known as the Tompkins Square Meeting was organized by the mayor to end the official period of mourning. It began with a procession and mass meeting in Tompkins Square Park. Muffled drums and a funeral dirge sounded as the procession formed on 13th Street and 2nd Avenue and then stepped off behind Marshal Henry Pfeiffer. The parade included representatives from every German Society in the city and was no less than 5,000 strong. Twice that number packed the park during the ceremony which followed. Inspector Schmittberger and 600 police attended, but were not needed.

On May 30, 1905, the Memorial Fountain was dedicated near the North West corner of the park. The eight-foot tall shaft of pink marble depicted a boy and a girl on a riverbank gazing after a steamboat as it headed out to sea. The inscription read, "They were the earth's purest children, loving and fair."

By the 1980's, the pink of the marble was enshrouded in graffiti and the fountain's font had been chipped and broken. In 1989, do-nations from New Yorkers and boating enthuse-iasts, collected by the Maritime Industry Mus-eum refurbished and restored the monument to a nearly new condition.

In November of 1904, a plaque was mounted on the side of Niederstein's Hotel (now a restaurant), adjacent to Lutheran Cemetery in

Tompkins Square c. 2003

Queens. This hotel was the headquarters of the Organization of the General Slocum Survivors and continues to the present as the place descendants meet after the annual memorial service in the cemetery. Shortly after its installation, the plaque was stolen. The vandals turned out to be neighborhood boys and the plaque was returned but, at this writing, it is not on the building and its whereabouts are unknown.

Also in 1904, the New York Chamber of Aldermen passed a resolution to commission a plaque. The whereabouts of this tribute is also unknown at present.

The citizens of France also presented a tribute to the congregants of St. Marks Church and the city of New York in the form of a bronze wreath, designed to be displayed on the church's façade. The wreath carried a banner across its center showing a river with a plank of wood representing the *General Slocum*. This tribute is also lost to time.

ଔ A Cross Too Great to Bear ଔ

That the Slocum survivors suffered as much as the victims of the tragedy is obvious in the ways that they coped with their grief, losses, and anger, often for the rest of their lives. Many moved away as soon as their dead were buried, or soon after. There were suicides and individuals who went quietly insane. Some were not so quiet!

September 2, 1904, The New York Times –Hugo Bertuch age 36, of 171 Brown Avenue, Bronx, lost several relatives on the General Slocum. Neighbors said he was never right afterwards. His landlord, John Garvey, discovered smoke coming from Hugo's apartment. He broke in and saw Mr. Bertuch's bed on fire and the renter heaping other furniture on the fire. After his arrest, Hugo told police he had had a telephone call from Heaven, during which he was told that Mrs. Garvey had compelled Captain Van Schaick to set fire to the General Slocum. God then ordered Mr. Bertuch to set fire to the house, Mrs. Garvey, and her family. Mr. Bertuch was transferred to Bellevue Hospital.

"... I myself, after a happy married life of twenty years with as good a woman as God ever made, with a beautiful daughter of nineteen years, am today suffering with the effects of their loss and my experience in that disaster. I cannot follow my business prospects any more, owing to nervousness and a general breakdown."

—George Wunner, Vice President, Slocum Survivors Association,
in testimony before Congress

"... You, who no doubt have happy homes and families of your own, while some of us have none. Our loved ones are gone, while you have your dear wives and sons and daughters. How happy it is for you to go to your homes and firesides. We cannot do so. We look at the four walls of our homes, which are cold and cheerless, as we have no families to greet and cheer us."

–Charles Dersch, President Slocum Survivors Association
in testimony before Congress

To this day, each June two wreaths of 61 carnations are laid at the General Slocum Disaster monuments; one at Lutheran Cemetery and the other at Tompkins Square Park.

The Mayor's Committee on Relief

Rev. F. Holter, from St. Paul's Lutheran Church, Jersey City, New Jersey arrived on the 15th to help in whatever capacity he might be needed at St. Mark's. He was momentarily stunned by the crush of people in the street near the church waiting for word of loved ones. Members of the congregation were manning tables in the church vestibule for this purpose, and many other clergy were already on hand. He entered the church; found a table and chair, then returned to the street where he set up a hastily worded sign announcing that he would accept donations from individuals wishing to help. He remained at his post without rest for almost 24 hours. Nearly fainting from fatigue, he was convinced to rest for a few hours, after which he returned to his task. As word spread, he was almost mobbed by the crowd forcing pennies, dimes, and dollars into his hands. On the first day he collected $1,000.00.

Meanwhile, the Reverend Mr. Feldman manned an information station in the vestibule of the church to provide information and comfort to the survivors and the mourning families. He collapsed after thirty-six continuous hours of work, and was put to bed under the watch of a physician, while other volunteers took over the work. Although well meaning and much needed in the early hours of the disaster, the obvious shock and lack of formal organization limited the effectiveness of these volunteers who struggled to keep up with events.

The following day, Commissioner McAdoo sent two policemen to the church to begin transmitting a flow of data to police headquarters. These were later augmented by the Mayor's Relief committeemen who were experienced in working with large events and groups of people. These gestures relieved not only the exhausted volunteers but, also the congregant families waiting for information. It also facilitated the Mayor's committee in assessing where and how great the needs of the survivors were. As the flow of data into official channels and news reports grew, many diverse organizations rose to volunteer their resources. Two such efforts illustrate the ecumenical nature of the aid given.

The Young Ladies' Auxiliary of the Talmud Torah gave an entertainment on June 28th at Beethoven Hall, 210 5th Street, for the benefit of the Slocum Disaster Survivors; it was well attended and raised about $200.00. In the same week that the Vicars General of the New York Archdiocese met and decided to send a contribution t the Mayor's Relief Fund.

The Mayor's Committee decided that Pastor Holter should continue to man the subscription table at St. Mark's Church. Mr. Hill, of the Charity Organization Society[1] was appointed Secretary of the Allied Committee and took personal charge of all relief work at the headquarters of the Society in the basement of St. Mark's church.

The committee organized a public memorial service, which was held at the Cooper Union. Although the crowd nearly exceeded the capacity of the building, and most were German, few were from the victims' families. 400 members of the German Singing Societies participated. Nathan Franks conducted the Metropolitan Opera Company Orchestra. The Mayor, Julius Harburger of the Mayor's Relief Fund, the Reverend John Heisman, George V. Von Skal, Editor of the Staats-Zeitung, and Judge Morgan J. O'Brien made remarks. At the conclusion of the service, the audience remained seated and still for several minutes. During that time the audience hushed anyone daring to speak or trying to leave.

By June 18th, $15,000.00 had been received in the Mayor's Office. Within the first ten days the fund reached $100,000.00 [by comparison, $1.00 in 1904 equals $30.00 in 2003]. Thinking that this rate of subscription would continue, the Mayor declined offers of assistance from large philanthropic organizations and individuals here and abroad but, his confidence was premature. The funds did not last long, as funeral and hospital bills mounted. Also, having made one donation, most contributors did not feel the need to continue in this particular philanthropy.

In contrast to the city's generosity, the German-Americans' pride complicated the committee's efforts to assist those in greatest need of relief. Mr. Ridder put out a plea to the newspapers for the public to assist the committee in locating those in need of help; "Ask the public to send us word of the needy. They won't come to us." Some of the less secure families found themselves going through their savings, plus whatever the fund could provide while many middle-class families declined outside help. As a result, those who declined but, who most needed the community's support, were in such distress that the traditional avenues of the family, the Mutual Aid Society, the church, and the city, were all depleted of funds at the same time. Many of this group had no choice but to declare bankruptcy. This was in spite of the fact that only an estimated three quarters of all eligible families and / or individuals actually came forward to accept assistance[2]. Among the middle-class, the options were not significantly better, plus they carried the additional stigma of not being able to "take care of their own".

District Superintendent of Schools, Gustave Staubenmuller, assisted the committee via his in-depth knowledge of the families and children of the affected area. He organized a sub-committee of teachers to canvass the neighborhood searching for families and individuals in need of assistance who had not come forward, either from grief or pride. According to Northup;

> "... so imbued are they with the feeling of independence and self respect that with the entire city ready to empty its pockets in token of sympathy, it has been extremely difficult to gain the consent to accept financial aid.
>
> "Mr. Ridder, chairman of the Relief Committee, was in despair early in the week. The committee was meeting daily and its chief business developed into, not trying to meet demands as might have been expected but, in trying to find people who would take the money. 'Ask the public,' Mr. Ridder said to the newspaper representatives, 'to send us word of the needy. They won't come to us.'
>
> "... Where young children have had no parent to resent the efforts of help, an aged grandmother, as poor as they, would come forward and forbid it. "'No, no', cried one, 'it would be thrown up to them always that they had taken charity. It must not be.' "I had enough saved to pay for one funeral, said another but, I did not think to have five at one time. I will pay it, though, if it takes me two or three years."

A cornucopia of goods and services were offered to the committee in the days immediately after the disaster. From private citizens to international organizations, it seemed as if everyone wanted to contribute. Western Union offered free telegrams to any person rescued from the wreck or any of their relatives in order to let family know they were all right. The same courtesy was extended to the families of the dead. A Mrs. Chaudler and Miss Mason, local society ladies, came before the committee and offered to pay for all mourning items wanted by the needy of the congregation who had lost relatives. They also offered to furnish clothing, food, and other provisions as needed and they would also "look after" the families who lost their breadwinner. Doctors Otto Maier, 212 E. 18th Street; Emil Hotzen, 67 7th Street; Eugene Bachmann, 312 W. 34th Street, all members of the German Society were required by the society to provide services gratis to all survivors and victims' families. Several organizations offered a home to the orphans of the disaster. The American Volunteers offered accommodations for 25-50 orphans at their Darien, Connecticut headquarters. Rev. Dr. Berkemeir, of the Warburg Lutheran Orphan Home, Mt. Vernon, New York, advised the committee that although full at the present time, he had made arrangements to convert buildings on the home's property into proper accommodations and would accept all orphans of the disaster.

Even the dead were remembered in the charity of the city. One unnamed undertaker offered to bury all unidentified victims free of charge. Of the various cemeteries of the city, several offered burial plots of extraordinary size. Superintendent Boystede, of Woodlawn Cemetery, reported that he was authorized to give such a plot on Martin Luther Avenue. Fresh Pond Cemetery of Long Island offered its services free to St. Mark's Church members.

Private and political organizations organized themselves to help. The 10th Assembly District Tammany club voted Julius Harburger the Chairman of a Committee for Relief Work in the district. This was St Mark's. district and many of the members were affected. The Jefferson Club of the 35th Assembly District was offered a resolution by Borough President Haffen to contribute to the Relief Fund. The resolution passed. Other areas of the country with large German populations joined the rush to help. Mayor William B. Hays, Pittsburgh, sent Mayor McClellan a note to Mayor McClellan that he was personally arranging a benefit for the Slocum Disaster sufferers. The benefit would be a minstrel show at the Nixon Theatre, Pittsburgh, on the next Friday night. Closer to home, Madam Lobel, of the New York Yiddish theater gave a benefit performance of "The Lost Paradise" for 1,500 people and realized $500.00 for the fund. J. Edelstene & B. Thomashiosky, management of the house, donated their location and the orchestra's services for the benefit. One "leading citizen" even gave Ocean View Cemetery a donation to buy plots for all the victims.

For over a week, the New York papers listed contributors and the amounts of their generosities in columns beside the listings of the dead. A few of the first donations received included:

St. Mark's congregation	$1,500.00
H. H. Rogers, Standard Oil Millionaire	$1,000.00
Jacob H. Schiff, philanthropist	$1,000.00
The New York Staats-Zeitung Newspaper	$1,000.00
10th district Tammany club	$ 600.00
President Theodore Roosevelt	$ 500.00
Bernard M. Baruch	$ 250.00

During this time, it was reported that an "unregistered contribution" from John D. Rockefeller put his entire fortune at the disposal of the committee for any deficiency after donations were exhausted. Mr. Ridder announced that the reports were premature and that nothing was definitely decided. Both Mr. Rockefeller's and Sir Thomas Lipton's [$1,000.00] offers were politely refused[3] as if taking its cue from the German community, the city announced that it could and would take care of its own.

At a special Meeting of the Board of Aldermen, the Board of Estimate & Appointment was authorized to issue $50,000.00 in bonds to meet the expenses of the Commissioner of Police, Commissioner of Health, and the Commissioner of Charities in assisting the survivors and relatives of the victims. It was decided that with the thanks of the Relief Fund Committee, this issuance would not be needed. At the same time it was announced that the Mayor's Committee had pledged $5,000.00 to undertakers and another $500.00 for drugs, medical services, and mourning clothes. When these funds were distributed, the Fund's expenses were largely cleared. Donations were closed at $108,504.04, which the committee felt was more than ample for any remaining claims. In 2003 dollars, this sum was close to $3,255,000.00.

Amid the confusion and the sheer magnitude of the catastrophe, Mayor McClellan might have been forgiven for overlooking a few seemingly trivial details. However, this was not the case. When it was brought to his attention, probably by Mrs. McClellan, who was on the board of the nurses' training school at Bellevue Hospital, and thus in daily contact with the human conditions of the disaster, that certain establishments were inflating the cost of mourning related staples, the Mayor publicly announced that no such thing would be tolerated and prices again settled at more reasonable levels.

Typical items and their costs, as reported in the New York Times:				
Black Ribbon	Formerly	$0.05/ yd.	Inflated to	$0.25/ yd.
Floral Wreaths		$2.00-3.00		$10.00-15.00

Gouging, by morticians hawking clients at the morgues, was immediately handled by police intervention. Schemes ranged from inflated pricing to the insistence that bankbooks or insurance policies be handed over as security of payment. Then suggestions of additional services (like bands, mourning trappings, many coaches, etc.) often followed, even when there were orphaned children whose care needed to be considered. In several cases, valuables were retrieved by the police for families who had handed over such collateral against their payment for services. This led to discussions, by the New York Association for the Improving of the Condition of the Poor concerning the need for the establishment of a company of undertakers which would help the needy and render aid without extravagant charges so that the poor would not be victimized by unscrupulous morticians in the future. In several other incidents concerned citizens, who suspected individuals presenting themselves as soliciting funds for the Mayor's Relief Fund or a private collection, called in the police. In each case, the citizen felt that something was amiss and that the solicitor had no intention of handing over the contributions he had collected. All such monies confiscated by the police did, indeed, find their way into the official charities for the benefit of the *Slocum* victims and survivors.

In August of 1904, the Mayor's Committee disbanded with an unspent balance of about $17,000, after setting up a trust fund for the children orphaned by the disaster and making some hardship grants. In the September report of the commission to the Mayor the following statistics were listed:

120	Lost their whole family
597 out of 956	Victims had been identified (359 left as missing/unidentified).
$200,000.00	Had been found in cash, jewels, and bankbooks on the dead
$224,205.80	Was collected by the joint fund; $20,000.00 remained unused and would be set aside for the future needs of those who were made to some extent, dependant by the Slocum disaster. This sum was turned over to the committee at St. Marks church who would hold it in trust. Out of this they received $1,500.00 stipulated for the relief of two Catholic families on the list.

On 14 February 1906, out of the $20,213.00 that was received from the city, St. Marks still held $9,769.00. The church committee filed suit in Albany, New York asking that, since the unnecessary funds could not be returned to the contributors, many of whom were anonymous or had given only partial addresses, the Supreme court should take in hand, for accounting purposes, the remainder and that the court should then pay over to St. Mark's the same sum to use in such ways, as they deemed proper, since the survivors were no longer drawing it down. Attorney General Meyer was made a party, representing the unascertained beneficiaries. This was a novel situation in New York law.

By 20 May 1906, the Trustees of the Committee of Lutheran Ministers who had charge of the Relief Committee Fund, with a remaining balance of $9,008.75, were divided. Some thought that the monies should go to St. Marks, since survivors were not drawing it down but, others were opposed. The trustees appealed to the State Supreme court and Justice Blanchard handed down his opinion. "Under a 1901 law, trustees must administer the fund for twenty-five years and the wants of sufferers must be relieved upon appeal at any time during that time. At the end of the twenty-five years, the remainder and any interest might be adjudicated on the filing of another application. Since the fund was a trust and subject to court control, it had to be held for twenty-five years. This balance was separate from the $125,000.00 collected by the Mayor's committee. The church had received over $20,000.00 directly and many of those donations were anonymous. Some donors stipulated payment to St. Mark's church (this aggregate equaled $7,691.25), and had already been distributed to the church. Direct payments to sufferers had amounted to $3,300.00 and no applicant had been denied."

Appointed by New York City Mayor, George B. McClellan::	
Herman Ridder	Chairman & Ex-Officio of the Sub-Committee for Disbursement of Funds
Jacob H. Schiff	Zionist & Philanthropist, Treasurer
H. B. Scharmann	Assistant Corporation Counsel, City of New York, (Temporarily assigned to the committee as chairman of the Sub-Committee for Disbursement of Funds)
Hubert Cillis *	President of the German Society
Thomas Mulry *	Erskine Hewitt *
Robert A. Van Cortlandt*	John Fox 8
John Weinacht	Morris K. Jesup
Charles A. Dickey	Joseph C. Hendrix
George Ehret	Julius Harburger
Isaac N. Seligman	Eugene A. Philbin
Louis C. Maegener	John Crane
Representatives of the Lutheran Congregations attending the Allied Committee;	
Rev. Dr. J. J. Heischmann	St. Peter's Lutheran church, Brooklyn, Chairman
Rev. Dr. Jacob Loch	Brookly;n
Rev. Dr. Emil Kraeling	Zion Lutheran Church, Brooklyn
Louis W. Kaufmann	A publisher

The Charities Department yearly report, to the City of New York, for 1904 revealed the following statistics concerning the *General Slocum* disaster:

- 172 people were actually from Germany, as opposed to sec-ond or third generation German-Americans
- 19 babies were recovered, only 2 survived
- 25% of children between 1 and 5 years of age survived
- Henry Gerdes, age 80 of Brooklyn, was the oldest man to die
- Mary Werter, an 83 year old widow from E. 12th St. was the oldest woman to die
- Margaret Herbolt, age 76 of 416 E. 5th St. was the oldest female survivor
- Stephen Erklin, age 7 weeks and Adella Liebenow, age 6 months were the youngest survivors
- $7,834.07 was spent to cover the expenses of the department in caring for victims of the disaster
- $ 950.00 was spent separately on the Special Committee's con tract with the Metro Equipment Co. to drape the city Hall (interior & exterior) in crepe for 30 days

Per Dr. Darlington, of the Health Department, 1,200 individuals were lost and of these, hundreds would not be recovered. Perhaps 100 victims had not been reported or identified, such as girls recently arrived here as domestics; husbands and wives who were childless and had no one to report them.

The Globe & Commercial Advertiser, on the 16th of June 1904, reported that perhaps fifty bodies were claimed before reports were made and so were not counted. All of these were removed from the Alexander Avenue Stationhouse. Another newspaper reported that bodies floated so thickly between the *Slocum* and the shore that they could be used as a bridge. This was somewhat of an exaggeration. But another report that one tenement building, which was divided into fifteen apartments, had seen fifteen families destroyed was much closer to the truth.

Recently completed research in the Wagner College's Lutheran Synod of New York Archive revealed a somewhat higher total of victims than the city's official number. According to the notes of the Ministerium of New York for 1904, 1,329 congregants had boarded the *General Slocum* and 922 had died. This pushes the total of passengers (congregants plus friends, other relatives, etc.) who boarded the ill-fated ship well over 1,500 and perhaps as high as 2,500.

As for any claims that the survivors' and the victims' families might have brought in the courts over the *General Slocum* disaster, there was virtually no hope. Per the United States Revised Statutes, the liability of the owners of the wrecked steamer was limited to the value of the hull, cargo, and engines after the wreck had been raised, *if the vessel was seaworthy*. If the vessel was not seaworthy, its owners were liable for the damages awarded by the courts. The *Slocum* was raised and the hull sold for $2,800.00 Out of this sum the wharfage and other charges had to be deducted first, leaving about $1,200.00 to $1,300.00 per claimant. According to the city, the total amount of claims was $1,475,673.00. The city divided the claims into three categories;

Claims for deaths	$1,361,152.00
Claims for personal injuries	80,000.00
Claims by the City of New York	34,521.35

Henry Weidemann, a barber from 79 E. Houston Street, filed a $50,000.00 claim for the loss of his wife, Caroline. His son also filed a suit for the loss of his wife, Helen. Henry's son-in-law, Emil Reichenbach, also filed for $5,000.00 in the loss of his 2 ½ year old baby. From this extended family, only one member who had been aboard, a daughter was saved. By the city's estimate, this family would have been entitled to about $3,900.00

---

* Members of the Sub-Committee for the Disbursement of Funds

[1] This organization was something like a private clearing house for charities, watching out for con artists and organizing joint efforts when needed.

[2] It has been estimated that of those considered "poor" among the congregation, ¾ accepted some aid; among the "middle class" survivors, only about ¼ of those who were eligible accepted assistance

[3] According to Northrup, Mayor McClellan cabled Sir Thomas Lipton, thanking him for his offer of a donation, but refusing on the grounds that the city would take care of its own. This seems to fly in the face of other donations accepted from other states and other countries.

THE GENERAL SLOCUM SURVIVORS' ASSOCIATION

In July of 1904, just fifteen days after the disaster, the Organization of the Slocum Survivors was formed. According to the Certificate of Incorporation filed in Albany, its mission was... "To assist and help one another in their trials and troubles," which in 1904 was interpreted to mean any kind of support including, if necessary, becoming the legal guardians of the orphans and those dependents deprived of support by the disaster. The Organization would "...commemorate the occurrence of the catastrophe annually; promote the welfare of and assist those who have suffered as a result thereof; to use all proper means and methods for the prevention of any future like occurrence; and in that behalf to petition the legislation for appropriate and proper legislation."

By the end of 1904, bi-weekly meetings were being held at Niederstein's Hotel [now Restaurant].* According to Munsey's Magazine, the association represented the dead and the seven hundred relatives of the dead; all were mourned by the three hundred members. Later, the group moved to Mozart Hall on 86th Street in Manhattan and met monthly.

Charles Dersch was elected the first president. (He had lost a wife and daughter. When he died in 1938 his only close relative was his brother Fred.) Other officers were similarly affected. The Secretary had lost his wife and two children, the Treasurer his wife and four little ones. In a rare public statement, Dersch expressed the position of the officers "We are three homeless men, but have suffered no more than many others." Concluding the same interview with Munsey's, Mr. Dersch then stated; "They call us good-natured Germans, but we have determined that our dead shall be avenged and that others shall not be destroyed as ours have been."

Within a few months of the group's incorporation, these three men, who lead the group for almost thirty years, began their campaign. First, they loudly backed Congressman Sulzer, in his denouncement, of the reinstatement of Edward Van Wart's License as Pilot by supervising Inspector Rodie and his request that the Department of Commerce and Labor investigate the matter. The officers of the organization remained so focused on their aims that at one time, after Captain Van Schaick's incarceration, when rumor spread that he was running a ferry from Sing Sing across the Hudson, for the prison, several representatives of the group traveled to Sing Sing to oppose his privilege, which, in fact, turned out to be false. (Sometime later, he was put in charge of a small boat called the *Bristol*, which carried convicts and provisions back and forth to Bear Mountain while a site there was being prepared for a new prison.) Ever vigilant, lest any individual or organization get by with a reduced penalty for their implication in the disaster, the association actively opposed the early release (parole) of Captain Van Schaick.

In 1910, Congress had passed a law providing a Board of Parole who could approve parole for any prisoner of good conduct after one third of his sentence had been served. Van Schaick would fall into this category in 1912. The week before his parole board meeting, Dersch and several members called on U. S. District Attorney Wise for his support and were met with a cool reception, even though he had prepared the case against Van Schaick. It seemed he now favored the parole and he intimated that Judge Thomas thought along similar lines. A letter and telegram were sent to R. V. Ladow, President of the Board of Parole, serving notice of the association's intent to oppose the parole. Then Dersch addressed the membership;

> "... it is not malice at all. This man was the only one punished for the crime that broke up so many of our homes. No one criticized our zeal for that. Now we insist that his sentence be gone through."

A copy of the Association's resolution was sent to Ladow and President Taft.

At a meeting in February, 1911, about 100 members of the Association officially adopted a resolution protesting the Captain's parole. He was the only person actually punished under the law; therefore, he should see no mercy and serve his full term in spite of efforts by his wife and the Association of Masters, Mates, and Pilots, who actively lobbied for his release and/ or pardon. President Dersch reminded the group of President Roosevelt's conviction that he would not help a justly convicted criminal escape punishment. On 11 August 1911, after three and one half years in prison, the captain was paroled.

Through the years, the meetings both social and civic continued. Those records which survive, at the New York Historical Society and the New York Public Library, are incomplete at best, but follow the group up to the 510th meeting on December 5, 1961. This was the last meeting of the original Association. Over the years, the membership dwindled until, in those final years, only a handful of members still attended.

In the cover of a notebook used to record the membership and the memorial dates for those who died, in an unidentified hand is the following, written as if the author feared that he might forget and so sought a more lasting personal recollection:

"Losses Personal
 Mother – sister – brother
 Relatives
 Oellrichs
  Aunt – Helen 19 mo.
  Eliz. 3 Fred 6 Wilh 4

                    Behakus
                     Anna 13
        Total – 9 in all 3 our family   6 relatives"

The author would like to add two more entries in the diary of survivors: Catherine Ullmeyer (Gallagher) Connelly, who died October 2002, age 109 and the last survivor of the *General Slocum* disaster, Adella (Liebenow) Wotherspoon, who turned 100 years young in 2003; she died in January of 2004 at the age of 100 years.

Both of these ladies were tireless in their strivings to keep the memories of the disaster alive. They faithfully attended the services in Queens and Manhattan and retold their families' stories to anyone who asked. Neither one dwelled on the past but, both created lives richly filled with family and happy memories.

The General Slocum Memorial Association, organized decades after the dissolution of the original association, was created in association with the Queens Historical Society. Primarily populated by the descendants of the disaster, the group continues the tradition of a memorial service in Middle Village, Queens, and a grave side wreath-laying at the monument in Lutheran All Faiths Cemetery. Another yearly service, held in Tompkins Square Park, Manhattan, at the monument dedicated to the children lost in 1904, is organized by a committee of descendants and interested citizens in conjunction with the city's Department of Parks and the Maritime Industry Museum.

A WORD ABOUT THE APPENDICES THAT FOLLOW

The story of the *General Slocum* did not end on June 15th, 1904. Nor do the preceding sections of this book adequately illustrate the contributions of the many hundreds of New Yorkers who cared enough to help in one way or another. Beyond the few family stories, which give us a glimpse into that day from the perspective of the ones who tried to help, there were many, many more which are only known by a line in the newspapers of the time. The police, fire, and medical personnel, divers, and private citizens have been grouped according to their association with the disaster in the appendices. It was estimated that fifty vessels from schooners and steamers to rowboats helped in the rescue and recovery work. Many fire men and police officers did round-the-clock duty for several days and all whose names could be located have been listed. Many individuals received official recognition for their acts of selfless courage. These folks, many of whom refused the gifts from grateful survivors' families or the official recognitions of various organizations, or the city as "just doing my job", deserve no less than the heartfelt thanks and remembrance of us, the descendants of those they labored to save. We can only wish that we knew all of their names.

For those readers who wish to delve deeper into the legal ramifications of the disaster, several sections of the appendices expand on the coroner's Inquest and the several trials, suits, etc. arising from the disaster, including the entire reformation of the Steamboat Inspection Service.

At the conclusion of the Centennial Voyage of Remembrance to North Brother Island, a wreath of 61 red carnations was set adrift over the place where the *General Slocum* grounded and burned. As the wreath floated seaward it was flanked by white carnations carrying the prayers and thoughts of several hundred descendants.

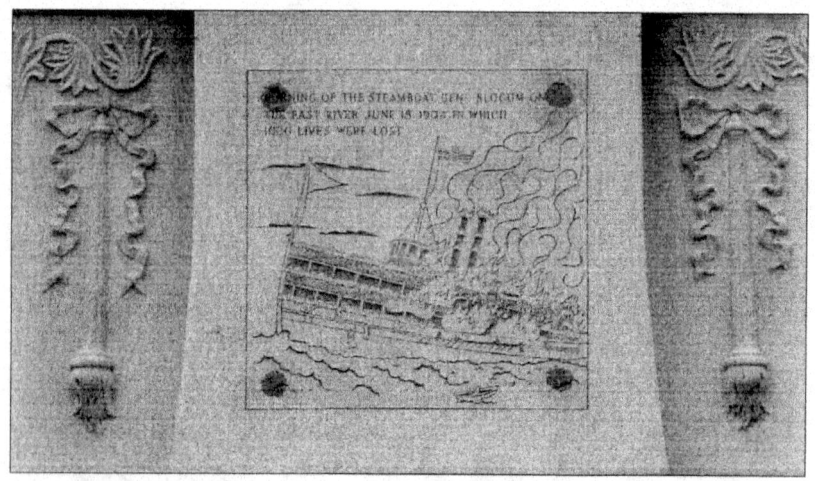

This engraving on the face of the General Slocum Monument, Lutheran Cemetery, Queeens, New York, was commissioned to replace the original bronze plaque after it had been stolen and returned. The scene is taken from the sketch of a survivor done soon after the disaster.

While technically inaccurate, it is no less emotionally accurate.

APPENDIX

Passing through Hell Gate
    The Navigation of Hell Gate   104
    Map of the Route of the *General Slocum*   145
    Timeline to Disaster   146
    The *General Slocum*   149
    The *Grand Republic*   151

The Work of Heroes
    Without a Thought for Themselves   151
    Vessels of the Rescue & Recovery   154
        Fireboats
        Ferry Boats
        Steamships
        Tug Boats
        Dredges
        Derrick
        Vessels of the 1st Battalion New York Naval Reserves
        Vessels of the U.S. Navy, Brooklyn Navy Yard
        Privately Owned Vessels
    Medical Facilities & Personnel and Other   168
        Departments and professions Assisting
        in the Rescue and Recovery
            Riverside Hospital
            North Brother Island Lighthouse
            Riker's Island Hospital/ Work House
            Bellevue Hospital
            Lincoln Hospital
            Lebanon Hospital
            Harlem Hospital
            Roosevelt Hospital
            New York Hospital
            St. Vincent's Hospital
            Other Hospitals Mentioned in the Press
            Department of Docks
            Merrit-Chapman Derrick & Wrecking
            Other Institutions and Professions
    New York's Finest   178
    New York's Bravest   182
    Private Citizens   183

Roll of Honors
    Department of the Treasury Life Saving Medal   185
    U.S. Volunteer Life Saving Service Medal of Honor   185
    New York City Police Department Medals   186
    Members of the Force to be Recognized by the Volunteer   187
    Life Saving Service and by the Department for the Roll
    of Honor
    New York City Fire Department Medals   189
    International Recognition   189
    Unsung Heroes, Stories from the New York Press   190
    Above & Beyond the Call of Duty   192

Legal Repercussions of the Disaster

Federal Officials	194
City Officials	195
Steamboat Inspection Service, New York	196
Officers of the Knickerbocker Steamboat Company	197
Crew of the *General Slocum* Including Concessionaires and Band	197
The Scales of Justice	200
The Coroner's Inquest	202
Proceedings of the Federal Grand Jury	215
The Trial of William H. Van Schaick	218
The *Slocum* Inspectors' Trial	221
The Vendors' Trials	223
Before the House of Representatives House Committee on Merchant Marine and fisheries	224
HR 4154 – Legal Aid & Families' Suit	226
United States commission of investigation Upon the disaster to the Steamer *General Slocum*	228
The Steamboat Inspection Service	232
Final Thoughts	242

Bibliography 249

Other Sources for Research 254
Organizations
Libraries
Collections
Web Sites
Cemeteries
German Lutheran Churches
Vital Records

Lists of Persons aboard the General Slocum June 15, 1904 264

Validated List of Victims/ Survivors/ Missing Persons 266

Non-Validated List of Victims/ Survivors/ Missing Persons 356

Navigation of Hell Gate

> "[Hell Gate] At half tide, roaring with might and main
> "like a bull bellowing for more drink" but when the tide
> is full sleeping as soundly as an alderman after dinner";
> in fact, resembling "a quarrelsome toper who is a peaceful
> fellow enough when he has no liquor at all or has a skinful,
> but who, when half-seas over, plays the very devil!"
> -History of New York Vol. III

Since the 1600's when the first European seamen navigated through Hell Gate, this confluence of the Harlem River, the East River, Long Island Sound, winds, and tides has wreaked havoc on anyone foolish enough or brave enough to test nature at her worst. Still today, when the tide turns at Hell Gate, it turns twice.

The East River is not a river; it has no natal origin of its own. The Harlem is not much better. Both of these watercourses are more like canals whose direction is dictated not by the earth's gravity but by the moon's tidal pull. With the turn of the incoming tide, the northward surge flows up along the eastern side of Manhattan for sixteen miles, winding along until it meets Long Island Sound near Throggs Neck. At the same time, the rising tide flows up the Hudson River along Manhattan's western shore and into the Harlem River, which flows out into the East River. Two hours after starting north, both currents meet in Hell Gate. This head-on collision causes eddies and rip currents that whirl around each other like water poured from a huge unseen pitcher. The basin of Hell Gate fills, which causes the Harlem River to reverse and run back toward the Hudson as the East River races up through the Bronx Kills, Little Hell Gate and the Bay of Brothers toward the Long Island Sound. While these twin tidal currents are running northward along the Manhattan shores, another tidal flow follows the northern shore of Long Island from east to west. It enters Long Island Sound near Orient Point and follows the coastline until it pushes into the East River, also at Hell Gate. Four hours after the tide's turn at the entrance of New York Harbor, the current from the Sound and the currents from Manhattan converge.

Then the Sound current runs south for the open Atlantic beyond New York following the East River in reverse. There is never calm water in Hell Gate; the modern name, a corruption of the Dutch, is an inadequate description of the complexity of tides running over and under each other each time the tides turn. The British navigators of the eighteenth century compared it to the Norwegian Whirlpool infamously referred to as the "Maelstrom." Even this is not adequate. Besides the currents, which reverse bihourly, rocks and reefs lie below the surface waiting for unlucky mariners. Nineteenth century navigation charts only hint at the madness of traversing Hell Gate and are peppered with names like Long Pot Rock, the Grid Iron, and Frying Pan Ledge. In the days of sail, hitting one of these underwater obstacles could destroy a ship in minutes. Once helpless, a ship could easily find itself in the grip of a tide pushing it several miles before it either grounded or sank.

As New York grew, so did the dangers. The occasional vessel of earlier times grew with the city's size and increasing commerce until at least 500 ships per year sailed through Hell Gate in the early 1800's. Steamboats, which initially drew rye smiles from New Yorkers as a novelty of the day, proved their value with their initial runs through "the Gate." Belching great clouds of black smoke in their wake, these ships could generate power at will and travel in a straight line through the East River channels. Power meant speed to forge through currents, without tacking in the wind, plus greater maneuverability around obstacles. With an even greater growth of commercial shipping on the East River in the 1830's, the city was outstripping its ability to reduce risk further. More ships in already tight sea-lanes began increasing the inherent dangers of the "Gate." Two ideas that competed for support were either to train a cohort of pilots to take all vessels through the passage, or to remove the underlying obstacles. Until that time, ships approaching New York waited in the Lower Bay off Sandy Hook, New Jersey for a rising tide to come into the port. Those ships wishing to hasten the completion of their crossing, entered off the eastern end of Long Island and followed the Sound to Hell Gate; then south on the East River. By mid-century, steamers had replaced a great many of the Clipper Ships thus reducing some of the risks of a sea voyage. But to risk any ship in so dangerous a passage as Hell Gate, was costly, furthermore it was impacting New York's claim as the premier port of the nation.

Finally, New York decided to tame the capricious channel. By the 1850's, upward of 1,000 vessels a year grounded in the "Gate." The deadly bottleneck would be destroyed once and for all. Surveys were done between the 1840's and 1851, when the first chart of the area was produced. What it showed would make even modern navigators quake in spite of their sophisticated electronics. At the eastern end, in mid-channel, lay a shelf of rock only eight feet below the surface and extending out over one-hundred and thirty feet into the waterway at cross-angles to the current. Where the flow followed the river at Hallett's Point, the rocky substratum created such a violent current that those vessels traversing the "Gate" had to go nearly up on the rocks of the point. Vessels moving in the opposite direction, toward Manhattan, had only a slightly easier time in the Middle Channel, which was the main ship channel, and which had only nine feet between the keel and another ledge. At flood, the currents ran through Hell Gate at ten knots.

Work began on the removal of the worst of the obstructions, but was halted until the end of the Civil War. In order to destroy the Hallett's Point rock ledge, galleries were drilled into the rock, the water was pumped out, and only thin supporting columns were left to support the upper layers of rock until the manmade caves could be packed with explosives and detonated. The resulting explosions, of the 700-foot by 300-foot ledge, sent a column of water fifty feet into the air. Between 1869 and 1876, the digging and blasting of smaller formations essentially eliminated Hallett's Point and reduced some of the hazards to navigation. But the "Gate" remained treacherous. The bottom had been lowered to only twelve to eighteen feet below the surface, but this was not uniform, outcrops still existed. And after all that work, the tides still drove ships down the Narrows and onto the remainder of Hallett's Point Reef or Flood Rock. In 1885, engineers turned their attention to Flood Rock, a nine-acre section of substrate, which showed only about two hundred square feet above the water. When the "greatest manmade explosion" went off, the ground moved with a "sickening jar" before waves generated by the explosion could reach the shore. Rocks rained down and then subsided. The years of work had been costly, but the loss of shipping during the same period had averaged two million dollars a year. The cost was worth it, navigation had been improved, but "Hell Gate" still had not been tamed.

In 1904, the currents of Hell Gate still ran one on top of the other in opposite directions with each tide turn. The waters still raced through with such violence that swirling rip tides were formed and rebounded off the neighboring islands. In that centennial period of heady exuberance over man's ability to best nature and bend the natural world to his desires, Hell Gate was still traversed with trepidation and at full throttle. Hell Gate pilots were known for their skill. To move cautiously was to give up what little control a pilot had to guide his ship around the thousand demons of the channel. Each knot of current speed had to be met with equal engine speed. The alternative was a splintered destruction among the shoals, rocks, and natural seawalls. On June 15th, 1904, the tide was running up the channel at full force toward an 11 a.m., high tide. The *General Slocum* was running at ten knots (11.5mph), which was something near her top speed. The headwind from the north may have slowed her somewhat, but she was doing nearly fifteen knots with the help of the tide as she entered the gate; and approximately 18 knots (20.7 mph) as she pulled clear. She must have shot out like a bullet. By the time the captain and pilots had breathed a sigh of relief, she was abreast of Sunken Meadow. Turning or stopping would have been difficult. A ship does not stop like an automobile; there are no brakes per se. To turn into a particular pier, island, etc., maneuvers including slowing the engines and throwing one into reverse, while the other remains in forward gear, must be completed to begin the turn. As the ship rounds the turn, the engines must again be precisely coordinated, only then can the helmsman slow her forward progress by throttling down. While making the turn, the ship may become unmanageable in the wind, heavy surf, or strong tides. For the *General Slocum*, turning broadside would probably have made her slow to respond to the helm and possibly spun her around as the tide ran north and the wind pushed south. She might have cracked her hull on submerged rocks in deep water, since she was fully loaded and drawing about an eight or eight and one half foot draught. There was no good solution, having committed to the channel; the only logical way was straight ahead until the way was clear to make for a place to beach.

After the tragedy, one of the newspapers chartered a boat and an independent but experienced pilot to follow the captain's course. Pilot Van Wart, who had been at the helm on the 15th, advised which course he had followed. Aboard were reporters and photographers to record any likely landing before North Brother Island. Lead lines were continually cast to check the water's depth against the most recent navigational charts and firemen worked to determine the speed and progress of the blaze. Their conclusions stunned many New Yorkers; the *General Slocum* had nowhere to go! Their estimate was that from the time the blaze was discovered until the captain had to assess the severity of the fire and set a course would have been no more than one and one half to two minutes. They stated that if the vessel was midway from Ward's Island and exiting the Gate, when the fire was discovered, she would have been at Sunken Meadow before the decision to beach on the Meadows could have been made. The news crew also determined that there was no place at all to ground on the starboard [right side, Queens shoreline] of the channel or near the end of the Sound closer than the Brother Islands. The commission impaneled by President Roosevelt came to other conclusions.

In a rare "years later" interview, with the New York Times, Albert F. Frese, a teenager in 1904 just starting his career with Funk & Wagnall Publishers, reiterated his experience that day. His is one of the more lucid accounts. This telling is from 1954, and it sounds like the plot of a nightmare.

> "I saw the flames leap from the forward deck. No one screamed or cried out. Not at first. Faces froze. Then the captain began pushing the boat. The whole boat vibrated and shook. A man ran by 'don't go down there, it's a furnace down there!' I never saw him again. People on Blackwell's Island ran to the beach with make-shift rescue gear hailing the *General Slocum*. I went over the rail in a front first dive. None of the gang was with me."1

Even in the telling of what seems a coherent story, there are discrepancies. Different witnesses claimed the first sight of smoke or fire from as far south as about 50th St. others as far north as 97th St. or, in one case, 102nd St. One passenger says that the boat began to vibrate and shake at 86th St, another at 145th St. There are many such stories, most far more incoherent or fragmented than the one above.

Determining exactly what happened and when during the ill-fated trip of June 15th, must have been extremely difficult for those whose job it was to illicit the truth from dazed and injured passengers, frantic crewmen (some of whom undoubtedly recognized their own jeopardy even before they were approached by the coroner or the police), and well-meaning passers-by who did not realize the full impact of what they were seeing. With so much going on, even trained observers would have found it difficult to determine precisely the actual sequence of events. It should also be noted that among the many testimonies are several from observers on the eastern Queens shoreline which state that up until the General Slocum turned toward North Brother Island from 138th Street, there was no smoke or flames to be seen. This is because they would have seen the starboard or right side of the vessel. The starboard side did not begin to burn until the Slocum was nearly on the rocks. In fact, some rescues were accomplished from this side of the ship in relative safety. These accounts have not been included below for the reasons stated here.

From the day of the disaster through the various legal proceedings of the following week, every paper and legal panel tried to answer the all-important questions; what did the captain and crew know and when did they know it? The two figures of primary concern, Captain Van Schaick and Edward Flanagan, instantly became the most infamous men in New York. To this day questions regarding the behavior of both men during the infamous hour leaves one with as many questions as answers. The conclusions of the Coroner's Inquest jury and the Grand Jury, which led to prosecutions, do not entirely or satisfactorily answer those questions. The captain changed his testimony once at the beginning of the legal proceedings. This may be do to his advanced age and the shock of the disaster plus his injuries. He had been severely injured and was clearly suffering during his testimony at the coroner's Inquest. These injuries would follow him for the next two decades as a limp from a broken ankle and heel, a painful back, and the loss of one eye from burns. On the other hand, Flanagan changed his story for each recital. He continually sought to shift blame from himself to anyone else at hand. His obvious performance so infuriated the jury of the Coroner's Inquest that they recommended that he be brought up on charges for perjury and cowardice, but he never was.

Secretary Cortelyou's commission also noted that overall there were discrepancies in the testimonies of those passengers on the Main and Hurricane Decks and the men in the pilothouse. Based on the speed of the vessel, they determined that this discrepancy in time was three to five minutes and covered about three quarters of a mile in distance between when each group first became aware of the fire.

In order to establish as closely as possible the actual time/ distance relationships along a timeline, the author has compiled the table below. All available accounts were laid out first by time (where noted by a witness); then by distance as measured by blocks along the Manhattan and Bronx shorelines. Because the "cross-streets" of Manhattan, and to a lesser degree in the Bronx, are laid out in a pattern of twenty blocks, moving south to north, equals one mile; the time and distance between witness accounts could be determined in a relatively accurate way. The progress of the fire, based on witnesses' accounts was then laid in to the table. Next, the captain's and the crew's accounts were added in. Finally, the whole table was compared to the accounts in the Report of the Commission on the Sinking of the General Slocum (Secretary Cortelyou's commission). Italicized notes between the time/street/event breakouts have been added to give an idea of what the probability of rescue for both passengers and ship was at a point in time. These represent only the author's opinions.

It is also the author's conclusion that if everything possible had been done properly and promptly, the outcome would have been much the same. Once the ship had passed Blackwell's Island, little remained to be done regardless of where the ship was beached.

June 15, 1904 Marine forecast: Low tide 5 am & 5 pm
                                 High tide 11 am & 11 pm

                                 The wind was from the Southward, a
                                 moderate breeze on the starboard
                                 quarter of the bow.

---

[1] He was rescued by a tug.

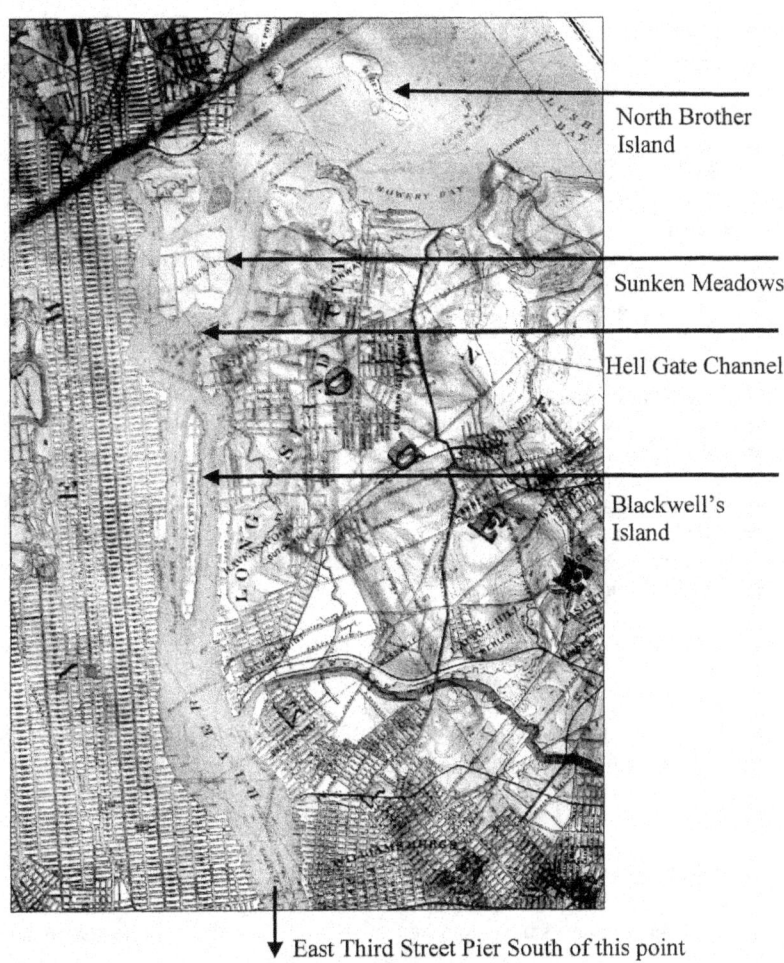

Timeline of the Disaster

9:20/ 9:30 North of 3rd St. 7 Mi. to NBI As the *General Slocum* backs out of the pier and begins to cruise north on the East River, Edward Van Wart is at the wheel; Edwin Weaver is also in the Pilothouse; Captain Van Schaick is in his cabin.

9:45 50th – 55th Sts. *Fire has started below decks, but all aboard are un-aware.* Deckhand McGrann tells Flanagan "There is fire forward and it's got a good start." **Charles H. Lang**, on board the *General Slocum* with his family, overhears two crewmen saying that there is a fire aboard. He looks at the Manhattan shore and knows they were between 50th and 55th Sts. by a brewery he recognizes. At this point he begins planning his family's es-cape.

9:57 83Rd ST. Passing Hallett's Pt. Lighthouse 3 Mi. to NBI *As the excursion moves north, a little smoke begins to curl out the portside porthole, which has been left open.* **Frank Perditsky,** From his vantage point forward of the pilothouse, on the hurricane deck, looks down and sees smoke coming from below the bow. He runs to tell the Captain about the smoke at 83rd St, he knew the street by recognizing a park "he knew well." The captain dismisses him, thinking his report of fire is a hoax. Van Schaick goes to the pilothouse between 83rd and 86th St. to take the boat through Hell Gate. The law mandates his presence in the pilothouse during the passage.

Passengers and ship could have been saved if the ship grounded or docked before entering Hell Gate.

9:58 86th- 89thSt. Passing Blackwell's Is. Captain Van Schaick starts to push the ship forward; the whole boat vibrates and shakes. People flock to the beach on Blackwell's Island with makeshift rescue gear, hailing the *Slocum* to pull in. The captain, concentrating on the water ahead, does not see them.

9:59 90th St. - 92nd St. Entrance to Hell Gate *As the Slocum enters Hell Gate, Fire is running along the portside toward the stairs to the upper decks.* **Jacob S. Jacobs** sees smoke coming from a cabin near 90th St. **John Engleman**, a tug man onboard the *General Slocum* with his family sees fire. After entering Hell Gate, Flanagan <u>thinks</u> he sees fire and immediately tells the pilothouse via the speaking tube.[1] Capt. Van Schaick has the wheel. A reliable witness on the Hurricane deck notices fire off the Marble Yard at 92nd St., one block from his home.

10:01 96th - 97th St. Exiting Hell Gate West of Negro Point Bluff, Ward's Is. *Fire is coming up the stairs to the upper decks and taking control of the Main Deck.* A man runs past **Albert Frese** shouting "Don't go down there, it's a furnace down there!" Albert never sees the man again. This may have been Flanagan. **Albert Frese** decides to jump. At a point opposite 97th St., several of the crew, on the Main Deck, see puffs of smoke coming through the seams in the flooring above the "second cabin."

It is unlikely that all the passengers and the ship could have been saved beyond this point

10:02 98th - 98th St. Ward's Is., six lengths eastward of Sunken Meadow *Fire is in control of the Main deck bow, the Promenade, and the Hurricane Decks*. The captain testified later, that he left the pilothouse and descended to the first deck to see how far advanced the fire was and then returned to the pilothouse. His estimate was that it took him one to two minutes.[2] Secretary Cortelyou's committee decided that he was out of the pilothouse no more than one minute at most, and so could not have gone to the main deck and back. It is the author's opinion that he may not have been able to get to the lower deck due to the smoke and heat billowing up the companionway and the throngs of passengers racing up the stairs in search of safety. Whether or not he makes it to the first deck and back, the captain has no more than one and one half minutes to determine the severity of the fire and his course of action. Several passengers from the Hurricane Deck testified that the fire was known and a message had been sent to the pilots before the ship was past the east end of Ward's Island. This places the ship about one fourth mile west of where the crew stated it was. Their estimate of the ship's location was west of the eastern end of Ward's Island. On north Brother Island, **LuLu McKibbons**, hospital switchboard operator, sees the *General Slocum* between Ward's Island and Sunken Meadow trailing smoke from her port side and making for the island. She begins calling every police and fire station on Manhattan for help.

10:03 102nd St. Exiting Hell Gate / Sunken Meadow The captain and crew stated that they knew of the fire from the speaking tube message about four to six lengths east of Sunken Meadow Spindle. Another account states that at about the same time, another boy, on the first deck, sees smoke and flames. He tells Coakley, who informs Flanagan, who calls Engineer Conklin and the Pilot Van Wart on the speaking tube. If the fire is as far advanced as noted above, these incidents could not have transpired on the main deck so late in the voyage. As the *General Slocum* turns toward North Brother Island, she is about 1,000 feet from the Manhattan shoreline. The commission felt that the men in the pilothouse had knowledge of the fire before the steamer passed the east end of Ward's Island, three fourths of a mile west of where the pilots said they knew, and therefore they could have beached either in Little Hell Gate to the west or in the Bronx Kills to the East of Sunken Meadows. Further, they could have gotten her stern into the wind thus pushing the fire off the bow and away from the passengers. However, Sunken Meadow is deep and the bottom is rocky. The ship would have to beach broadside to the swift current, which was pushing north and east toward the Sound. It is entirely possible that grounding here might have resulted in higher numbers of casualties and higher numbers of bodies being carried out to sea than at North Brother Island. **Patrolmen Kelk** and **Van Tassel** begin herding the passengers up to and toward the stern of the Hurricane Deck, they then try to control the panic with little success.

The ship is now passing the point of no return

10:04 104th St. 1 ½ Mi. to NBI Captain Van Schaick states that he learned of the fire about this point. [2]

10:05 Abt. 110th St., Lawrence Pt., Astoria By 110th St. Slocum is totally ablaze, passengers had twenty minutes to live. Flanagan calls the captain to report the fire. **Captain William Alloway**, on a dredge off Lawrence Pt., Astoria, sees smoke and sounds his whistle to warn the passing ship.

10:11 125th St. – 129th St., Casino Beach, L.I. 1 Mi. to NBI The flames rise to ten feet above the decks. The ship is enveloped in smoke; two thirds of her 250 feet are ablaze. The *Abram S. Hewitt* responds to an alarm of ship on fire from her Brooklyn Pier. The Bronx and Queens piers are lined with kerosene and oil tanks; by docking there, the *Slocum* might have set the docks on fire.

10:13 130th St. – 132nd St. Health Dept. Pier The *Franklin Edson* vacates her berth at the pier so *Slocum* can come in, but the captain seems not to have noticed. The *Slocum* is at her closest point to land (300 feet from the Bronx shore). Women and children are holding on below the smoke and fire." **Sampel Johnson** of the Astoria Hell Gate Branch of the U. S. Life Saving Corps. sees the *General Slocum* pass on fire and puts out in pursuit in a rowboat.

10:20 / 10:25 138th St., Bronx The captain states that he tried to come into a dock, but was waved off by a tug. Two units of the NYPD, Engine 60, and Hook and Ladder 17 deploy on the dock and wait for the ship to turn in so they can cover evacuating passengers with spray, but the *General Slocum* steams on. Above 138th St., the captain is observed fighting to keep the ship on course as if the helm is unresponsive. A man on a float at the foot of 138th St. describes the scene to the press; "She came up in a cloud of smoke and whistle screaming, the stern deck was black with people. The stanchions were burning through one by one. As she turned toward North Brother Island, she showed her starboard side. There was a line of hawser from the stern to the paddle box below the lower deck. It was hung with women and children from end to end.

The Bronx piers are the Slocum's last hope of rescue

Abt. 10:25/ 10:30 138th –140th St. North Brother Is., southern end, James J. Owens, a mason working on the laundry building, sees the flaming hulk making for the island. The following armada of boats begins picking up about one tenth of the passengers in the water. Captain Van Schaick testifies that he felt the ship buck and vibrate (possibly the first explosion) as he steers for North Brother Island. She speeds past the sand beach and rounds the point of the island where she grounds in a rocky cove. She is twenty-five feet from the seawall and nearly at the dock of Riverside Hospital. The *Slocum* beaches with her bow in seven feet of water. She is about twenty feet from shore; her stern is in thirty feet of water and is angled out to about fifty feet from shore. Almost all passengers' watches stop at this time, probably from impact with the water. The port rail collapses and catapults hundreds into the water. A muffled explosion is heard; red clouds of sparks erupt from below and the Hurricane Deck falls in. As the *Slocum* comes to rest, many more victims are thrown into a pit of flames amidships. The water surrounding the flaming hulk boils. Flames reach twenty to thirty feet above the top of the stacks.

Neither the passengers nor the ship can be saved

10:30 – 11:30 Abt. 143rd St. North Brother Is., northern end The captain is seen in the wheelhouse to the very end. His hat is ablaze before he jumps onto the submerged rocks of the island, injuring his foot, back, and ankle. Another four to six hundred victims die after the grounding; screams are heard on the far Bronx shore. All thoughts of rescue have changed to recovery. No one else is alive to save.

    There were many versions of these scenarios published before, during, and after the Coroner's Inquest. Both the captain and Flanagan changed their testimonies. The captain's reports may have been influenced by his age and the shock of his injuries, and were only changed once; however, it is clear when period accounts are read, that Flanagan changed his testimony for each recital. The Coroner's Inquest jury noted in its recommendations that he should be brought up on charges for perjury as well as cowardice, but he never was.

    The commission felt that overall there were discrepancies in the testimonies of those passengers on the Main and the Hurricane Decks and the men in the pilothouse. Based on speed, this discrepancy in time was
three to five minutes and covered about three quarters of a mile in distance between when each group first became aware of the fire.

---

[1]   Note how Flanagan's accounts change throughout this timeline.
[2]   This is the one inconsistency in the captain's testimonies –at what point did he descend from the pilothouse in order to evaluate the fire and his course of action.

The General Slocum
The Youngster of the Line

Launched April 18, 1891; 250 feet long; 70 foot beam, 7 foot draft / 8-8 ½   feet fully loaded
Hull : Devine Burtis Shipyard
Walking Beam & Engines: W. & A. Fletcher Co., Hoboken, N.J.
Upper Works: John E. Hoffmire & Sons, 808 5th St., New York, N.Y.

The *General Slocum* was built on the same 1878 plans as the *Grand Republic* although more modern and possibly safer options were available; The *General Slocum* conformed in all specifications, including safety features to her older sister. Passengers boarded on the Main Deck, which was primarily enclosed as an extended cabin; took refreshments on the second or Promenade Deck (this is where the band was seated on June 15th, 1904), then preceded up the stairs to the top or Hurricane Deck. Her only apparent modification was her seventy-foot beam. This measurement included a thirty-eight foot main deck amidships with an additional thirty-two feet added for an expanded passenger promenade area. She was longer and wider than other paddle wheelers in the harbor and was rumored to be one of the fastest ships around with twin thirty-one foot side-wheels capable of 18 knots (apx. 15 mph).

Originally built to carry luxury passengers to off shore races at Sandy Hook, the *General Slocum* had done so for her first several years in commission. But time was beginning to show. Her velvet appointments in the main deck salon were no longer bright and endless repainting on top of older paints gave her a dulled ambiance in spite of new varnish and well-polished brass. By 1900, luxury charters no longer requested her. Instead she had become the darling of the day-trip set. Church groups like St. Marks German Lutheran Church Sunday School loved her space; everyone could bring their entire family plus friends. Between May and the end of June, the *Slocum* carried charters nearly every day. For the rest of the summer she made ferry runs to various beachfronts around the city.

Over the years, in an effort to make up for the lost luxury business, the Knickerbocker Steamboat Company had requested and received several amendments to their seasonally renewable license, which increased their allowable limits to 2,500 passengers but occasionally she carried up to 4,700 without the authorities taking notice. Loaded as she was for the St. Mark's excursion, her draft would have been about eight and one half feet; over one full foot deeper than normal.

When an inspector was aboard, he confined his attention to the hull and the boilers. The *General Slocum's* eight lifeboats, two life rafts, hoses, buckets, and approximately 2,500 life preservers were generally given only a visual inspection. Perhaps one or two might be pulled down for a better look, but unless one disintegrated in the inspector's hands, it was doubtful that any negative comment would be made about their condition. Inspectors were paid by the number of ships they certified not by how well they inspected them.

THE GRAND REPUBLIC
A Grand Old Lady of Steam with 36 years in Commission

In an eerie case of history trying to repeat itself, on July 6, 1910, the *Grand Republic*, now owned by the Iron Steamboat Company, nearly mimicked her sister ship, the *General Slocum*. As she returned from the Rockaways on an early afternoon run, smoke began billowing up from the galley. Whistle screaming, she plowed through the Narrows at top speed. Her master, Captain Edward Carman, had decided to make for shore as swiftly as possible. She put in at the Crescent Athletic Club's pier in Bay Ridge, Brooklyn, with a small flotilla of would-be rescuers falling in behind. Even with the help of a flood tide and southerly wind pushing her in the right direction; it took six minutes to land where the fire department was waiting.

Unlike 1904, the crew had been drilled on fire and safety procedures, knew their stations and equipment, and were primarily concerned with the welfare of their passengers. Fate had also dealt a kinder hand on this trip. The *Grand Republic*, which was licensed to carry 4,000 passengers, only had about 50 aboard, having dropped her charter at the beach in the Rockaways.

One of the vessels, which came to the rescue, was the *Cygnus*, another side-wheel steamboat owned by the Iron Steamboat Company and mastered by Captain Van Schaick's son, also William H., known along the The waterfront as "Captain Billy".

Without a Thought for Themselves

In the first year after the *Slocum* tragedy, several authors brought out books which exploited the horrors of that awful day. They dwelled on the lines of bodies along the beach and neatly arranged on the hospital's front lawn. They included pictures of traumatized survivors huddled in blankets waiting for a ferry ride home. They emoted over the throngs of grieving relatives and friends at the doorstep of St. Marks. Over the years, other authors filled their pages with their versions of what happened, based on the newspaper accounts, which were often inaccurate at best. Although mentioned, none ever spent much book space on the rescuers. This is a pity. New York at the dawn of the twentieth century was not thought of as a very friendly place. From the pickpockets down town to the "Robber Baron" industrialists uptown, there seemed to be cut throats of every sort on every corner. The out-of-towner needed to be wary!

But on June 15, 1904, New York opened her heart, her arms, and her purse to her neighbors. While it was easy for a society matron or the neighborhood grocer to support the relief fund, one for the "bragging rights" the other because he knew the families involved, it was not so easy for the average man in the street to toss aside his daily business and at risk to life and limb rush to the aid of another. But that is exactly what the city did. From the boatmen in the harbor, who seeing smoke pursued the flaming wreck, to the cop on his beat who ran to the water, commandeered a row boat, and pulled for the inferno; each had but one thought –to help an-

The *Patrol*, the *Massasoit*, and the *Zophar Mills* work the rescue with a New York, New Haven & Hartford tug

other in need. Seldom has any city worked so quickly or so intensely for so long to accomplish what turned out to be so little. The stories abound of the rescuers who spent the rest of their lives trying to live down the nightmares, the pain of what we would refer to as traumatic shock from seeing so much carnage and feeling so helpless to halt the inevitable finale. When it was over, many were rewarded, some very handsomely for their efforts, some modestly refused all accolades. Just as the list of victims may never be reliably complete, so also will the Rolls of Honor for the heroes never be complete.

Each individual deserved the finest the city and in many cases the nation could give. Hundreds were awarded ribbons and medals; some received jewelry or cash from the grateful families of those they saved. Hundreds of citations were handed out at vast recognition ceremonies. Bands played and politicians shook the heroes' hands. Huge audiences looked on in admiration. The atmosphere was charged for months with a feeling of pride over the communal achievements. The outpouring of gratitude from the city to its citizens, and from both national and international organizations was overwhelming. But true to the hero ethic, almost to a one, they faded back into their daily routines, normal activities, and ordinary jobs, carrying on as if their greatest glory was helping a child to cross a street. They had done nothing, only what was right. But in an odd twist of fate, many of these same Good Samaritans would find fate less kind farther down the road. Several of the police officers, who risked their lives in flimsy row boats, or in the water, would die the victim of out of control drivers in the street. And far down the road, in the 1930's, many of the golden medals, so precious to the families of the heroes, were sold to put food on the table.

Today, the records of these worthy individuals' exploits are often difficult to find. Precincts and firehouses have come and gone. Unit designations change. Musty old books are boxed and archived somewhere in the city and no one repository can list every medal recipient or what kind of service each performed.

As the *General Slocum* steamed up river, trailing smoke from a porthole below her deck line, a fleet of water craft large and small were taking note and falling in behind. Many could not approach her speed and so held back, choosing to pick up the early jumpers. Others pursued at a closer range, first picking up the jumpers who were carried past them on the tide, and then rushing in to collect the living rain of children thrown over the rail by panicked parents and parents who jumped hoping to remain near their little ones. As back decks swamped and bows blistered in the heat, these mighty workhorses of the river saved many dozens at a time. As they moved in and out in time to the rhythm of the rescue, skiffs and row boats worked around them like packs of hounds snapping up a few living and a few dead on each pass, then racing off to unburden their cargoes on the tugs or on the shore before returning to their hunt. Amid the smoke and confusion, there were the scoundrels who tried to charge living victims before allowing them in their boats, and the ghouls who cut off fingers for the golden bands or ripped the earl lobes of the dead for diamond studs, but these were only a few among the hundreds who sought to save.

On the island, the heroics of doctors, nurses, orderlies, and "ward helpers", some of whom were hardly more than children themselves, became overnight legends; like the young girl who suddenly learned how to swim *as she swam out* to catch babies thrown overboard by mothers at the point of incineration. Everyone took part. All able hands, and many barely recovered from the sick bed themselves took part in the rescues. So heart wrenching was the scene unfolding on their doorstep, to the contagious disease patients, that they had to be restrained by locking the wards lest in their desire to help they infect everyone with whom they came in contact.

    Virtually every police precinct and fire house in the Bronx and Manhattan was represented. Officers on their day off or completing their twenty four hour tour on reserve in their stations came by whatever means could be found and worked for days, around the clock without rest. Below are the lists of river craft, medical, police, fire, and citizens who responded to the call and have been found among the scattered records of the city.

Vessels of the Rescue & Recovery

Fireboats:

Zophar Mills
Built in 1882,
Engine Co. 51
120' long x 25' beam, 12' draft, and could pump 6,000 gal/ min. She was named for a 19th century volunteer fireman who initiated the fireman's Fund for widows and orphans. When she

was commissioned, the *Zophar Mills* was designated the first floating fire engine. The *Mills* was stationed across from North Brother Island at either 90th or 96th Street. As the *Mills* arrived on station, one of the firemen aboard saw Mrs. Christina Gessman (27) of 114 E. 4th Street go overboard from the *General Slocum*. Without stripping off his gear, he dove in and grabbed her but, before he could return to the fireboat, she clenched him in such a grip on his head and neck that they both sank. When they surfaced again, she would not loosen her grasp and was about to drown them both, the fireman struck her in the face to render her unconscious. This opened a small wound on her cheek. He then returned to the boat with her and she was revived and taken to the hospital. Later the wound turned to erysipelas, causing her death.

When all chance of finding survivors was gone, Captain Conley ordered the *General Slocum* be towed off the shore and out of the way. Working with several tugs, the Mills hosed the flaming wreck as it was towed to an area one half mile from Hunt's Point in the Bronx.

Crew:

Thomas F. Conley, Captain
> The first fire company to arrive at the foot of 138th Street phoned for the Zophar Mills to pick them up. Captain Conley came whistling to the rescue from Ninety-ninth Street, where she was docked but, seeing the *General Slocum* already beached, she turned around and headed for 138th Street, where she took aboard Captain Geohegen and the reserves from the Alexander Ave. Police Station, plus Engine co. 60 and Ladder Co. 17 who had deployed on the pier in expectation of the *General Slocum* tuning into their location for evacuation. The *Mills* arrived on scene gathering victims and bodies along the way. These were picked up from both the water and other vessels. She then began pumping four streams of water on the fire, which was already out of control. The *Mills* then turned her hoses on the *Wade* to keep her from catching fire while lashed to the *Slocum*.

Fireman Joseph Mooney, Jr. had just transferred to the *Mills* from the *William L. Strong* the same week as the disaster and was credited with three rescues.

Fireman George Lawler saw a woman go overboard from the *Slocum* and begin to drift away. He swam after and caught her, then returned.

Abram S. Hewitt
Built in 1903,
Launched Nov. 7, 1903. She was 117' long x 25' beam x 9.5' draft, and could Pump 7,000 gal./ min. she had originally been put into service as Engine 77 and was stationed at Brooklyn.

Hewitt received the call shortly before noon to respond to the burning steamer. Before going to the scene she was detailed to meet Deputy Fire Commissioner Thomas W. Churchill, Chief of the Department Edward G. Croker[1] at E. 67th Street. The *Hewitt* ferried these officials to the disaster, arriving after the fire was out, and was ordered back to her berth.

Crew: No crewmen have been identified.

William L. Strong built in 1898, the *Strong* was 100' long x 24' beam, 12' draft and able to pump 6,500 gal./ min. and designated Engine 66.

She was named for the 90th mayor of New York, who took office immediately after the consolidation of Greater New York. Her station was at Grand Street, Corlear's Hook, on the East River just below the Williamsburg Bridge. The *Strong* worked alongside her sister ships in the recovery.

Her powerful searchlights may have been used to continue work after dark.
Crew: No crewmen have been identified.

Ferryboats:

Massasoit (Dept. of Corrections Ferry)
She was moored on the far side of the Brother Islands, and waited for the *General Slocum* to come to her. Because of her deep draft, she stood off the grounding and loosed small boats to the scene. After the rescue concluded, she ferried bodies to the Charity Pier morgue for 24 hours, 36 coffins stacked on her deck on each trip. The *Massasoit* earned four medals from Congress. Coroner O'Gorman prior to the Inquest interviewed her entire crew.
Crew:
Fredrick W. Parkinson*, Captain / Pilot (1872-1938), 430 Wales Ave.,New York, was a
 commander for 45 years and had been trained by his uncle, Captain Henry
 Rick.(see *Franklin Edson*) He was a seasoned Hell Gate pilot and before his career
 concluded, he had worked for both city departments and private firms and had
 supervised the construction of the *Correction*, the *Rikers Island*, and the *Hart
 Island* for the City of New York. On scene, at the *General* disaster, the captain
 directed *Massasoit's* efforts alternately from the smoldering inferno of the
 pilothouse and from a loop of rope swung over the side to facilitate his bringing up
 victims from the water. On her second trip from North Brother Island to the
 Charity Pier morgue, she encountered people clinging to flotsam in the river and
 overturned lifeboats. All of these were taken aboard. The captain and crew saved
 forty souls in all.
James J. Duane*, Mate, 329 E. 39th Street, Manhattan, launched a lifeboat which saved
 about seven victims and recovered eight bodies on its first trip. He could
 remain beside the *Slocum* because the deck crew kept a fire hose trained on him.
Albert Rappaport, Coxswain,* 108 E. 114th Street, "Redhead", jumped in and rescued eight
 persons, five girls, two boys, and one woman. While rescuing a teenage
 boy, Rappaport's underwear slid down around his ankles. The combination of
 this hobble and the bulk of the boy, nearly kept him from reaching a lifeline and
 being pulled back aboard. Two of his rescues were Minna Well of 1235 3rd Ave.,
 Manhattan and William Weel of 29 1st Ave., Manhattan.
William M. Hatch, Chief Engineer, 219 E. 88th Street, New York, received Volunteer Life
 Saving Service Medal for his work at the disaster.
Nicholas Ryan, Assistant Engineer, 418 Clinton Street, Brooklyn
James Farrelly, 170 Edgecombe Ave.
James Caffrey, Stoker, 26th Street, East River
John Corcoran, Stoker, 321 West Street, Manhattan
John Lynch, Boatman, 428 2nd Ave., Manhattan
John A. Cunningham, Deckhand, 822 3rd Ave., Manhattan

Franklin Edson (Dept. of Health ferry to No. Brother Is.) This vessel was named for a former mayor of the city from the 1880's. She was berthed at Mulberry Street, Manhattan and usually sailed from the 16th Street pier carrying patients from Riverside Receiving

Hospital to North Brother Island. On the morning of the 15th of June, she was tied up at the 132nd Street pier (Health Dept.).Captain Rick was just going ashore when the urgent blasts of her whistle brought him back on the run. As they steamed to the scene, the hose lines were made ready for duty; but on scene, the crew could see that hoses would be useless and began immediate hand-to-hand rescues. The *Edson* came alongside allowing passengers to jump onto her deck. Her paint blistered for thirty feet back from her bow and her forward windows cracked from the heat. Coroner O'Gorman interviewed her entire crew prior to the Inquest.[2]

Crew:

Henry Rick, Captain. (58) was a twenty-two year veteran pilot for the Health Department and the Department of Charities. The captain not only directed the boat's rescue efforts, he went into the water to save victims. Fifty victims were rescued and ten bodies recovered by him. When the city wished to give the Captain a medal for his work in the rescue effort; he refused because picking up those in distress was just part of his job. Captain Rick left the *Franklin Edson* in July of 1904.

Charles Johnson, Mate

Samuel J. Mills, Fireman (hired 26 Oct. 1897)

William D. Palmer, Fireman (hired 1 Mar. 1904)

Andrew Andrews, Deckhand (hired 10 Aug. 1903

Frank Lagato, Deckhand (hired 23 Oct. 1894)

George W. Johnson, Deckhand (hired 1 Mar. 1904), helped his friend, James J. Owen, rescue victims along the shore of North Brother Island. The two are credited with saving eighteen people.

George A. Palmer, Engineer, (hired 7 July 1890) is also listed as a crewman on the Civil List for 1904, but he does not appear in contemporary accounts.

Fidelity (Dept. of Charities Ferry, aka "Potter's Field Boat") She carried men and supplies to and from Manhattan, Riker's Island, North Brother, and other city islands. Not only assisted in rescue, but continued ferrying bodies to the temporary morgue at the Charity Pier for several days, as did her sister ships *Gilroy, Wickham,* and *Companion.*

Crew:

Edward McEvoy, Captain (1858-1957) Once held the title of "Commodore of the Fleet", which was voted him by the skippers in service to the city. He was also the founder of the Harlem Yacht Club

Bronx

The *Bronx* ran to North Beach from 133rd Street; she passed within several hundred yards of the *Slocum*, but did not stop to assist. She was probably too slow to catch up with the side-wheeler.

Crew:
John H. Monforte, Captain.
No other crewmen have been identified.

Easy Time (Car-boat [possibly a ferry for automobiles]) other accounts call her a tug boat. She came up on the *General Slocum*, whistle sounding the alarm and towing several small boats for survivors to climb into; she carried Captain Van Schaick, the pilots, first mate, and engineers to the 134th Street pier.
Crew:
Mr. Churchill, Captain
No other crewmen have been identified.

Unnamed Ferry
Crew:
Willard Palmer, Captain. (1880-1946), won commendations for his efforts. He saw the
    General Slocum burst into flames and remained on scene all day,
        personally diving in 17 times. His vessel brought 145 bodies to the Charity Pier.
No other crewmen have been identified

Companion
Gilroy
Wickham
Haarlaem (Astoria Ferry)

Steamships:

Patrol (Police Boat) This steel hulled, twin screw, police launch was 135' long 118 tons and fitted out as a fire boat, she also carried a small deck gun. She was built in 1893 at Sparrow's Point, Maryland and her berth was at Pier A, North River. Commissioner McAdoo later described the patrol as; "... practically a small gunship or revenue cutter in type... Patrol is fitted as a fire boat and can put four streams of water on a fire and was called out on fire alarms as well as police calls." In the Year End Report to the city, Captain Albertson described Patrol's service at the disaster as follows; the steamer Patrol attended and assisted at the fire on the excursion steamer *General Slocum*, June 15, 1904, standing by the burned vessel until all bodies had been recovered from the wreck and surrounding waters, when she accompanied the burned hull from where it was beached to the Erie basin on June 26, a period of twelve days". This service included aiding in the overnight recovery efforts with her searchlights. And after the hulk was moved to Erie basin it was impounded for inspection by the coroner, fire officials, and federal authorities.

Crew:
Mr. Dean, Captain.
Mr. McGovern, Captain.
No other crewmen have been identified

Graylark (Police Launch) The launch carried the Mayoral Party to the rescue/ recovery site.

Priscilla, Steamboat (Fall River Line), passed the recovery sight so fast that her wake caused rescuers to lose their hold on victims and the bodies were lost to the tide. Several other unnamed vessels were accused of doing the same thing.

Kills (Docks Department Launch) The launch was off 138th Street when the *Slocum* was sighted. She rescued eighteen victims and recovered twenty-one dead, which were taken to Riker's Island. (see Lembach, Albertina in Families ID section)
Crew:
Mr. Halloran, Captain had received the Congressional Medal when he was in the Lifesaving Service.
No other crewmen have been identified

Gloria (Steam Launch) This boat belonged to Steinway, LI. Boathouse. She brought Police to the island and picked up twenty bodies.
Crew:
William Henry Muff, Owner
No other crewmen have been identified

Hudson (Revenue Cutter)
Crew:
James Bradley, Captain
No other crewmen have been identified

Tug Boats:

John L. Wade (John L. Wade Tug Co.) The *Wade* happened to be tied up at the pier on North Brother Island, taking on water, when word came in concerning the *Slocum*. She cast off and moved out into the stream, then waited for the *Slocum* to approach. As soon as she was spotted, between Sunken

Tugs like this, photographed on the Hudson, could be seen all along the New York Waterfront

Meadows and Wards Island, trailing smoke, the *Wade* began moving in to position to render assistance. The Wade followed the *General Slocum* onto the beach (*Wade* on the shoreward side), nosing the steamer in tighter, then lashed to the flaming ship. During this maneuver, her propeller fouled on a rope, which ended her ability to back away. She was later pulled off the island by the tug *Golden Rod* while the *Zophar Mills* hosed her down to prevent her bursting into flames. Besides one-on-one water rescues, the tug also put over all her life vests and life boats for the victims in the river. Coroner O'Gorman prior to the Inquest interviewed the entire crew. Victims rescued by the *Wade*, when they discovered that the ship was Jack Wade's entire livelihood, raised enough money on their own to cover all repairs necessitated by his remaining at the rescue for so long, but he refused to use the funds for that purpose and donated them to the Relief Fund. Two victims, Mrs. Elusca and Mrs. Anna Sickman, rescued by the *Wade*, wrote many letters and raised enough money to replenish all of the *Wade's* life preservers and lifeboats.

Crew:
Jack (John) Leonard Wade, Engineer & Owner received severe burns by the tug's proximity to the *Slocum*. He noted that the whole of the *Slocum* Was on fire in one minute. The *Wade* lashed to the stern of the *General* and was immediately pelted with flaming victims jumping from the upper decks. As they hit, the crew doused them with buckets of water and pushed them toward the front of the vessel to keep her stern from swamping. Several of the *Slocum's* crew also came over and helped with the rescues. The exception was Flanagan who tried to cast off the lashing line and separate the two ships. Jack Wade quickly put him off (into the water) and continued his rescue efforts. The crews' shirts were burned off their backs. When advised that the ship might go up in flames at any minute, he reportedly announced "Damn the Tug! Let her burn! Let her stay where she is. What's a tugboat to a human life?" This was not just his paycheck he was sacrificing; he was the sole owner of the tug. This vessel is credited with the rescue of 155 people. Jack was one of the first people to bring up the condition of the life preservers as a reason

for the loss of life; he also was quoted as estimating the *Slocum's* passengers at 2500. He began his own relief fund for the victims and was one of a handful of rescuers to receive a Congressional medal for his heroism. On January 27, 1912, the daughter of Mrs. T. Ladmore notified Jack Wade that he had saved her mother and she had remembered him in her will. He received a bequest of ten acres of land in Farmingdale, Long Island and an undisclosed number of shares of stock in an oil company. When interviewed, Jack Wade said he did not remember the lady.

Edward "Ruddy" McCarroll, Deckhand / Fireman was among the first rescuers in the water . Before climbing back aboard the *Wade*, he had rescued at least four people; then was nearly drowned by frantic victims in the water. He barely got aboard before one of his rescues pushed him back in to save her son. At one time he had three children holding onto him at one time, which prevented his swimming back to the *Wade*, and he had to let one drift away.

Antonio (Tony) Marcetti, Deckhand, was credited with several in-water rescues.
John McConnell / McDonnel, Deckhand
Bob Brannigan, Deckhand

Hustler (derrick tug) (Merritt-Chapman Derrick and Wrecking Co.)

Hustler set up off Hunt's Point, having towed in three dive barges. She was used as the recovery divers' diving platform. Dives were made on the low tide. She also assisted in moving the hulk of the *General Slocum* to Robbins' Dock in Erie Basin after the disaster. She was lashed to port and *Champion* to starboard. Together they guided the hull out of Flushing Flats.

Crew:
C. F. Risedorf / Risdore, Captain.
Peter Gilligan, Diver aboard (of the Dock Department), 173 E. 117th Street
John M. Rice, Diver aboard (of the Dock Department), Bklyn.
David Tullock, Diver aboard (of the Dock Department), 223 Madison Ave., Bklyn.
Albert Greenberg / Blumberg } Divers aboard and first divers down on the
Henry Heyer[3] }16th. These two men may have been the
} ones ordered by the starboard Chief
} Wrecker, Captain Devlin to retrieve the
} seacock to the pumps standpipe and to
check if it was open.

William E. Chapman (Wrecker Tug) (Merritt-Chapman Derrick and Wrecking Co.)

She came from Hunt's Point and assisted in towing the *Slocum* to the location where she finally sank. *Chapman* then got a crew aboard the *Slocum* almost immediately after the tow was dropped. She later carried Superintendent Inspector Rodie to the scene.

Crew:
Wm. L. Chapman, Company owner
Mr. Turner, Captain
Harry Heyer, Diver aboard [3]
Albert Blumberg, diver aboard
No other crewmen have been identified

Director
On the first pass along the stern of the *General Slocum*, before she was beached, the *Director* had so many passengers jump aboard, that her stern deck was nearly awash and her bow was out of the water. She ran close alongside and picked up seventy-five people from the water, then made a pass close to the bow. Many more jumped aboard including one unnamed woman who landed on the roof of the pilothouse.
Crew:
John McAlister, Captain, Quay & Franklin Sts., Brooklyn
No other crewmen have been identified

Goldenrod (White Star Towing Co.) aka "Towboat" By the captain's count, he and his crew saved 250 victims. He testified at the Inquest that he considered the life preservers in good shape and volunteered to go into the river wearing one. He said that he saw people clinging to other people who were wearing life preservers, and that in each case both were afloat. *Goldenrod* carried a deckhand who had been a gunner at the Battle of **Santiago, during the Spanish American War.**
Crew:
William Hillery (Holley), Captain, 117 Freeman Street, Brooklyn
No other crewmen have been identified

Albert F. Bennett (Dalzell Towing Co.)
Crew:
Albert F. Bennett, Captain (1886-?), he took superintendent Barnaby of the Knickerbocker
    Co. to the scene of the disaster and was on scene near the island for many hours during the rescue. Known for his tight-lipped attitude, he had become a pilot on his 21st birthday and later a Master. He retired after 50 years on the water. His brother, William, was the captain of the tug *J. Fred Lohman* of the same company.
No other crewmen have been identified

Monarch (Merritt-Chapman Derrick and Wrecking Co.) This tug had a 3,000 hp winch that was used to raise the hulk of the *Slocum* off the bottom in a cradle of chains, so it could be towed off North Brother Island and may have been the vessel that brought divers to the wreck on the night of the 15th.

A.C. Sumner (Gauthier Towing Co.)
Crew:
Joseph Rea, Captain (1855-1932), was later the captain of the tug *Red Ash*.
    He won a Knights of Columbus Medal, for his rescue of 125 victims, which he later donated to the Jersey City, New Jersey Library.
No other crewmen have been identified

Arnott (Keeler Transportation) Caught fire from *Slocum* three times
Crew:
Mr. Van Elton, Captain
Mr. Anderson, Fireman- Dove in and saved six women and two children.

Mr. Olsen, Engineer- Made two dives; rescued three children under six years of age.
No other crewmen have been identified

Briggs The *Briggs* assisted in moving the hulk of the *General Slocum* to Robbins' Dock in Erie Basin after the disaster. The *Briggs* and the *Unique* took over from *Hustler* and *Champion* for the trip down river behind the police steamer *Patrol* and a docks department launch.

Champion This tug assisted in moving the hulk of the *General Slocum* to Robbins' Dock in Erie Basin after the disaster. *Hustler* was lashed to port and *Champion* to the starboard side of the *General Slocum*. Together they guided the hull out of Flushing Flats.

Summer[4] (New York, New Haven & Hartford Railroad Co) Responding to the signals of passengers, the *Summer* pulled in alongside the flaming paddle-box of the *Slocum*, and took off several people. Then she docked near the beach and picked up people as they floated by.
Crew:
Mr.Van Gelder (Gilder), Captain
No other crewmen have been identified

Walter Tracy Unlike other vessels, she was able to come alongside the *Slocum* before she grounded, and was pelted with children. Before she was full, she was smoldering. She continued to follow the *Slocum*, picking up survivors from the water, then sent lifeboats to the grounding and took victims to Port Morris, Bronx.
Crew:
Tom Flannery, Captain
No other crewmen have been identified

William H. Gautier The *Gautier* tried to extinguish the flames of the beached side-wheeler with her fire hose, but to no avail as the vessel was totally engulfed.

Manhattan (Department of Docks and Ferries) On the 16th she carried Mayor McClellan from the Fulton Street dock to North Brother Island.

Quigley This vessel dragged the East River with grappling irons, while steaming up the East River, opposite Randall's Island, and retrieved several bodies.

Theodore (Theo) William Major of the *Theo* picked up Lucy Hencken of 169 South 2nd Street, who fell off the third deck.

Unique *Unique* assisted in moving the hulk of the *General Slocum* to Robbins' Dock in Erie Basin after the disaster. The *Briggs* and the *Unique* took over from *Hustler* and *Champion* for the trip down river behind the police steamer *Patrol* and a docks department launch.

Unnamed (Daley Co.)
This captain ran his vessel alongside the beached *Slocum* and picked up two survivors and ten victims.

Unnamed New York Central Tug #7
Crew:
Edward Stokes, Captain.
No other crewmen have been identified

Unnamed New York Central Tug #17 This tug cut loose her tow to get in closer to the rescue.

Unnamed New York, New Haven and Hartford Tug # 14 This tug anchored near the beach and picked up people as they floated by.

Unnamed Tug from Union Gas Works This tug was credited with saving twelve victims from the Sound.

Wheeler came from Port Morris, towing a scow and fell in behind the *Slocum* picking up victims.

The Sun
E. A. Bayliss (See Sloop Bayliss,)
E. Levy
Elsa
Margaret

Dredge:

Unnamed Dredge Was working off Astoria and picked up twelve victims in its steam launch, *Mosquito*.
Crew:
Mr. Dean, Captain
No other crewmen have been identified

Unnamed Dredge William Alloway, Captain, Was working about one quarter mile beyond Broadway, Astoria; he saw a burst of smoke and set off four blasts on his whistle as warning.
No other crewmen have been identified

Derrick:

Monarch She followed in tow, behind the hulk of the *General Slocum* as it was moved to the Erie Basin, after the disaster.

Vessels of the 1st Battalion of New York Naval Reserves
These two vessels arrived from the foot of East Twenty-Fourth Street and

were put at the disposal of Commissioner McAdoo. One was sent to aid the Harbor Police recovering bodies along the shore, while the other was used as a ferry between the island and the foot of East 136th Street:

Launch Oneida *Oneida* ferried divers to North Brother Island on the evening of June 15th.

Launch Franklin Under the direction of Commander Franklin; and the command of Lieutenant Barnard with a crew picked from the ranks of the *New Hampshire*.

Vessels of the U.S. Navy, Brooklyn Navy Yard
(unnamed) Navy Tug, which was fitted with searchlights, assisted.

Privately Owned Vessels:

Sloop (Schooner) Bayliss (see Tug E. A. Bayliss,) *Bayliss*' a yawl (no name) was dispatched from the foot of East 137th Street with several policemen and two crewmen, to the disaster. They maneuvered into the paddle-box and rescued several people. In another account, Policeman Herbert C. Farrell, from the Alexander Avenue Station, grabbed several strangers and commandeered the small boat from the tug *E. A. Bayliss*. They rescued twenty two victims and retrieved sixteen bodies.
Crew:
Olaf Jensen[5], Master
Samuel Patchen, a Negro steward
No other crewmen have been identified

Unnamed (Naptha Launch) This launch was crossing from Little Hell Gate to the East River. It followed the *Slocum* until it was able to run in at her starboard paddle-box and rescued three children. Although burned in the process, the sea man on the launch then landed them on North Brother Island, and took a position on the seawall where he pulled in forty more survivors.
Crew:
Peter Jensen, [6] Owner
No other crewmen have been identified

Launch Peter
Crew:
Policeman Andrew Woods, Peter Jansen [6] and John Rau rescued twenty-three victims and retrieved thirty-eight bodies.
No other crewmen have been identified

Yacht Candida (New York Yacht Club) J. Bond of 86th Street, North River wrote to the New York Times to correct observations that this vessel did not assist. The yacht had sent its lifeboat to assist in the rescue with the Mate in charge. They assisted for over an hour until no more living victims could be found.

Schooner Allison Coming up river in tow, she cast off her tow, then picked up eighty five victims on the first trip to the Bronx; eighty five more on her second trip, to Manhattan; then, fifteen to twenty more on her third trip, also to Manhattan.

Sloop Easy Time
Note: Per Chas. C. Bothur *Easy Time* was a motor yacht, with James C. Ward; Pilot but, news accounts list her as a sloop. There may have been two pleasure craft with the same name.
Crew:
Mr. Churchill, Captain.
No other crewmen have been identified

Surprise Sloop /Launch
She followed the *General Slocum* up stream, picking up victims. The crew reported seeing women and children sink in life preservers.
Crew:
Reuben A. Tudor, Captain
Granville Gibbons, Crewman
No other crewmen have been identified

Elsie (Auxiliary Sloop)
This vessel landed sixty two victims and thirty two bodies at North Brother Island. The Barclays were credited with fifty of the rescues.
Crew:
H. Burgi, Owner
Charles Wetzel, Club Steward
Rudolph Zimmerman
Frank Barky
Robert Start
Fred and Charles Barclay

Unnamed Barge (Lone Star Boat Club)
On the 16th, as this vessel was passing, the crew saw a body in the paddle-box, which they retrieved and passed off at North Brother Island.

Unnamed Row Boat Henry Berg worked grapples along the island's shore to bring up bodies.

Unnamed small boat Adolph Walter (1854 - 1932) received a Congressional medal for valor for his rescue of fifteen people from the decks of the *General Slocum*. He pulled his small boat in alongside where larger vessels couldn't go.

Unnamed rowboat Samuel Johnson of Astoria, a member of the Hell Gate Branch of the U.S. Life Saving Corp, was at work at the foot of 138th Street, when the *Slocum* passed on fire. He put out in a rowboat with James Gray. They picked up twenty-two victims in the wake of the steamer.

Bronx Yacht Club, at the foot of Willow Street and Bronx Kills. Eleven members in three small launches, over a half mile area, saved 110 drowning victims. Within six hours, they recovered 127 bodies from the water and surrounding beaches. (The *General Slocum* passed their facility on the way to North Brother Island)

Metropolitan Rowing Club Several members in a four-oar barge (rowboat), near Riker's Island, found one body and was towing it in when they were hailed by a launch which took them and their charge in tow. The barge and the body were dropped at North Brother Island.

Stuyvesant Yacht Club Members put out in a motorized sloop to help. At one point at least 50 small boats (row boats, small skiffs, etc.) assisted in the rescue at North Brother Island.

---

[1] Based on his experience with the *General Slocum* disaster, the Hoboken Pier fire, and general marine duty, Chief Croker began redesigning fireboats for the City of New York in 1906 and moved fireboat units out of the regular battalion structure and into their own marine units.

[2] This entire crew was listed as living aboard the *Franklin Edson*, on the 1904 Civil Lists.

[3] This diver may have worked with two different vessels during the recovery.

[4] Other accounts list her as a Starin Co. barge in tow of the tug *E. Levy*, but that is unlikely.

[5] Olaf Jensen, master of one of the several vessels grouped together here, was awarded a medal for his work in the rescue, but no record has been found which substantiates which vessel he actually sailed.

[6] Due to conflicting accounts, the two individuals referenced by this number may be the same person.

Medical Facilities & Personnel, Other Municipal Departments and Professions Assisting in the Rescue & Recovery

> "The news of the disaster spread rapidly and, from every hospital in the city, ambulances, physicians, and nurses were sent to the scene, and prompt and efficient service was rendered to the injured that were taken from the wreck."
> –Harpers Weekly, June 25, 1904

Ferries and private craft, even rowboats loaded to swamping, brought relatives and friends from Harlem to the island in search of their families. By noon, close to 3,000 people –hospital staff, victims, police, and friend and relatives were on the island. Police maintained a surprising degree of control over the crowds, keeping them out of the way.

The Department of Charities sent doctors and nurses from the hospitals on Randall's Island, Metropolitan, Infants, and the Alms House Hospitals. Doctors and nurses came from all over the city; from private practice as well as public and from as far away as Newark and Paterson, New Jersey. Ambulances were dispatched to the shoreline on both sides of the river seeking those who might swim to shore or float in on the tide. Although all hospitals above 23rd St. sent staff to North Brother Island, these were reported as receiving victims of the disaster:

Riverside Hospital, North Brother Island_(The Isolation Ward, Contagious Diseases, Tuberculosis Ward, and Fever Pavilion were all divisions of Riverside hospital a reported in the news accounts). This facility would later be the home for "Typhoid Mary" when she was finally and permanently confined for life.

The original Riverside Hospital, now only a vague memory in the mind of New Yorkers had been located on Blackwell's Island (Now Roosevelt Is.), but had been re-established on North Brother Island decades before 1904. This hump of ground had originally belonged to the Bronx, but had recently been transferred to the borough of Queens.

The island had also undergone extensive landfill, which doubled its usable building area. Unless they had had family or friends detained there for measles, scarlet fever, or typhus, most of the passers by on the 138th St. ferries or other vessels had no idea what type of institution was located on the island as they sailed by. The broad green lawns and red brick buildings along the shore gave no hint of the various pavilions and contagious disease wards secreted behind shrubbery down garden paths.

When called on to deliver patients, either the ferry or the Health Department's own vessel the Franklin Edson, which sailed from the dock nearest the Reception Hospital on 16th St. those who worked on the island, seldom left, sometimes remaining for years without even an excursion to the mainland. The hospital was connected to the city by phone and cable. At the Boiler House, near the ferry dock, on the west side of the island, there was excellent fire equipment. When the *General Slocum* was still one mile away, the fire crew was set up with pumps going. As the steamer passed their position, they followed her along the shore, so as to be at the right place when she grounded.

Eventually, victims of the disaster filled every ward bed, doctor's and nurses' quarters, every stable and shed (excepting only the contagious disease wards). Nearly 100 nurses and patients assisted in the rescue and recovery efforts.

Dr. George Taylor Stewart, Superintendent of hospitals, City Health Department, and Head of North Brother Hospital Corp, was the Chief Administrative Officer of Bellevue Hospital In 1902. After midnight of the 15-16th, Dr. Stewart had to order his personnel and patients to bed before they were struck by total collapse. Only the carpentry crew stayed up making coffins.

Dr. Charles Roberts, of the City Health Department. The mayor, Deputy Counsel Breckenridge, and Commissioner Darling called at his home on the island to congratulate him on the work of his staff.

Dr. Samuel P. Watson, was the Head of the Isolation Hospital, North Brother Island. His patients were so upset by the scenes outside their windows that they rioted, wanting to help, and had to be calmed. It took 50 staff to restore some measure of order. Dr. Watson first locked down the entrances to the Contagious Disease Wards, and then ordered monitors be placed at each door, to keep patients in and others out, he then ran to the beach to help direct rescue efforts. When an unknown man was seen trying to steal valuables from a body, he assigned personnel to guard the bodies from looters.

Dr. William T. Cannon, Intern, Assistant to Dr. Watson

Dr. Wharton B. Mc Loughlin, Intern, Corp of Physicians North Brother Island, was in charge of the TB Ward, took rowboat out and rescued 6 victims, then attended those on shore. Once the recovery was complete, he began assessing the condition of those patients who had helped and decided that none were worse off for their efforts.

Dr. Max Weissman, an intern, and several others listed just below, worked the entire day at reviving rescued excursionists.

Dr. William D. Lord   } These doctors first took rowboats out to the grounding
Dr. Bruno Horowitz   } and brought in victims, then treated survivors on the
Dr. Helgeson   } shore.

Nurses and Tuberculosis patients formed human chains stretching out from the shore until they were nearly neck deep to catch and pass back to land both living and dead victims as they drifted by.

Mrs. Kate L. White, Matron, was the Head of Cooks and Waiters. She triaged on the beach, brought supplies then went in and continued rescuing victims using 35 foot wooden ladders as extensions for victims to grasp and be pulled ashore. Her staff fed all rescue and recovery workers several times on the 1st day and then, as scheduled, with the wards.

Miss Kate O'Donnell, Assistant Matron, "Domestic", had not been taught to swim but taught herself, by sheer determination, at the scene of the rescue efforts. She recovered fifteen people, seven dead and eight living.

Cassie McMannus, Nurse, "Domestic" saved seven. She received a commendation from the City and another from Congress along with a Bronze Medal. In 1906 she fell ill and, having exhausted all her resources, became a charity patient at Bellevue Hospital. When she died her body was nearly sent to "Potter's Field" and would have been buried among the penniless and unknown of the city, except that a reporter for the New York World heard of it and came forward to pay for her burial at Mt. Olivet.

Mary Ann McCann, Scarlet Fever patient & Ward Helper, "Domestic" from Ireland, and an excellent swimmer; she was tall with brown hair and blue eyes and all of sixteen years old. She rescued nine victims. The staff of the hospital was so moved by her courage and ability that they brought her situation to the attention of senior staff. The teenager had only been in the United States for about four weeks, and most of that time had been spent on the Island recuperating. She was still supposed to be bedridden when the disaster struck, and the exertions and exposure could have cost her life. District Attorney Gavin took her in (due to her age and lack of family, she had become a ward of the city when she became ill) pledging to provide a home and education for her. She was later graduated as a trained nurse from the Florence Crittenden Training School, Washington, D.C. The hospital staff recommended her for the Carnegie Hero Medal, but it was denied. Although she tried to escape public attention, she was located as an essential witness to the various legal proceedings which followed the disaster. The jury of the Coroner's Inquest was so impressed with her story that they wanted to give her a gold medal and "gorgeous purse" as a reward.

Mary Maher, Helper on the Measles Ward, "Domestic", saved three boys and a woman.

Margaret Lawrence, Helper, "Domestic", saved ten victims.

Pauline Peutz, Nurse (Waitress), was a former lifesaver from Asbury Park, NJ. The eighteen-Year old saved five children and five women. Pauline Peutz, Nurse

Lulu F. McGibbon, Hospital Clerk and Switchboard Operator, First telephoned every hospital, police station, and firehouse in Manhattan for help, then ran to the beach and rescued several people including two infants. She was a good swimmer.

Kate Harding (poss. Nurse) waded out up to her chest and pulled in struggling victims.

Henry N. Malabar, Chief Clerk at North Brother Island

Mrs. Allen, a workwoman at North Brother Island, swam out from the pier to save several women.

George Doorley, Superintendent of Out of Door Work, Contagious Disease Landing (the island ferry pier), helped in carrying bodies and patients from the water to the hospital lawn. He is also credited with identifying and pursuing a thief trying to remove valuables from a body.

Joseph S. Gaffney, Chief Engineer, he first set up hoses to pump from shore, but when that was of no use, he waded in and worked in deep water pulling victims in with a long fire hook until exhausted. At some point, he persuaded five men to make a human chain in the water, with himself as the outer most man, and rescued twenty people by passing them back along the chain to shore. He was credited with rescuing twenty living and fifty dead passengers. From where he was working, Mr. Gaffney saw Mary McCann's repeated rescues and, when she fainted from exhaustion in shallow water, he carried her and her rescue to shore. It was Gaffney, an unnamed surgeon and, several others who related Mary's story as a hero to the press.

James J. Owens, a mason working on North Brother Island, notified Chief Engineer Gaffney of the island's pump house, of the impending disaster, and the two put out in a skiff just as the *Slocum* was grounding. Owens repeatedly swam between the boat and the steamer rescuing victims. Another account puts James in a boat with George W. Johnson, a friend from the *Franklin Edson*. Either way, Owens first encouraged victims to jump from the flaming excursion vessel, then maneuvered them into a small boat. He is generally credited with 18 rescues. At one point, the small boat was so full it had to put into shore while Johnson remained behind in the water with a woman on his back and a boy on his arm. When Owens returned, he transferred these, then found two more children before accepting a ride in for himself.

Others listed:

Mary Clark, Chambermaid
Nellie Duffy, Cook "Domestic"
Catherine Hanley, Cook "Domestic"
Florence Denning, TB. Hospital
Hattie Walker
Miss Edith V. Smith
Clara Anna Lay, Nurse
Miss S. C. Wolfstenholm

Miss Sloan
Alice Meinhart, Kitchen Maid "Domestic"
Eleanor Wrenn, Nurse
Florence Rhodes, Nurse
Miss Atkins
Agnes Lamb, Nurse
Martha Rutledge, Nurse

Frank Margate, Handyman, a Civil War Veteran who had fought at Gettysburg, also noted as helping Coroner O'Gorman who interviewed all of Riverside's staff prior to the inquest.

North Brother Island Lighthouse

Commissioned in 1869; decommissioned 1953 Originally had a 6th Order Fresnel lens; after 1900, a 4th Order lens was installed. What is left of the house is still visable from the north eastern shore of Randall's Island, but the light tower is gone.

J.T.P. Jacobs, Lighthouse keeper

The light, which stood on the southern tip of North Brother Island, rose above the river and sound on a small promontory edged in boulders between the sea grasses above and the rocky beach. The tower of the light was white and equipped with both light and fog bell. In an earlier time, the light keeper's wife had made a few dollars making up picnic fare for relatives ferrying over to see the hospital's patients, but since the introduction of isolation wards as necessary to contain contagious diseases, her province had become limited to the small green lawn which ran around the base of the lighthouse / keeper's cottage, all of which was separated from the hospital's grounds by a fence.

On June 15th, the keeper could not have missed what disaster was unfolding in his neighborhood. In order to run into the beach or pier on the western side of the island, the *General Slocum* had to come past the lighthouse. With her whistling flotilla of would-be rescuers in tow, it must have been an unforgettable sight. With his mission to save those imperiled at sea, the keeper, though not mentioned by name in the accounts, must have been one of the first to run toward the inferno.

Rikers Island Hospital / Workhouse

Before the disaster was over, bodies had arrived at Rikers Island, by boat and by current, which were held for transportation to the temporary morgue at 138th St. or the pier near Bellevue Hospital at 26th St. Some living victims were also cared for there. According to the New York Times, Dr. Nathan E. Broder, resident physician, of 45 Eldridge St., was on duty that morning, supervising several prisoners of the workhouse. When the *Slocum* was sighted, fully ablaze, covered in smoke and heading for North Brother Island, prisoners, John Merther and Dan Casey, literally compelled the doctor to take them to the scene of the rescue efforts. The three put out in a small rowboat and worked as a team to haul in bodies and living victims. They put in at North Brother Island to hand over their load and then returned to deep water for more. This was repeated until all were too tired to continue. They then returned to Rikers Island and their original assignment. By the end of the day, the police estimate showed 90 bodies on Riker's Island.

In another instance, The Evening Sun, that night, reported that the bodies of seven women and two children were picked up by a rowboat along with one woman and two small boys, who were still alive. Was this the same boat manned by Dr. Broder and the prisoners, we may never know. I suspect it was another team of heroes who could not remain on the shore safe while others perished.

According to the New York Corrections History Society, there were no permanent structures on Riker's Island at this time and anyone in need of assistance would have been helped in a tent on the island.

Dr. Nathan E. Broder, of the Workhouse (Res. 45 Eldridge St.), was instructed to attend with every physician he could spare.
John Merther, Prisoner

Dan Casey, Prisoner

Bellevue Hospital 27th St., Manhattan
This venerable institution was located behind the 26th St. pier. Among its facilities was a large

morgue adjacent to the pier. Because of their proximity, the pier and morgue became the primary destinations for families searching for missing loved ones. As a result, the hospital supplied personnel to the morgue and pier and catering for the various workers involved in both locations for a full week.

Dr. Michael J. Rickard, Assistant Superintendent of Bellevue Hospital, took charge of the pier as regarded the landing and disposition of bodies as they were received. registrars and clerks from the hospital assisted him. After working twenty-four hours straight on the 15th, and twenty-hour days for the next week, he took a one-week leave of absence to recuperate from his strenuous labors, this he spent on City Island.

Miss Jane A. Delano, Superintendent of Training Schools (1902-7), (Res.120 E. 26th St.), and later Assistant Superintendent of Bellevue Hospital, helped in coordinating staff activities at the hospital. In a twist of irony, Bellevue Hospital nurses who had earlier watched the lively throngs pass by as the *General Slocum* sailed north, now helped the mourning families in the hospital and at the morgue. Miss Delano sent nurses in twos and threes out along 26th street all that day and night, to inform families of new arrivals at the morgue, and to render whatever service was needed by the crowds.

Dr. Brooks, with some of his personnel, men from the City Lodging House, and 2 prisoners from Rikers Island, pulled bodies from the water and set them in coffins for transport to the morgue. This is probably P. Bellows Brooks from Norwich, NY. He interned at Bellevue 1903-1905 in the 3rd Surgical Division, rising from 1st Junior Assistant to Senior Assistant and at the end of his internship was offered a position as House Surgeon in 1905. He later belonged to the Chenango County Medical Society and the New York State Medical Society.

W.D. Howard, Nurse, was assigned to the Charity Pier to help prevent suicides, since no record of this person could be found among the Bellevue Hospital records, he may have been a member of the staff of either the Bellevue Hospital or the New York City Morgue, 3rd St.

Dr. Ayre, arrived at the foot of 138th St. in his automobile with three other doctors he had picked up along the way.

June 22, 1904 – In his report to the Board, Acting Superintendent Michael J. Rickard, noted that the hospital had received 845 bodies from the *Slocum* and that during the first three to four days after the incident, the hospital could not follow its usual custom of transferring bodies from Bellevue to the City Morgue [Dept. of Charities] because of the overcrowding, but by the date of the report, this had been relieved. He further noted that his entire attention had been given over to the matter of the *Slocum* in identifying and disposing of bodies brought to the City Morgue. The hospital had also sent food and supplies to the employees of the morgue and served meals at the hospital for 100 workers per day during the first week.

Lincoln Hospital 141st St. & Southern Blvd., Bronx This facility was originally founded on 65th St. to care for newly freed slaves coming north after the Civil War. They received the majority of the *Slocum* victims.

Superintendent William Daub,
  President of Lincoln Hospital
Hannah Daub, the Superintendent's
  daughter

Lebanon Hospital, Caldwell & Westchester Avenue & 151st St., Bronx This hospital probably received the second largest number of victims. Police were needed the first day to help handle the crowds.

Supt. of Nurses, Grace Mary Spratt, upon hearing the first news of the disaster, assembled a squad of eight trained nurses and left for the Island. Once there, she organized their efforts in support of the Riverside Hospital Staff.
Nurse Anna E. Malloy (1881-1958) worked on scene at the rescue. She received a Medal of Honor from Lebanon Hospital and a Certificate of Honor from the U.S. Volunteer Life Saving Corp.
Nurse McCallum, was in charge of Children's Ward.

Harlem Hospital 103rd Street, Manhattan
The morgue facilities at both North Brother Island and at Bellevue were overfilled almost immediately and about 200 victims' remains were sent to the Harlem Morgue. This hospital also received survivors.

Dr. Krauskopf rode ambulances to and from the docks picking up survivors.
Dr. John Donovan (1879-1950) had only graduated from NYU in 1902, and in 1904 was an intern at Harlem Hospital. He received a medal from the city for his rescues.

Roosevelt Hospital
12 doctors, 2 ambulances, 10 nurses were sent to North Brother Island.

New York Hospital
Many doctors, ambulances, and nurses arrived at the Island.

St. Vincent's Hospital
Four doctors and eight nurses responded to the disaster

In The Bronx
Dr. David E. Alexander, a 1900 Columbia graduate; an obstetrician and general practitioner, saw flames from a medical call he was attending in the Bronx. He ran to the waterfront, jumped into a rowboat, and headed out into the river where he picked up several survivors on his way to the island. Once there he helped to administer first aid.

Other hospitals listed in the press;

German Hospital, (became Lenox Hill) Between 67th and 68th Sts. And Lexington and 4th Avenues, Manhattan, was built in the late 1860's; at a cost of one million dollars. All funds were raised by the New York German Community.

Flower Hospital was associated with New York Medical College (downtown) The Mayor ordered all Police surgeons, ambulances, and patrol wagons to the scene. But his suggestions had already been implemented. Nine Police surgeons arrived on the Island to aid in any way possible.

Dr. Darlington ordered all health Department boats to begin patrolling from the Battery north to North Brother Island looking for bodies.

Department of Docks the department of Docks divers were on the wreck from shortly after the grounding until 11PM on the 16th, and then began again at 7AM on the 17th. Although most became emotionally as well as physically exhausted by the enormity of the task, none asked to be relieved.

Bob Russell,	Department of Docks Diver
Mr. Edwards,	Department of Docks Diver
Peter Gilligan,	Department of Docks Diver (working with the tug *Hustler*)

GHASTLY SIGHTS CAUSED SICKNESS

GEORGE J. SMUNCK OF THIS CITY WITNESSED RESCUING OF BODIES OF THE SLOCUM.

SUCCUMBS TO STRANGE DISEASE

Was Student at Board of Pharmacy Institute—Case Puzzles Hospital Physicians.

The remains of George J. Schmunck, who died yesterday at Mount Sinai hospital at New York, were brought to Syracuse early this morning and taken to the home of his parents, Mr. and Mrs. George Schmunck, at 915 Townsend-st. The funeral will be held at 3:30 o'clock Sunday afternoon and burial will be at Woodlawn.

The death of the young man was rather a sad one. He was 20 years old and after graduating from the Syracuse public schools entered the employ of Druggist William Muench in North Salina-st. For five years he remained with Mr. Muench and until about three months ago when he went to New York to enter as a student the Board of Pharmacy Institute.

When the work of rescuing the bodies from the ill-fated steamer General Slocum was begun in New York harbor, Schmunck was one of the first at the scene. He was at North Brother island and the morgue almost constantly from the time the first bodies were recovered. Some time ago he began to complain of feeling ill, and it was supposed that he had been overworking. Then his skin began to turn a ghastly yellow hue, and he took treatment for jaundice. He grew rapidly worse, however, and two days ago was taken to the hospital in an unconscious condition. The doctors at once pronounced it a case of the rare disease of acute yellow atrophy of the liver.

This disease, which is a complete breaking down of the functions of the liver and a shrinking of that organ, has as its preliminary symptom an apparently light attack of the jaundice. From that its progress is rapid till it soon gets beyond medical aid.

Relatives of the dead man say they believe he contracted the disease while witnessing the rescue of bodies from the wreck of the General Slocum and while working about the morgue at that time.

The physicians at the hospital were unwilling to say anything positive about the cause of the disease. They did not deny the possibility of its being contracted in the way suggested. One opinion ventured was that the system might have been in a weakened condition at the time, and the unusual strain and unhealthful surroundings might have brought on the disease.

Dr. Daniel Lewis was asked for an opinion on the case last night, and said he was not prepared to say anything authoritively about it as it was an unusual disease, and little understood by him. He, however, said he did not see very well how it could have been contracted directly from the bodies.

David Tullock,	Department of Docks Diver (working with the tug *Hustler*)
Thomas Caulic,	Department of Docks Diver
John E. Ronan,	Department of Docks Diver
Charles P. Everett	(Employed by the city for this recovery only); 1898 he dived the *Maine* in Havana Harbor for the Federal Government to determine if damage was a mine
Nicholas Belzer	

Merrit-Chapman Derrick & Wrecking Co.

Merrit-Chapman was in charge of raising the wreck but, before doing so, ran dragnets over and around the hull hoping to bring up more bodies. Even in the first hours that the divers were on scene; their mission was to recover forensic evidence along with the bodies. Sections of hose and standpipe were retrieved in order to determine if the hose had been attached properly and if the valve on the pipe was open, allowing water to go through. One report also claims that a lifeboat was also recovered, which had several bow seams parted, and another opening between the planks about one-foot long. This could have been one of the many commandeered craft, which rowed out to the rescue, or it may be the only life boat from the *Slocum* reported as having been lowered.

Captain Tom Kivlin, Chief Wrecker of Merrit-Chapman, noted that the hull could only be made into a barge due to the substantial damage sustained.

John M. Rice Diver

Albert Blumberg,	Diver
John M. Rice,	Diver
Henry Heyer,	Diver

Other Institutions & Professions

The news accounts were full of stories about normal people who felt the call and performed extraordinarily. Here are some who were named.

William Henry Muff, Hotel and Boathouse owner at Steinway, L.I., Olaf Jensen, Sailor and Sam Patchen, Negro porter, assisted Police from the 16th St. Station to rescue twenty-six victims.

Sailors Scard & Wood, were credited with pulling twenty victims into boats.

The marble works opposite North Brother Island, on the Manhattan side, was shut down for The day; the boss and his crew of 150 men crossed on any vessel they could find to help.

Thomas McQuade, recovered thirty-six bodies off North Brother Island by grappling from a boat.

Alderman John H,. Dougherty, of the Bronx arrived on scene with Coroner O'Gorman and helped with the rescue, working with a Harbor Patrol unit The Superintendent for the Metropolitan Street Railway, at Commissioner McAdoo's request, sent six gasoline flare lights used for overnight street construction, so that the efforts would not stop at dark.

John H. Fitzpatrick, who managed a staff of embalmers, worked Midnight to 6 AM each night at the Charity Pier.

Philip Wagner,[1] Undertaker, 138 2nd Ave. Or Ave. A } These two men were brothers
Jacob Wagner,[1] Undertaker, 508 6th St.   } and members of St. Mark's.
Philip handled thirty-one and Jacob fifty-one bodies. This was their entire capacity. Stolts, an undertaker from the Bowery, sent 250 coffins and tons of ice for the first wave of victims arriving at the Charity Pier's temporary morgue.

A diver ascends from the wreck having seen the face of death at arm's length

The 2nd Battery of the New York State Militia (Bronx Field-Artillery Unit) sent two cannons which were loaded on a railroad barge and towed to a point near where the *General Slocum* was seen to catch fire. They were then slowly towed up toward North Brother Island, continuously firing blanks. It was hoped they would dislodge any victims stuck in the mud. Thirty were dislodged in this manner.

---

[1] Other reports listed these two men's surnames as Herrlich

New York's Finest

In his 1906 book, Guarding a Great City, Commissioner McAdoo had the following comments regarding his force's service on the day of the *Slocum* disaster; "It is a fine service, taken altogether, and those who saw these brave, big, tender-hearted fellows the black day of the *Slocum* disaster, standing for hours in the water of the Sound, and working tirelessly to recover the dead and render service to the living, felt proud of them and of the city they serve."

Police Commissioner   William Gibbs McAdoo
Deputy Commissioner Waldo Rhinelander
Acting Superintendent of Detectives Rickard

Chief Inspector Cortright met with Insp. Max Schmittberger and Captain McDermott, of the 5th Street Station, at Police Headquarters, to discuss police arrangements for the funerals.
Inspector Brooks was in charge of 50 officers on duty at the Bellevue Morgue, helping visitors locate and identify relatives and friends.
Inspector Kane was in charge of 170 men detailed to handle the funeral crowds at Lutheran Cemetery
Inspector McLaughlin worked with Captain Shire was in charge of the 26th St. Morgue
Inspector Max F. Schmittberger, 1st District, had overall responsibility for all police at Funerals; commanded 23 sergeants; 10 roundsmen; 400 policemen organized into squads of 10 men and one officer. Each funeral was accompanied by one squad. No night sticks were carried by this detail. (He personally commanded Police Guard at Pastor Haas' family funeral) Later, he headed McAdoo's squad of 50 German speaking officers who attempted to discover every survivor and every victim's family in order to correct errors in the news coverage and compile an accurate list. These objectives were not accomplished entirely.
Inspector Elbert O. Smith, of the Harbor Squad, formerly in command of the Marine Dept. of the police force. (He had trained Kelk & Van Tassel)
Inspector Richard Walsh, West 65th Street Station is mentioned in news accounts.
Captain Charles L. Albertson, of the West 152nd Street Station, sent boats with fifty reserve officers to North Brother Island. He was also in charge of 300 police & divers at the Island.
Captain Becker, of the Hamburg Avenue Station had to callout reserves for crowd control at the funeral at the Blohm residence
Captain Carrigan, of the North Beach police supervised round-the-clock beach patrols for bodies between Astoria and Whitestone
Captain William Dean, of the Harbor Squad (a.k.a. Steamboat Squad, 42nd Precinct), Commanded the police patrol launch *Patrol* on duty at North Brother Island.
Captain Joseph C. Geoghegan, of the Alexander Avenue Station, 35th Precinct, undoubtedly saw more of the disaster within the walls of his own precinct house, than most officers saw on scene. Bodies were laid out wherever there was floor space, and families moved through looking for both the missing and the dead.
Captain Stephen McDermott, of the 5th St. Station, which was the "Little Germany" Precinct, was heard to comment that his best people had died.
Captain Shire, who is variously listed at the East 35th Street Station, the Alexander Avenue Station, and the 21st Precinct, worked with Inspector McLaughlin.

These two men formed lines outside of the morgue pier to facilitate the procession of relatives and friends coming to identify the dead. The line formed west of First Avenue, with a flanking cordon of officers on either side of East 26th Street, thereby allowing the searchers to enter by the south sidewalk and leave by the north.

Captain Murphy, from the Hamilton Avenue Station, called the attention of the Health Commissioner to unsanitary conditions of the Slocum, after the disaster, noting that holding the wreck in city limits was a public health hazard.

Captain White, assigned to the East 35th St. Station was in charge of 2 sergeants, 2 roundsmen, 80 patrolmen on the Charity Pier.

Captain Gallagher }
Captain Burfiend } These two were also mentioned in the press.

Lieutenant Wilson, of the 19th Precinct, was mentioned in the press.

Detective Sergeant Kensler } on duty as aids to Coroner O'Gorman, were noted in a letter by
Detective Sergeant Postoff } him to Commissioner McAdoo for their nearly continuous on-duty status June 16th – 26th.

Detective Prundy, from station house near St. Marks Church, was in charge of a plain clothes detail that handled crowd control outside St. Marks Church and mourners' homes.

Detective Ross, who was on Inspector Schmittberger's squad, worked the crowds at the Funerals looking for pickpockets. He was responsible for collaring Benjamin Lieberman (17) in the act of grabbing Mrs. Rosie Fischer's purse. Lieberman then had to be protected from the crowd who wanted to "lynch" him on the spot.

Detective Edward P. Mulrooney, of the Harbor Squad, was detailed to go to Lincoln Hospital, Bronx, and put Captain Van Schaick under arrest. Mulrooney later became New York's police commissioner.

Dr. E. T. Higgins, Medical Representative of the Police Dept., was stationed at the morgue.

Sergeant Lawrence Hines was in charge of the temporary telegraph unit at 132nd St. and the East River, and sent communiqués to headquarters.

Sergeant Lonergan was in charge of the temporary telegraph facility at North Brother Island, this outpost and Hine's kept superior officers informed as circumstances developed and changed throughout the rescue and recovery.

Sergeant Lane, from the West 152nd Street Station, arrived at North Brother Island with 50 reserves by police boat.

Patrolman Gilderman, of the Harbor Squad and Patrick Nolan of E. 138th Street found the daughter of Ex-Alderman Smith, 12th District, on shore and returned her to her family.

Officer John A. Scheuing, Alexander Ave. Station Saw the Slocum while on his beat near the waterfront at 138th St. He commandeered a soda water wagon, in which he was driven to the foot of E. 141st St., where he took a boat and pushed into the stream. Scheuing rowed directly into the port side of the vessel although wind whipped flames were flying overhead. Within 100' of the ship, his skin began to blister. He soaked his blouse to protect his head and shoulders, and rowed on. Under the paddle-box he found five victims. He wedged a foot into the wheel works and helped all five to the safety of the boat. After relieving himself of his load at a barge, he returned for thirteen bodies, then put ashore at North Brother Island and helped the coroner tag 171 bodies over the course of the day. His determination shone as a beacon to others who followed his lead in spite of the incalculable risk. Officer Freese arrested John Davis in Williamsburg, Brooklyn for soliciting money for the Relief Fund without intention of turning it over to St. Marks or the city.

Policemen Gick from the Alexander Avenue Station, was at the foot of 138th St. as the *Slocum* passed.

Patrolmen Zuk }
        Holt } of the 136th St. Station commandeered a yawl and rescued 26 victims

Policemen James Collins }   Farrell, worked with Officer J. Scheuing on the *Zophar*
    Herbert C. Farrell[1] }   *Mills*, pulling people from the water. Another account places both officers, of the Alexander Avenue Station, 35th Precinct, on the *Bayliss* with Olaf Jensen, where they rescued 24 people under the paddle box.

Roundsmen Nicholas Klute }   From the Harbor Squad; they recovered 217 bodies after
        Giloon }   arriving at the Island on the Naval launch *Oneida*.

Policemen George A. Mott }   These were among the first policemen to reach the *Slocum*.
  (poss.) Peter Murphey}   They remained on scene for the rest of the day working with
      Skelley }   the.Chapman Co. crew and the Naval Reserve crew.
  Wm. A. Grey (Groh) }   This whole group of bracketed officers is also listed as from
     John Healey }   the 42, Sub.[Station]

Roundsman Daniel Ryan   } Working together,
Policemen   Corbett   } these five recovered 107 bodies
          Franklin   }
          Powers   }
          McKeown }

Policeman James Murtha first assisted in rescuing several people then helped recover bodies from the water.

Roundsman Wood, of the 35th Precinct, was in charge of the Harbor Squad detail at North Brother Island. Inspector Albertson singled him out for his devotion to duty during the rescue and recovery, his aid in identifying bodies, and for the many victims he had saved. This may be Patrolman Andrew Wood of the Alexander Avenue Station who was reported working a fire on the shore, saw the *General Slocum* come by, grabbed two civilians and ran to the Launch *Peter*. He was credited with 23 rescues and 38 bodies retrieved.

Bicycle Patrolman Rush Webster saved Otto Meinhardt and others.

Policeman Thomas Cooney was assigned to North Brother Island. He rescued eleven, then drowned trying to save a twelfth victim.

Patrolman John Hines, from the Mulberry Street Station } responded to the call for help,
    Edward Schnitzler, hid beat was on 102nd St.   } not knowing that this was the excursion vessel carrying their families.

Policeman Edward P. Norton[2,] 19th Precinct, worked at the recovery site for twelve hours. He arrived having picked up volunteers along the way.

Policeman Essig was on the 3rd Street pier as the hulk of the *General Slocum* passed the 3rd Street pier, on the 27th, the officer ordered "Hats Off!" To which a Crowd estimated at hundreds, responded by men doffing headwear, and women bowing their heads in silent prayer.

Also mentioned in the press accounts;
    Policeman John A. Darrow took a small rowboat and saved several people.

Policeman John Leonare, from the 19th Precinct
Policeman George Young
Policeman Daniel Sullivan
Patrolman Louis M. Schauberger
Patrolman Patrick Kenny
Roundsman Pugh from the Canarsie Station
Bicycle Patrolman Schneider

Onboard the *General Slocum*; [from the 42nd or Harbor Police Precinct]
The Police Board only awarded two bronze medals, which went to Kelk and Van Tassell, because they actually saved lives while risking their won.
Officer Charles Kelk, from the 5th St. Station, stationed himself at the stern rail, as the vessel was heading for North Brother Island. Kelk picked up and hurled many children to the following steamers. He also saved a woman whose skirt was ablaze by rolling her on the deck in a baby's blanket.
Officer Abel R. Van Tassel, from the 5th St. Station, was credited with saving fourteen people. The "Portly" officer stayed aboard until the vessel grounded, then passed children to safety. He was hit by falling debris[3] from an upper deck, which struck him on the back of the neck rendering him unconscious. He was revived by the shock of hitting the cold water; he turned on his back to float to shore, allowing as many women and children as possible to hold onto his arms and legs like a raft. James J. Owen, Bricklayer, and patient at North Brother Island, pulled Patrolman Van Tassel out of the water. Both Van Tassel and Kelk had trained under Inspector Elbert O. Smith, who formerly commanded the marine department.

Officers Haslinger } Were detailed to the 3rd Street Recreation Pier the morning of the St.
       Lang   } Marks excursion to maintain order during boarding.

    Several other precincts, whose personnel were not named but, whose stations were mentioned included the Tremont Avenue Station, the W. 152nd Street Station, the W. 125th Street Station, the Bronx Park Station, the High Bridge Station, and the W. 100th Street Station. Each of them sent wagons and reserves.
    Police officers and civilians, hired by the Police Department, patrolled the beaches on both sides of the East River for five miles north and five miles south of the wreck in search of bodies. Before it was all over, nearly 100 policemen were reported as working on North Brother Island. This is probably a low estimate.
    By the latter part of the week of June 20th, fully one third of all valuable (jewelry, cash, bank books, etc.) had been claimed, at an estimated value of $25,000.00. After the Inquest, if no living relative had come forward, these items went to the Public Administrator. Where there was reason to assume that living relatives might be found, property was held by the coroner's office for a "reasonable period", then turned over to the courts for disposal.
    Although other officers are mentioned in various news accounts, no other evidence that their names and/ or exploits were actual. Therefore, they have not been included.

---

[1] In another account, Ptlmn. Herbert C. Farrell, from the Alexander Avenue Station was reported as at a fire on shore, saw the General Slocum pass, grabbed several strangers and commandeered the small boat from the tug *E. A. Bayliss*; he was credited with 22 rescues and 16 bodies retrieved.

[2] Notes in cover of book formerly owned by Officer E. P. Norton; "Norton; worked on this steamer 12 noon until midnight E. 138th St. dock –Alexander Ave. Precinct. Was notified about fire and received orders from Hdqtrs. by Officer Beasley to proceed to ship and pick up volunteers. 1500 dead missing unknown 3,000 on board."

[3] Another account says that the debris was a woman jumping from the upper deck.

James Collins, Herbert C. Farrell from the Alexander St. Station, and Olaf Jensen worked together to save lives.

New York's Bravest

Chief Edward F. Croker was a commander from 1899-1911; he was a twenty-eight year veteran at the time of the *General Slocum*. He was instrumental in the evaluation of the events of June 15th for the Inquest and Grand Jury.

Fireman Thomas J. Cahill, Engine Co. 38, Manhattan, arrived with Coroner O'Gorman on North Brother Island, and in spite of it being his day off, worked until midnight on the recovery.

Fireman Joseph J. Mooney from Ladder Co. 16, was already a decorated hero of the force[1] and had recently transferred to the fireboat *Zophar Mills*. He first dove in and saved two women, while wearing full gear. He then nearly lost his life when a third very heavy victim deranged, by panic, nearly drowned him. They were both rescued by another fireman who threw a rope to them. He returned to the deck and resumed his normal duties.

Fred Hoffman, Company unknown, was aboard the *General Slocum* as a passenger with his family and survived.

Fireman George Lawlor, Engine Co. 46, 451 E. 176th St., was assigned to the *Zophar Mills*; he rescued a woman being born away by the currents by swimming after her and bringing her back to a tug.

Fireman Ernest Plate, (1866-1935), had been a fireman since 1890. He jumped from an assisting tug and rescued many of the non-swimmers in the water. At one point a "heavyset woman" nearly drowned him but, his buddies rescued him. He was credited with saving twelve victims and received $2,800.00 and medals from seven organizations.

Fireman Eusebius Murphy, (1873- 1927), from Hook & Ladder 19 of the Bronx, took a small boat and several people. He was given the Fire Department Medal.

Ladder Co. 17 (143 rd St.) & Engine Co. 60, (352 E. 137th St.) 138th St. & Locust Ave. Both of these companies were from the Bronx. Several of their members took a small boat and rescued jumpers while the rest of the land based units could only watch [2]

---

[1] June 6, 1903 –William L. Strong Medal
[2] Tom Guldner www.MarineFires2aol.com.

Private Citizens

Hundreds of civilian heroes go unknown to this day. These include a lady from the neighborhood near the 138th Street docks who opened her heart to a survivor. The woman had survived a jump into the river, was picked up by a rescue boat and brought to the Island. Her clothing was almost burned off, so the hospital staff gave her a blanket. She was put aboard a boat ferrying people back to 138th Street and dropped off there to find transportation home. Dazed and without her purse, she was approached by a well-dressed matron who took her in hand. The survivor was escorted to this lady's home, dressed from the woman's closet, and sent home in the family's carriage. As the crowds swelled at the uptown train stations, conductors boarded tattered and filthy survivors for free. Other strangers pressed train or cab fare into grimy hands to help strangers get home.

Two interesting sidebars to the heroic efforts expended during the rescues and recovery of the *General Slocum* played out within a year of the disaster. On June 17th, 1904, Edward Bannigan, of the Bronx Health Department, a resident of Long Island City, was brought before Magistrate Connorton. He had assaulted a police officer, James Stanton, at North beach on June 12th. Alderman Dougherty happened to be in court that day and told the magistrate that Edward had been a rescuer at North Brother Island on the 15th and that he had been brave at the risk of his own life, and therefore should be given consideration at sentencing. The charge was reduced to Disorderly Conduct and his sentence was suspended. The audience applauded.

Many survivors found their way home on the "El" Train

In June, 1905, almost a year to the day of the disaster, Adam Rickenback of 27th Street, Camden, New Jersey, decided to row out and have a look at the hulk, which had been towed around the toe of New Jersey and was riding at anchor in the Delaware River awaiting refitting as a barge. He moored beside the charred hull and climbed aboard. The lower section of the hull was carpeted in sand from Long Island Sound. This had not been pumped out before the Slocum's trip down from New York City because it acted as ballast. Frederick Creamer, her new owner, had not considered where the sand had come from. With the changing flow of tidal currents first at North Brother Island and then at Hunt's Point, the sand had washed in and out several times before settling. Unknown to Mr. Creamer it contained an unexpected surprise. As Mr. Rickenback walked about, his foot hit an unusual resistance. Buried in the sand were unidentifiable human bones!

Roll of Honors

Various accounts, from 1904, note that a total of nine "Congressional Medals" were awarded for the *General Slocum* Disaster. The only surviving list, which was located by a congressional research staffer, lists all recipients whose names are followed by dates. The others were either listed in the press accounts or contributed by descendants.

Department of the Treasury Life Saving Medal

These medals were referred to as "Congressional Medals" in the press. This is a misnomer as the actual Congressional Medal [of Honor] is only awarded to military personnel for acts of bravery while in service. The Department of the Treasury Medal was originally authorized by Congress in 1874 specifically for civilian acts of bravery, in peril of life, in the water. Although such a medal may be ordered by a member of congress {thus the misnomer), it is struck by the Treasury Department.

Department of the Treasury Life Saving Gold Medal
  Patrolman John A. Scheuing, Presented 08/23/1907
  Patrolman Herbert C. Farrell, presented 12/01/1906

Department of the Treasury Life Saving Silver Medal
  Mary McCann, presented 03/18/1909

Department of the Treasury Life Saving Bronze Medal
  Bicycle Patrolman Rush Adelbert Webster
  Adolph Walker
  Fireman Patrick J. Lynch
  John L. Wade presented 09/08/1906
  Patrolman James Collins presented 12/01/1906

U. S. Volunteer Life Saving Service Medal of Honor

On the evening of December 7, 1904, Mayor McClellan, Congressman Joseph Goulden, New York University Chancellor MacCracken, Police Inspector Albertson, Commissioner Lantry of the Department of Charities and Corrections, and Colonel J. Weatley Jones of the Life Saving Service presided over a ceremony at the 2nd Battery Armory, Bathgate Avenue and 77th Street. In an event as bright as June 15th had been dark, the U. S. Volunteer Lifesaving Service awarded four hundred medals to the most celebrated of the *Slocum* disaster's heroes and heroines while two thousand survivors, relatives of victims, and notable New Yorkers cheered from the audience. Twenty-four of the medals and eighty certificates of recognition went to members of the New York City Police Force.

Gold Medal
Patrick J. Lynch, also received American League of Honor Medal.

Medal Degree Unknown
Capt. Samuel J. Berg, was a member of the Volunteer Life Saving Corp of New York. News accounts make note that he was a powerful swimmer and a professional lifesaver. He was credited with saving fifty victims.
Kate Harding, walked into deep water and helped rescue many victims.
Captain Parkinson, of the steamer *Massasoit, assisted* and/ or rescued many people including Hannah Ludemann of White Plains who presented him with his medal.
E. J. McCann, This unknown hero's medal[1] was displayed on the Internet.
Police Sgt. Michael J. Fitzgerald, a Medal and Certificate were awarded
James C. Hennis, received a Certificate of Honor.
Patrolman William Krossett, was given a commendation by the U. S. Life Saving Corp.
Edward F. Otto
Charles C. Ward
Mary McCann

New York City Police Department Medals

Due to the ongoing evolution of medals and the criteria for awarding them in 1904, the author has not satisfactorily identified the correct form of the Police Department Medal for illustration here. It should be noted though, that in 1904, as today, when an officer received any of the commendations listed below, he also received the Legion of Honor.

In 1904, the Department's Medal of Honors could only be given for an exhibition of physical courage and were ranked as follows:

1. The Departmental Medal   (M)
2. Honorable mention (with or without the medal)   (H)
3. Commendation   (C)

Patrolman John A. Scheuing   (H)
Patrolman Charles Kelk, Police Department Medal of Honor
Corrections Officer James Duane, Bronze Medal
Policeman James J. Collins, Police Department Life Saving Medal
Policeman Michael J. O'Connor, Ploice Department Honor Legion
Patrolman William Krossett

See the following pages for other police officers who received honors but, for whom no record of their service has been found.

Members of the Force to be Recognized by the Volunteer Life Saving Service for their work in assisting and rescuing victims of the *General Slocum*, June 15, 1904 and members of the Force recognized by the Police Department for the Roll of Honor for meritorious service at the same disaster[2]

N.Y.P.D. Special Order #54 March 30, 1905 states "…for heroic work at the *Slocum* disaster, where he rendered services at personal risk in the rescue and care of the living, and in the recovery of bodies and care of the dead."

On this list, names with an L were recognized by the Life Saving Service; those with a C, H, or M indicate the Police Department award received.

NAME/ RANK	PRECINCT	NAME/ RANK	PRECINCT
CAPT.		**PATROLMN.**	
Charles L. Albertson	C	Peter J. Hunt	4 L
Charles Kensler	6, District L	James P. Morrisson	4 L
Joseph C. Gehegan	35 C L	William J. Finigan	4 L
Simon Blumel	6, District L	John O'Keefe	4 L
Patrick Byrne	36 C L	Edward D. Ehlers	24 L
Frederick Wohlfarth	76 C	Charles A. Schultz	24 L
Cahrles C Wendell	34 C	Louis M. Schauberger	24 L
James B. Ferris	37 C	John A. Coleman	25 L
William Dean	42 C	John V. Ahearn	28 L
Surgeon Edward T. Higgins	C	Martin Early	16 L
Mark F. Horrigan	16 C	Nicholas Brooks	C
		Thomas J. Carmody	16 C L
INSPECTOR		Elbert O. Smith	C
William B. Reilly	17 H	Martin Early	16 C
Edward Brady	16 C L		
ROUNDSMN.			
Joseph W. Cooney	17 C	Richard O'Conner	6, District L
James Wren	19 C L	Cornelius O'Donnell	17 C L
Benjamin Merrit	19 L	Edward D. Hoffman	29 C L
James Cooney	19 L	Andrew Wood	35 H L
James P. Flanagan	22 L	John A. Park	36 C
Joseph L. Ferten	22 L	John Kenny	37 L
John J. Dust	21 C	John A. Kenny	41 L
Thomas Ryan	21 C L	Harry Dobert	42 C
Charles F. Morris	21 C L	John F. Dwyer	42 C
Joseph L. Naughton	21 C	Daniel M. Gilloon	42 C
Patrick J. Reid	22 L	James W. Hallock	42 C
William F. O'Connell	21 C L	Nicholas Klute	42, Sub. C L
Antone A. Strausser	21 C L	William J. Morris	42, Sub C L
Albert E. Blythe	21 C L	Daniel Dillon	42, Sub L
James P. Flanagan	21 C	Daniel Ryan	42, Sub. C L
Louis J. F. Riedell	21 C	Michael J. Mulhall	42 C
Michael Kenny	29 L	Patrick J. Riordan	42 C

David P. Lawlor	29 C L	William Wettlaufer	42 C
Stephen W. O'Brian	29 C L	Hugh Quinn	Cent. Off C
John P. Nehill	29 C L	James Frawley	34 C
William Baumbach	29 C L	Isaac Milheuser	74 C L
John Meischlen	29 L	Charles S. Sands	29 C L

PATROLMN		PATROLMN	
John McDermot	30 L	David J. Goff	35 H
Jacob Zoranna	3 L	James J. Collins	35 H
Louis Redman	30 L	George P. Young	35 H
Charles Fischer	3 L	Charles Hefferman	35 H
William Table	31 L	Arthur G. Coulter	35 C
William Irving	31 L	Rush A. Webster	36 H
Theodore D. Kess	4 L	Charles B. Zeek	36 H
Charles Coughlin	33 L	Daniel Sullivan	35 H
Frank Pineau	32 L	Michael Fitzgerald	36 L
James F. Shevlin	32 L	Richard Holt	36 H
George H. Griffin	33 H	Thomas M. Reilly	36 C L
Thomas J. McManus	24 L	Gustavus Gick	35 H
Michael Voght	33 L	Hubert C. Farrell	35 H
Hugh Cribben	33 L	Patrick McCarthy	36 C
William Krossett	34 L	Patrick Clynes	36 C
Burtus Thompson	35 L	William C. Rice	36 C
John A. Scheuing	35 H	Christopher J. Lyne	36 C
Thomas Calligan	36 C	Thomas C. Back	36 C
William McCarthy	37 L	John T. Kelly	42, Sub.L
George A. Mott	42, Sub L	Peter Murphy	42, Sub.L
John Healy	42, Sub L	William Groh	42, Sub.L
Abel R. Van Tassell	42 M	Charles Kelk	42 M
Thomas Larids	74 L		
Ludwig Schmidt	74 L		
Joseph B. Fanning	76 L		
John B. Sasson	76 L		
William Higher	74 L		

HOUSE OF REPRESENTATIVES.
COMMITTEE ON ACCOUNTS,
WASHINGTON.   March 9 1909

Rush A Webster
794 East 145th Street
New York City

My dear,

I respectfully request that you come to the Morris High School, on Friday Evening, March 12th, at 8 O'Clock Sharp, when the Congressional Life Saving Medal will be awarded you.

Kindly report to me when you reach the hall.

Very truly yours,
J. A. Goulden

218 Nostrand Ave, B'klyn.
March 18, 1908

Mr. Rush A. Webster
Dear Sir:-

Your farm of to-day at hand and will the greatest pleasure will do you this little favor knowing what a brave deed you did for me and for all my life shall not cease to thank you for it.

Yours Truly
Otto Meinhardt

P.S. Inclosed please find the papers I signed for you.

These letters, sent to Officer Rush A. Webster, were from Otto Meinhardt, saved by the officer and from Representative Goulden, inviting him to a medals awards ceremony.

New York City Fire Department Medals

Unfortunately, in 1904, there was no one medal for valor in the New York City Fire Department. Several different medals could be issued for heroic service. Surely, this list is far from complete, but only these few individuals appear in news items or have been supplied by descendants.

Fireman Patrick J. Lynch, Fire Department Roll of Merit, class A
Fireman Ernest Plate, received multiple medals
Fireman Eusebius Murphy, unknown medal
Fireman James Mooney of the *Zophar Mills*, unknown medal
Fireman George Lawler of the *Zophar Mills*, unknown medal
Fireman Michael F. Nugent, unknown medal

INTERNATIONAL RECOGNITION

Word of the disaster flashed around the world. But nowhere did it have a greater impact than in Germany. Although the vast majority of passengers aboard the *General Slocum* and their families, were second or third generation Americans, family ties were still nurtured in Europe. While German immigration to the United States had fallen off sharply over the closing years of the nineteenth century, many still dreamed of making the trip or were in the process of saving for passage. No greater show of this familial bridge could be found than the expression of the German Imperial Family. The Kaiser, sent expressions of sympathy and support to both the city of New York and St. Marks congregation. The Empress, Augusta Victoria, wife of Kaiser Wilhelm, made her own awards to fifty-one nurses in "Recognition of Service" for their "prompt response and for rescue out of danger to human life". The Empress was considered a champion of "the new woman", and had received Susan B. Anthony when she toured the continent. The award took the form of a diploma bordered in gold with the Empress's likeness and signature.

The presentation was made on the 24th of April 1905, At Riverside Hospital's Nurses' Residence by the Acting Consul General Gneist, of the German Embassy. A delegation, which included the Consul General, Dr. Darlington of the Health Department and other officials, was picked up at the foot of 116th Street, by the *Franklin Edson*, for the trip to North Brother Island.

Supervisor of Nurses, Miss Edith V. D. Smith, received a gold brooch, which included the German Coat of Arms and was set with pearls and emeralds.

Those receiving diplomas were;

Lou F. McKibbon	Anne Denning	Annie Connor
Jane Morryle	Mary Frances Rigney	Iva C. Youmans
Kate L. White	Martha Rutledge	Nellie Brown

Annie Ledurek	Clara A. Lay	Rose Anne Sullivan
Pauline Puetz	Mary L. Hurley	C.S. Wolsienholme
Molly Shinnick	Alice Giles	Julia Harrington
Nellie O'Donnell	Lillian Bottomley	Sarah Simpson
Halley Walker	Mary Giddings	Agnes M. Lamb
Annie Dodd	Delia Connelly	Celia Brown
Mary A. McCann	Eileen Palmer	Cassie McManus
Mary J. Blaney	Rose Mahon	Annabella Reynolds
Mary Sullivan	Eleanor Wren	Mary E. Canning
Lillian B. Woodrow	Mary Clark	Mary Maher
Tessie Murphy	Mary Baxter	Ann Mures
Emma O'Connell	Gertrude Laiberg	A. Lillian Sloan
Mamie Taylor	Josie Burke	

Unsung Heroes
Stories from the New York Press

Charles Schwartz, Jr. age eighteen, and a machinist's apprentice saved twenty-two souls. His heroism was attained not only at risk to him-self, but also in the midst of his own grief. Charles had learned to swim at an early age. When faced with the dangers of the *General Slocum*, he first fitted his ten year old brother, Louis, with a life preserver. When panic erupted, Charles tossed his brother into the water, checked to see that he surfaced in good shape, and then headed down to the lower decks to see if he could help extinguish the fire, but realized that the boat was already lost. He was pushed back against a railing as it collapsed, and fell into the river. "The first person that I saw was Mrs. Addicks, who keeps a candy store at No. 53 Avenue A, and she called me by name, and I went over and helped her by keeping her chin above water and towing her a little. She got to shore all right and was not much hurt."

Charles returned to the water again and helped Miss Emma Haas, sister of the pastor, until a boat came and picked her up. "Then I saw my mother and grandmother. They were floating face downward. I got them both ashore and helped the doctors with them on the lawn. 'It's no use', said the doctor, 'we can't do anything for your people, my boy.'" As Charles sat sorrowing, he saw a man in a small boat. Charles swam out to join him. He repeatedly dove in and returned with victims while managing to keep from being dragged under. Each time he and the boatman returned to the beach, they brought in four or five more victims until the final count was twenty-two. On every trip the youngster saw his mother and grandmother on the beach. "If I had been a stronger fellow I might have done a great deal more, but I'm light. I weigh only one hundred and twenty three pounds stripped. Rather too light, don't you think?"

The unknown man; His identity was never discovered by the press, so they called him the Unknown Hero of the *Slocum* disaster. Tugboat hands saw him near the shore of North Brother Island with three women clinging to him. Between whatever water skills he had and one of the few good life preservers, the little group was floating into shore when a fourth woman clutched at him. The excess weight was too much and he began to sink. "Don't, I can't swim if you do that" he cried, but the woman would not let go. "Alright", he sputtered, "we shall all go down together!" But within seconds a boat picked them up and deposited all safely ashore.

Peter Wingerter, age thirteen, found four babies on the upper deck abandoned by the adults. He remained onboard with them until a tug got close enough that he could pass two over. He then slid down a stanchion to the main deck, carrying the other two infants, and passed these to a man in a rowboat. Nearby, a woman tossed her baby into the water; he dove in after it. Another boat man, who supposed that Peter was drowning, pulled him and the infant he had just retrieved out of the water. He fought his would be rescuer to let him return to the water, but was held fast.

William McGaffrey, only fourteen years old, tossed a dazed girl aboard a tug then swam to shore himself. He swam out again and rescued three exhausted men who were about to drown in the shallows.

Arthur Link was on the upper deck when he noticed a woman about to jump into the water with her baby. "If you can't swim, give me that baby!" he said. She did and then jumped. The boy put the infant on a chair, which he braced against a stanchion, so that the child would not be crushed in the crowd. When he felt the deck beginning to give way, he tucked the baby under his arm and jumped. He was swimming for shore, holding the baby up, but was about to sink when a man in a small boat came along side. "I can keep up all right. Take care of the baby." He told the man as he handed up the child. Arthur swam to shore under his own power.

There are hundreds more stories like these, some recorded in local papers outside the city. Brooklyn, Queens, Staten Island, and the Bronx all had their own newspapers as did New Jersey and Connecticut. Even more stories are only known to the families of unnamed rescuers. Many of the Good Samaritans and professional responders were so traumatized by their experiences at the disaster, that they could never tell their families what they saw. Others felt that to describe what they had done was immodest. Whatever the reason, many of the noblest of the stories have undoubtedly been lost to the ages and we must hope that they have been collected in the great Eternal Book of Life.

---

[1] His medal, pictured beside his name, is engraved with his name and date of heroic service, but no written record of his bravery has been located. No descendant has been located to provide the details of his bravery.

[2] This list without the letter codes for which citation was awarded, was found in the General Slocum Scrapbook at The New York Historical Society. The letter codes were added from the Civil Lists for 1904-1905 on file at the Municipal Archives.

Above and Beyond the Call of Duty

While one might assume that the employees of the Knickerbocker Steamship Company would be responsible for the well-being and safety of the passengers, especially during a disaster, the line is not so easily drawn for those only hired for the day. This group included the galley staff, ice cream and candy vendors and the band. Although the band had been hired to play throughout the day, they were not steamship company employees and might have been excused if they had first sought to save themselves and their relatives. In fact, they were also working at a reduced rate as a favor to Mary Abendschein, who had coordinated the arrangements for the trip. But, none of those conditions seems to have entered the minds of those intrepid musicians.

George Mauer, a locally known band leader was hard at work on the second or Promenade Deck of the *General Slocum* when fire was discovered. The little ensemble was entertaining the passengers with German tunes and marches. Amongst the crowd gathered to listen were Mauer's wife, Margaretha and their children Mathilda and Clara. There may have been a third child, Julius, who was reported in the papers as missing but, no death certificate was issued for him. The children's uncle, Julius Woll, also a band member, and his wife Freda were also aboard with their daughter, Freda. The Wolls were happily awaiting the arrival of another child.

When the headlong rush of passengers, panicked by the fire and smoke, began charging past the band and up to the top deck stairway, George kept the band playing, hoping to calm the situation. Soon it appeared that the musicians might be trampled, although they probably were still uncertain as to why. They abandoned their chairs and stood together with their families beside them and played on. Soon it became apparent that no music could calm the growing beast of fear that ran this way and that all around them.

As the crush of passengers grew, it became physically impossible for any of the bandsmen to play their instruments. By this time they must have known what the danger was, smoke was roaring up from the first deck and all hands were engaged in the procurement of life preservers from the overhead wire cages. Again, the band as one tried to help. They pulled down and passed out what life preservers they could disengage. It was not until each had recognized his undeniable peril, that they relinquished their thoughts of duty and charity and sought to save their families.

Several of the band's members and members of their families were rescued from the river but, the Maurer family was not so lucky. Mrs. Maurer was injured and succumbed to shock and pneumonia. George and the girls were able to jump overboard but, another passenger, who jumped at the same time, landed on top of the band leader. When his body was recovered the unmistakable mark of a man's heel evidenced the blow that killed him. His daughters died at the scene and his little son, Julius remains a mystery to this day. Some reports said he returned home, others that he was never found.

Another Maurer daughter, Mrs. Charles Geminn, who was not aboard, later told the New York Times "Mother told me that they saw some life preservers that appeared to be in good condition but, they were wired down and they couldn't get at them."

Federal Officials

President Theodore Roosevelt — Not only expressed his condolences, but then took an active interest in the investigations which followed the tragic event. He also ordered the immediate inspection of all excursion steamers and made public his intention to closely follow the inquest, and trial if it should be convened.

George Bruce Cortelyou, — Secretary of the Department of Commerce & Labor (1903-4) took over the investigation begun by Supervising Inspector Rodie and organized a committee, at Washington, D.C., composed of:

Chairman, Lawrence O. Murray, Acting Ass't. Secretary of the Department of Commerce & Labor Headed Steamboat inspectors' Investigation for Secy. Cortelyou.

Secretary, Herbert Knox Smith Deputy Commander of Corps. In the Department of Labor & Commerce

John M. Wilson, U.S. Army Major General [Ret.][1]
Cameron McR. Winslow, U.S. Navy Commander [1]
George Uhler, Supervising Inspector General, Steamboat Inspection Service, Washington, D.C.
Robert S. Rodie, Supervising Inspector Steam Vessels, 2nd District, New York, Albany
Thomas H. Barret, Inspector General, 2nd District, New York
James A. Dumont, Inspector General, 2nd District, New York

Secretary Cortelyou further ordered that a Federal Inquiry be conducted by the Steamboat Inspection Service under the direction of:

George Uhler,[2] Supervising Inspector General, Steamboat Inspection Service, Washington, D.C.
Robert S. Rodie,[2] Supervising Inspector Steam Vessels, 2nd District, New York, Albany
Thomas H. Barret, Inspector General, 2nd District, New York
James A. Dumont, Inspector General, 2nd District, New York ; Barret and Dumont supervised the inquiry

Joseph Aloysius Goulden, (1844-1915) Mr. Goulden was the Democratic Congressman from New York for the 18th District (1903-1911). He was born and raised in Pennsylvania and served in the Union Marine Corp during the Civil War. He moved to New York City and held the positions of Commissioner and Trustee for the public schools for ten years. His interest in things military continue with his involvement on the Board of Trustees of the Soldiers' Home in Bath, New York and as a member of the commission that erected the Soldiers' and Sailors' Monument on Riverside Drive in New York City.

Edward M. Bassett, (1863-1948) Mr. Bassett was the Democratic representative from New York (1903-5); born in Brooklyn, New York; attended the public schools in Brooklyn and Watertown, New York, and Hamilton College, Clinton, New York. He graduated from Amherst College in 1844 and from Columbia Law School in 1886. He was admitted to the bar and opened a practice in Buffalo in 1886, then moved to New York City in 1892, where he continued his practice. He was a member of the Brooklyn School Board until his election to Congress. In 1904 he declined renomination and resumed his private practice. He continued in various city positions into the 1920's. He wrote extensively on bankruptcy, eminent domain, and police power.

City of New York Officials

George B. McClellan, Jr, Mayor of the City of New York. He was the son of General George Brinton McClellan (1826-1885), and was born in Dresden, Saxony, Germany of American parents. He graduated from Princeton in 1886; worked as a reporter and editor for several New York newspapers. He was the U. S. Representative from New York's 12th District from 1895-1903. He resigned this office to become the Mayor of New York City (1903-1910). Mrs. McClellan was on the Board of the Bellevue Training School for Nurses 1904-11 and 1918-24. She undoubtedly added to both the Mayor's interest in and hand-ling of the *General Slocum* disaster, since she was seeing much of the story unfold literally in front of her eyes daily as she walked the halls of Bellevue.

John F. Ahearn, Manhattan Borough President
Louis F. Hoffer, Bronx Borough President
Martin W. Littleton, Brooklyn Borough President
Joseph Cassidy, Queens Borough President
George Cromwell, Richmond Borough President (Staten Island)
Jeremiah Fahey, Chief Clerk
Charles V. Fornes, President of the Board of Alderman City of New York
John H. Doherty, Alderman of the Bronx
Mr. Breckenridge, Assistant Corporation Counsel
William Gibbs McAdoo, Police Commissioner of the City of New York.

He was born near Ramelton, County Donegal, Ireland on October 25, 1853 and Immigrated, with his parents to Jersey City, New Jersey in 1865. He was admitted to the bar in 1874 and set up a law practice. Between 1870 and 1875 he tried his hand as a reporter before running for the State House of Assembly in 1882. In 1893, he was elected to Congress and served there until 1891. During this tenure, he held the position of chairman for the Committee on the Militia. After losing a bid for re-election in 1890, he moved to New York City and in 1892 resumed the practice of law. He was appointed Assistant Secretary of the Navy by President Grover Cleveland and served in this capacity between March 1893 and April 1897. In 1904, he was appointed Police Commissioner for the City of New York and served until 1905. By all accounts, McAdoo did an excellent job of managing the myriad details of the rescue, recovery, and crowd control throughout the week of the *Slocum* disaster At the end of 1905, he returned to the law. This was precipitated by the mayor's impression that McAdoo was reluctant in dangerous situations. In December of 1905 he asked for and received McAdoo's resignation. In 1910, Mayor Gaynor appointed McAdoo Chief Magistrate of the Magistrates' Court, 1st Division, where he served until his death in 1930.

Nicholas Hayes, Fire Commissioner

Thomas F. Freel, Former Fire Marshall City of New York. Mr. Freel was detailed by District Attorney W. T. Jerome to investigate the circumstances surrounding the origins of the fire Chief Charles Croker, New York Fire Department. Based on his experience with fire boats, he moved existing fire boat units out of the regular battalion structure and created the Marine battalion. He also, recommended design changes to fire boats which substantially improved their effectiveness and efficiency. He also was in attendance at the Hoboken Pier Fire.

Mr. Flammer, Magistrate
MR. Featherson, Commissioner of Docks
Dr. Thomas Darlington, MD., Commissioner / President of the Board of Health
Mr. Tully, Commissioner Dept. of Charities [180 N. 6th St., Brooklyn]
James Dougherty, Deputy Commissioner of Charities [1131 Crotona Park North]
Daniel Doyle, Department of Charities Employee [302 Humbolt St., Brooklyn]
Andrew Hughes, Department of Charities Employee [150 7th Avenue, Brooklyn]
Bernard Lamb, Department of Charities Employee [948 Manhattan, Brooklyn]
William Hogan, Department of Charities Employee [100 North 7th St., Brooklyn]
William Flanagan, Department of Charities Employee
William Lee, Department of Charities Employee [209 East 21st. St,. Brooklyn]
Hermann Weisbauer, Department of Charities Employee [1977 Madison, Brooklyn]
Dr. Maxwell, Superintendent Board of Education
Mr. Rogers, President of the School Board
Charles Roberts, Principal, PS 25, 5th St. near 1st Avenue
William Travers Jerome, District Attorney
Francis P. Garvan, Assistant District Attorney [44 W. 44th St.]
Joseph I. Berry, Bronx Coroner (1868- 1952)
William J. O'Gorman, Coroner of the Bronx
Dr. Curtin, Listed as Coroner's Physician, aka Medical Examiner
William Mahoney, Listed with the O'Gorman party at North Brother Island, function unknown
Mr. Ruloff, Queens Coroner
Martin Major, Chief Clerk, Queens Coroner's Office [Middle Village]
Dr. Gustave Scholer, Coroner of Manhattan [311 W. 48th St.]
Dr. Goldenkranz, Coroner [98 St. Mark's Place]
Dr. Thomas H. Creighton, Assistant to the Coroner
Dr. John H. Riegelman, Assistant to the Coroner
Mr. Fane, Superintendent of Morgues (tagged bodies at the Charity Pier)
Dr. Michael J. Rickard, Acting Superintendent of Bellevue
Walter Watson, Morgue Attendant at 26th St.
Timothy McCarthy, Morgue Attendant, 608 Water St.
P. J. Scully, Clerk, City of New York
Mr. Grafling, Superintendent Casino Beach Gas Works

Steamboat Inspection Service, New York

James A. Dumont, Supervising Inspector General, 2nd District, New York
John W. Fleming, Assistant Inspector of Boilers, Local Board of the Steamboat Inspectors
Henry W. Lundberg, Assistant Inspector of Hulls, Local Board of the Steamboat Inspectors

Officers of the Knickerbocker Steamboat Company

Frank A. Brnaby, President
James K. Atkinson, Secretary
F. H. Dexter, Treasurer
John Pease, Captain of *Grand Republic*, Commodore of the Line
Chrles E. Hill
C. Delacy Evans
Robert Storey
Floyd S. Corbin

Crew of the General Slocum

Officers:

William H. Van Schaick,	Captain (60/61) born in Cohoes, NY; He was raised in West Troy, NY; both in and out of season, he lived aboard the *General Slocum*. He was a pilot and master in New York harbor for forty years and was anticipating retirement at the time of the disaster.
Edward L. Van Wart,	Captain and pilot. (62/64) he lived at 331 W. 21st St. and was described as "scrawny and hollow cheeked" with sandy side whiskers and a little shorter than Captain Van Schaick. He was at the wheel as the *General Slocum* navigated Hell Gate and steamed for North Brother Island. One account notes that it might have been his error which caused the vessel to miss the sandy bottomed beach of the island and ground on the rocky area, thus keeping her stern from reaching the safety of the shallows. His pilot's license was lifted after the disaster but, was later reinstated.
Edward N. Weaver,	2nd Pilot. (28)12th St., Troy, New York; heavy of build; in season, lived aboard the ship.
Edward Flanagan,	1st Mate. (27) In the off season he was employed as an Ironworker. He regarded himself as a mate but, did not have the required license.
James Corcoran,	2nd Mate. (20) At the Inquest, he defined himself as a "sort of Head Deckhand."
Benjamin F Conkling,	Chief Engineer. (69) He lived at Catskill, New York. His License was lifted after the disaster, but was reinstated. Had been a Civil War Dispatch boatman & had been Chief Engineer on the *Grand Republic* before the *Slocum*.
Arthur B. Bell,	Engineer. (40) After the disaster, he gave up the sea for a career as a stationary engineer.
Everett Brandow,	2nd / Asst. Engineer. (34) He lived at Catskill, New York he was the only crewman Recognized for bravery

Deckhands:

John J. (T.) Coakley,	(30)
Thomas Collins,	(25) He had been a crewman for only four days
Daniel O'Neill,	(24)
William W. Trembley,	(33) He came from California and lived aboard the ship. Employed only a few weeks, he described himself as not accustomed to the water. He was cool of manner and repeatedly swam ashore with victims.

Martin Kraljich, On the last trip, he brought in 3 babies (one by holding its
John N. Brennan, clothing in his teeth), and passed out on shore. He was
taken to the Alexander Avenue Station House where he
related his story then fell asleep on the floor.
(19)
(48)

Oilers:
Elbert J. Gaffga, (20)

Fireman:
John Tyson, (39)
Frank Silveria, (32)
Michael Lee, (46)
Patrick Mullen, (24)

Stewards:
Albert (Walter) Payne, He was also listed as a porter. [Negro]
Michael McGann, (48) 2161 8th Ave. He is the only crewman known to have died during the disaster. He may have jumped overboard carrying the day's receipts in coin bags, which weighted him down and caused his death. He considered himself the ship's purser.
Jenny E. Freeman, (33) Stewardess [Negro]
Bessie Smith, (35) Stewardess [Negro]

Waiters:
Thomas Ryan, (28)
Charles Wicker, (54)
William Wubber, (34)
Morris Hubschman, (39)
Paul C. Leonard, (24)
James Plintin, (22) Captain's Waiter [Negro]

Concessionaires and Band
(not Knickerbocker Steamship Co. employees)

Galley men:
Henry Canfield, (46 /53) Cook 421 10th Ave. He swam from boat to boat saving many victims. [Negro]
Edwin Robinson, 2nd Cook (19) 414 W. 39th St. [Negro]
George Owens, Steward (39) He was in charge of the chowder stand
Edward Smith, (68) Pantry man Died [Negro]
Robert Lucas, (32) Coffee man Died
James Woods, (45) Dishwasher Died 337 9th Avenue
Lewis Potar, (39) Bartender
George Volze, (39) 2nd Bartender
August Lutjens, Volunteer Bartender

Ice Cream Stand:
William A.(E.) Ortman

Candy Stand:
Jacob S. Jacobs

Band:
Seated on the upper deck when the fire was discovered, the band remained at their posts and played to help calm the passengers until they were nearly rushed by panicked passengers. They stood and continued to play, so as not to be trampled in the surging mob. Finally the crush was so great they could not use their instruments any longer and they stopped and began first handing out life preservers, and then their own evacuation of the inferno.

George Maurer,	Band Leader Died. Wife and three daughters aboard
Isaac Abraham,	Musician Died
August Schneider,	Coronet He had talked the band into taking the day at a reduced fee for his friend, Mary Abendschein. He later identified her body. His wife and three children were aboard
John Buhl,	Brother-in-law of Agnes Zimmerman
George Dillemuth,	Violin He grabbed a life-preserver and threw it around his neck, then jumped before the steamer beached. He was picked up by a following vessel and was heard to repeatedly mourn for his lost instrument.
Julius Woll,	Clarinet
William Zimmerman,	Died
"Ikey,"	Drummer Boy Died

Most of the General Slocum's crew and several of the concessionaires were called to testify at the coroner's Inquest, the Grand Jury, and various trials. For more on the legal ramifications of the disaster see following sections.

---

[1] Appointed by the President at the request of Secretary Cortelyou
[2] Mapped out the course of the investigation

Scales of Justice

As soon as the magnitude of the disaster began to become apparent, Supervising Inspector Rodie was asked to send inspectors to check the *Grand Republic*, in order to allay the public's concerns as to her safety and her compliance to existing standards. Inspector Rodie declined. When pressed he stated that there was too much red tape involved and that "...the request would not be heeded anyway." It had to be received from only one source, the ship's master or owner, and then had to await possibly several reviews before action could be taken. The public thus discovered that the Steamboat Inspection Service was a red-tape monster, which could not move on its own accord.

That same day, Coroner O'Gorman stated in the press, that Supervising Inspector Rodie had gone to North Brother Island during the recovery efforts and tried to remove evidence of the neglect by his department. Further, O'Gorman said that if Inspector Dumont had requested evidence, it would have been generously provided.

On the evening of the 15th at North Brother Island, Police Commissioner McAdoo announced that he would construct a plan with the Dock Department to inspect every steamer in the harbor. The police would carry out these inspections. During the days preceding the Coroner's Inquest, Collector Stranahan and the Surveyor General of the Port of New York, Mr. Clarkson were thinking along the same lines, but chose instead of direct actions to sponsor a Conference to determine if better methods could be devised which would insure more thorough examinations of the vessels in New York Harbor. No results of this conference have been found in published sources to date. The New York press for the most part, followed a substantially different line of thinking.

Shortly after the disaster, a trade paper, American Siren and Shipping, published an astonishing article, which was later reproduced in the Literary Digest of June 25, 1904. This excerpt is unusual in the modernity of its thinking. All of these measures are now considered mandatory on all commercial shipping, save the use of asbestos, which has been replaced by other materials.

"Upon every deck of an excursion-steamer like the *General Slocum* passengers are seated as closely as in a theater. During a theater performance, however, firemen are stationed at all the principal points to take the most prompt measures for extinguishing combustion, to aid in preventing panic, and in saving life as well as property; there is also an asbestos fireproof curtain to exclude the stage from the auditorium. Upon one of our modern excursion steamers there is none of these things. Yet there is no reason why the superstructure of one of these craft should not be almost wholly fireproof. Light steel could take the place of wood in every particular, and the asbestos coverings now available are a protection which it is almost criminal not to use. There is no reason, either, why every deck of such a craft should not be divisible by fireproof sliding doors or asbestos curtains to prevent or at least retard the sweeping of flames from one end of the vessel to the other. These are things which must come in time, just as certainly as the theaters have been safeguarded. The *Slocum* catastrophe is liable to work damage to the patronage of excursion steamers and all similarly constructed vessels until such a reform is affected. To have such a loss of life from fire on any ocean steamer to-day is absolutely impossible, because no such conditions inviting disaster are permitted. Bulkheads, fireproof material, abundant fire-fighting apparatus, and the prevention of over-crowded decks, as well as the equipment of well-disciplined crews characterize all modern liners in the ocean trade. This is not to say that all ocean steamships are as yet perfect, but that our excursion-steamers of the type referred to are culpably imperfect."

Equally concerned with protecting passengers while aboard, and looking at the same model of commercial shipping as the American Siren and Shipping, the Brooklyn Eagle, also asked why other modern life-saving methods employed on ocean liners were not mandated on excursion vessels;

"The ocean liner, made to carry a thousand, is supplied with abundance of boats and rafts, fire-hose and grenades, her crew is well drilled, her hull is divided into water-tight compartments, so that unless she strikes a rock with a tremendous shock it is possible to close these compartments and preserve a large measure of her buoyancy. The iron partitions that will keep out water will also keep out flame and smoke, and the passengers can readily be gathered into the uninjured divisions of the ships. Yet we permit companies of people twice and three times as large as these liners would carry to put to sea in wooden cockle shells that a careless smoker may set on fire. It must be a salutary revolution in marine architecture that will be induced by this burning of the *Slocum*. We must have boats that, if they will burn at all, will burn so slowly that the passengers can be removed in safety. The tragedy of yesterday must never be repeated."

Volunteer rescuers cluster around the wreck in hopes of finding more victims alive

Neither of these publications, nor the hundreds of letters to the editor received in the week following the tragedy could anticipate just how far-reaching the changes they cried for would actually be.

Coroner's Inquest
2nd Battery Armory
177th St. & Bathgate Avenue, New York, New York
June 20 - 28, 1904

In charge of the inquest were;	
Joseph L. Berry, Presiding	Coroner, city of New York
William John O'Gorman, Jr.	Coroner for the Bronx
Francis P. Garvin	Deputy Assistant District Attorney. He was put in charge when the District Attorney, William J. Travers, was detained at Lakeville, Ct.
Mr. Burnett	U.S. District Attorney
Mr. Trunbull	Assistant District Attorney
Abram Dittenhoefer[1]	Ex-Judge, Personal counsel for Frank A. Barnaby
Terence J. McManus [1]	Counsel for the Knickerbocker Steamship co., from Black, Olcott, Gruber, Bonynge
Julius Mayer [1]	Ex-Judge, Court of Special Sessions, Represented Insp. Lundberg
Mr. A. S. Gilbert[1]	Also represented Lundberg
Captain Benjamin f. Perkins	Secretary of the Pilot' Association. Offered to assist with the Inquest
A panel of fifty-three talesmen was called from which to pick a jury of fifteen; a Mr. thorn was elected Foreman.	

The Coroner's Inquest has been reproduced here as a series of abstracts of testimonies taken from the various New York papers. To the best of the author's knowledge, no actual copy of the inquest exists today. This is unfortunate since it set the tone and laid the foundation for all further investigations, trials, and assumptions as to participants' personalities and expertise. Later judicial actions revealed little or no new or countermanding data. Still, some of the information, which might have been laid out in this proceeding, might have possibly answered some of the still perplexing questions associated with this horrendous event. As it was, this initial investigation into the causes of the disaster became a *cause célèbre*.

Even before the fire was out, New York wanted answers. How could this happen? Who was to blame? In 1900, few individuals considered a lawsuit as a normal recourse for death or damages. It was unheard of for individual to sue corporations, and when the rare case was attempted, the corporations won.

Since the steamboat company did not seem the most likely candidate for official blame, New York turned to the next most culpable figure. This they found in the *General Slocum's* captain, William Henry Van Schaick. Although a captain and pilot for half a century, his record became the immediate fodder of the press. His was the first and last name presented to the public. They clamored for his head on the proverbial silver platter. What the public never heard above the din of public outcry was the positive validations of his marine abilities by watermen who had watched his comings and goings daily for decades under conditions of every sort. Compounding the public's erroneous impressions was the somewhat romantic idea that a captain is always the "master of his own ship". On the high seas, and especially in military situations, this may be true. Distance to land and lack of communication of necessity made ship captains little Caesars. Decisions could not wit months for landfall! But, in inland waters, a ship captain was more like an office manager. He could and did make decisions about the running of his vessel but, he was also at the mercy of the company from whom he drew his pay. All of this would come to bear on the inquest and trials that followed.

  June in New York City was usually quiet, politically speaking, and the press coverage of the disaster brought out the politicians and city officials to "be seen' and interviewed on their way into and out from the inquest. Among the spectators on the first day were many notables; Congressman William Sulzer, from St. Mark's district, who would remain in the forefront of the survivors' struggles for nearly a decade; Representatives of Frank W. Higgins, Governor of New York; representatives of Commissioner Cortelyou, the Mayor, George B. McClellan, Jr., Police Commissioner William G. McAdoo, and Nicholas J. Hayes, the fire commissioner.

  The coroner recognized the need to move quickly and convened the inquest just five days after the disaster. Throughout the recovery, O'Gorman had directed the divers and police to collect items to be used as evidence. He had requested a retired fire marshal to investigate the fire's origin. Now he would show the world what had happened. The evidence included a section of the standpipe from the *General Slocum*, which showed no attachments. This meant that no effort had been made to use a fire hose on the starboard side of the ship (the side away from the fire). There was also a section of hose from the wreck and deplorably mangled and shredded life preservers showing the stamp "1891" signifying the year they had been inspected. Documents entered in evidence included a copy of the 1904 Steamboat License Inspection for the *General Slocum* and a letter from the New York Belting and Packing Company documenting that hose had been bought at $0.40/Ft. with a 60% discount; the final cost was less than the cheapest garden hose, which was available for $0.16/Ft.

  If negligence on the part of the company could be proven, prior to sailing, then the company would be responsible for the disaster and company property could be attached as well as the stockholders' assets. Mr. Barnaby and Captain Pease were both stockholders. Pease also owned company bonds. In fact, of all the principals of the company, only Van Schaick did not own stock or bonds in the firm. Pease was also listed as the head of the Operations Department and passed all procurement bills to the office for payment. He had made application for the inspection in May but, made no comment to Mr. Barnaby on the condition of the life preservers or hose.

Called to testify:

President Barnaby was the chief officer of the Knickerbocker Steamboat Company and a stockholder. He testified that he was not a steamboat man and so relied implicitly on Captains Pease and Van Schaick. If they had mentioned anything, he would have corrected the problem immediately. According to Barnaby, $12,000.00 was spent in February 1904 to get the *General Slocum* into top shape for the upcoming season. The *Slocum* had cost $165,000.00 and carried $70,000.00 in insurance. The insurance company considered her a good risk. Although asked to bring the company's books, he first stalled and then produced extracts of the books. Eventually he was compelled to bring the bills back to 1902 when 350 life preservers had been bought. The bills had been changed by the use of acid[2], to say *"General Slocum"*. He was then told to bring the bookkeeper for the day's afternoon session. She finally appeared the following day.

Miss S. C. Hall, the bookkeeper of the Knickerbocker Steamship Co. testified that the changed notations from *Grand Republic* to *General Slocum* were done by her. She made the adjustments when Captain Pease told her how they were to be deployed.

Captain John A. Pease, Captain of the *Grand Republic*; stated that he had put the *Slocum* in order that spring. He had nothing to do with the life preservers on board, and he did not know whether any of them were stuffed with bull rushes. He had purchased three hundred and fifty life preservers for the *Grand Republic* this year. He further claimed that he had overhauled the *Slocum* this spring, but had only supervised the hull and machinery. Captain Van Schaick had looked after the life saving and fire apparatus. He had positively never had any conversation with Miss Hall about life preservers for the *Slocum*. When asked about the cost of hose, he said good hose could not be bought for $0.16/ Ft. (He had not heard the earlier discussion concerning the cost of hose.)

Secretary Atkinson testified that Captain Van Schaick was under Captain Pease's supervision.

Oscar Kahnweiler, a principal of the firm which sold 2,250 life preservers to the Knickerbocker Steamboat Company for the *General Slocum* in 1891, testified that since that date, they had billed no more to the *General Slocum*. Furthermore, he testified that his life preservers when Properly stored, should last 20 years; poorly stored, as on the *General Slocum*, exposed to the weather and salt, no more than six years. When shown one from the *Slocum*, and asked if he would trust it with his hands and feet tied, he said "Yes."

Miss Reba Goldberg the bookkeeper, for David Kahnweiler & Sons. (life preserver Manufacturer) testified for which boat life preservers had been ordered. Pease had ordered all life preservers and had received all on board the *Grand Republic*.

John J. Coakley was the deckhand, who may have been the first person to see the fire, which he stated was discovered just after passing Blackwell's Island and the beacon light. He had been employed for 18 days at a rate of $6.25/week plus board. He had previously worked on the *General Slocum* in the 1890's. He stated that there were no fires drills or instruction and that the lamp room was not locked. When a small boy alerted him to the fire, he was having a beer at the bar. He went into the cabin to find something to smother the flames, but there was nothing there. Then he called three deckhands. He also stated that he felt the fire had started in the lamp room, where lots of hay from two barrels of glasses had been stored. He and Flanagan had gotten out a fire hose, which was badly kinked before it burst.

With no time to straighten it out, little water could be forced through. Also, a coupling had gotten loose and before that could be tightened, the crowd had panicked. In any case, the hose, he was sure, was not rotten. When the hose failed, he went to the top deck and loosed life vests. Another deckhand gave him a baby, but he couldn't help the passengers due to their panic. He got a boat down [in the water] but, it was swamped.

Walter Payne testified that there were many types of inflammables in the cabin which was Never locked. He was near the gangway forward with a shoeshine kit. Someone said something was wrong up front, so he and the mate ran forward, but were stopped by the smoke. The mate then called up to the pilothouse. The hose kinked, they tried to clear it but, the water only went through three or four feet and then burst. Payne then started pulling down life Preservers, and did not keep one for himself. He couldn't swim, and was later rescued from the paddle wheel.

Edward Flanagan the 1st Mate of the *General Slocum*, was on the witness stand for half a day. He reported that he was depressed the ordeal had unnerved him. He testified that he was a steamboat master in summer and a foundry mechanic in winter. It was his second season as Mate. He hired the seven deckhands and was in charge of the lamp room (forward cabin), but no charcoal or kerosene was in that cabin and only one barrel of mineral oil. He had not seen barrels of glassware, but thought the passengers might have brought bananas packed in barrels. He noted that the forward cabin door was always open. The fire hose was never used, not even to wash the deck, which was done from pier side outlets. He had first noticed smoke when Coakley approached him. Flanagan then went forward and looked down the companionway, returned, and called the captain on the speaking tube. He then went to the engine room and asked the second engineer to turn the water on. But when he tried to use it, the hose burst and the coupling flew off. He never saw the captain on the lower deck. Flanagan had been aboard when the season's inspection was made, which included the life preservers. The inspection constituted poking a few in the overhead with a stick or cane. All those that he remembered were twenty stamped 1891, the *General Slocum's* launch year and as far as he knew, only ten to life preservers were rejected by the inspectors before the season began. He could not remember any test of the standpipes. After the grounding, he jumped overboard. He had not seen the porter on the morning of the fire. At various times, he had lighted matches in the forward cabin.

Edwin N. Weaver the 2nd Pilot of the *General Slocum*, (statement taken aboard the police boat *Patrol*) He had been licensed since May 1900, and on the *General Slocum* five or six years. He never saw a fire drill. He was sent to buy one hundred feet of hose at the start of the 1904 season, and paid $0.16/ foot. He, Van Wart, and Van Schaick were in the pilot house and were past the black spar buoy on Sunken Meadow and three lengths past Sunken Meadow (about one hundred feet southeast of the Bronx Kills, between the spar buoy and the signal box) and the wind was blowing when Flanagan called on the tube. They were moving at full speed and with the flood tide. The captain went to see the fire and returned in one minute calling "put her on North Brother Island quick as you can". Weaver then started and continued to blow the whistle for assistance. If they had beached at Locust Point (129th St.), it would have taken five to eight minutes with a wrenching turn, but North Brother was a gradual change of course, and better for the passengers to escape. Van Wart had the wheel, and they went straight up the course. No one came to the pilothouse to advise them of the danger.

James Corcoran, self styled "Head Deckhand" of the *General Slocum*, had worked aboard for four years. In the off-season, he drove a newspaper delivery wagon. He had been standing just behind the mate on the left side [port] deck around 9:40 a.m. and saw fire on the mid-ship gangway, about 25 feet from the forward cabin. He yelled to the mate, then he and another deckhand tried to stretch out the hose, but it kinked. Water ran in it for about five minutes, but no pressure developed. By this time the fire was all over the entire bow end of the forward main deck cabin. Corcoran started for the top deck between 105-106th Street. He also testified that Conkling was among the first to leave the ship after the grounding.

Charles A. Long was a passenger and at about 57th Street, he overheard someone saying "There's something doing up forward"; he saw fire at 94th Street.

Benjamin F. Conkling had been Chief Engineer of the *General Slocum* for twelve years. There was no standpipe and valve to the forward cabin for fire control. The *General Slocum* was running slowly, [perhaps 6 mph] and had stopped several times for passing vessels. Her speed was not increased in Hell Gate. The fire swept from stem to stern in a maximum of 15 minutes; the deck fell in as the ship grounded.

Frank Perditsky (age 14) was a survivor; he had gone with his mother. He was near the pilothouse when he saw smoke coming up from below. He told the captain, but was told to "mind your own business," the ship was at about 83rd Street, near a park he knew well. [Later] he couldn't find his mother, so jumped overboard and swam to North Brother Island.

Thomas Collins also a deckhand, he had worked the *Slocum* for four days and had had neither drill nor instruction in procedures. He had been detailed to stand near the forward gangway at the time when the fire started. His first notice was a woman who was screaming. He took a hose down and ran the nozzle to the door, but it burst. Then all hands ran to the upper deck. He did not jump until they had grounded. He had seen no deficient life preservers.

Everett Brandow was an assistant engineer on the *General Slocum* for six months. He testified that Flanagan had reported the fire at about 10:30. He could release steam only into the fire room [boiler room] but, no other compartments were equipped. The *General Slocum* ran at full speed through Hell Gate; as she exited, Van Wart signaled for "slow" as passengers did not want to get to the grove before 1 pm. After the fire was reported, Van Wart called for "full ahead" then "stop" then "go ahead, slow, full ahead". All these orders came almost all at once. When they beached, Brandow stopped the engines. Conkling had gone to the donkey engine at the first alarm, and Brandow didn't see him again until they were both on the island. Brandow jumped ahead of the wheels and helped rescue a girl in the shallows.

Mrs. Marie Behrends of 88 3rd Street testified that they had just passed Blackwell's Island when she heard someone yell "fire." She found a life belt and put it on her daughter, Annie; it saved her life. She found her children and hung on to the rail until rescued (one daughter was saved, two others died). The officers and crew did nothing for the passengers.

August Lutjens He saw smoke coming from the fore cabin at about 90th Street and moments later saw the 92nd Street ferry. He watched the hose being uncoiled but, did not see any water come from it.

Paul Liebenow realized that there was a fire between 90- 92nd Street and heard others implore the captain to put in at Sunken Meadows. He cut his hand on the wire cages supporting the life belts as he tried to get them down.

John Engleman was aboard with his wife and family. He noted the fire at 92nd Street. Since he had worked for the New York, New Haven, & Hartford Tug Co., he knew the river's landmarks.

Henry Hardekopf of 343 Rivngton St., stated that his mother saw the fire at the north end of Blackwell's Island, she died.

William W. Trembley had been a waiter on shore until May of 1904. Now a deckhand, he Testified that he had assisted the engineer in pulling down the hose and tried to attach the nozzle without success. The hose was too defective to allow water to pass through it. By this time, they were past Blackwell's Island. He then pulled down thirty to forty life preservers noting that some had holes. Before the 15th he had called these holes to the Mate's attention and was told to remove them. (Preceding the inquest, he had become so overwhelmed with emotion that he was taken from an initial interview with Coroner Berry to the hospital.)

Daniel O'Neill (age 24) was working his first ship's job and had been on the *Slocum* since April of 1904. He had helped load five barrels of glasses Tuesday evening but, he didn't know where they went. He had not been in the forward cabin on the morning of the 15th. He was on the port gangway amidships when he heard a shout and saw smoke. Flanagan came yelling and O'Neill helped take down a hose and turn on the water. He saw the water gush and split the hose. There was a call for another hose at about the same time. He thought they were through Hell Gate. He and Corcoran waved to a passing tug with lighters, and then tried to pacify the crowd. He saw a rowboat coming alongside and he jumped into it to help the man at the oars, it capsized and O'Neill swam to shore[3].

Martin Kralijich was the youngest deckhand on the *General Slocum* at just 19 years of age, and testified to the lack of fire drills. Further, he stated that he was on the Hurricane Deck when he saw the fire as he was watching the children. He pulled down about a dozen life preservers. He tried to prevent panic but, as soon as the boat beached, he put on a life preserver and jumped overboard. No one had given notice that there was fire.

The Rev. Julius Schulz from St. Luke's Lutheran Church, Erie, Pa. was a guest of Pastor Haas'. He stated that his first warning was flames in the gangway. He saw a deckhand run from the cabin, brush aside inquiring passengers gruffly, and jump overboard. Children started jumping, he tried to restrain them and tell them to wait for a tug to rescue them.

John Van Gelder was considered an expert witness, having been a Pilot for twenty-seven years. He piloted Weaver and the investigators over the *Slocum's* course in the police launch *Patrol* If the wind, tide, and position of the *General Slocum* had been as stated, the captain could have put in against the bank of Port Morris Point between Locust Avenue and 129th or 130th St., in the Bronx; the wind would have held her against the bank and blown the fire forward giving the passengers a better chance to escape. When asked if he felt that the captain had done wrong, he said no, but he might have lost his head.

Andrew Stebbins a patrolman, detailed to Randall's Island, testified that he was on the Southeast side of the island when he saw the *General Slocum* about five hundred feet below the spar buoy and saw smoke from the main deck. One to two minutes later, the ship was opposite the Kills and there were flames from all parts of the main deck midway between the paddle box and the portside [stern].

Two unnamed policemen first testified that the Captain could have put in at 129th St. but, later admitted that they knew nothing of navigation, water and wind conditions, or the difficulties of turning so large a boat.

Patrolman Herbert C. Farrell agreed with another policeman, Thomas Collins, that the *General Slocum* should have beached on a mud bank on the West shore at 129th Street (Farrell had sailed the East River on his own for 20 years.)

The jury was taken by the Police boat *Patrol* to tour the wreck and was then taken up the East River along the identical course that the *General Slocum* had followed the previous Wednesday morning. The jury toured the re-floated hulk, while the Fire Commissioner's theory on how the fire started was graphically explained. Of the superstructure, only the iron stairs, rods, and walking beam survived. Passengers, who died on the deck, had nearly been incinerated and their jewelry had melted. Approximately half a bushel of jewelry and gold was tagged and removed after the fire by the police. Below the main deck level, the forward cabin where the fire started was comparatively well preserved and showed little damage. Barrels of oil stood only a few feet from the one where the fire started, and showed little scorching. These characteristics were repeated for other flammables, such as hay and paint. By contrast, tugs rendering assistance during the rescue by coming alongside the *General Slocum* were badly scorched; their paint blistered off; windows were broken from the heat and in several cases, superstructures nearly burst into flames and had to be doused by the fireboats. This was because the fire had been pushed up the companionway by the wind and away from the combustible materials.

Thomas F. M. Freel Captain of Engine Co. # 8, and recipient of the 1903 Stephenson Medal; Former Fire Marshal and expert witness was retained to assist Assistant D.A. Garven. He took the Jury through the wreck explaining how and where the fire had started.

Captain Edward Van Wart, Pilot of the *General Slocum*, limped into the inquest, having hurt his hip landing in very shallow water at North Brother Island. He had been a pilot nearly thirty years, and on the *Slocum* since 1892. He testified that twenty five life preservers had been bought in 1895, when her permit was increased from 2,300 to 2,750 passengers. He did not know if the inspection at the start of the season included the life preservers. The lifeboats were in good condition and had needed no repairs. They were only lowered when painted. He had not seen any lowered during the disaster. He had not personally inspected the hose.

Ida Hayden The step daughter of John Holthusen, Superintendent of the St. Mark's Parish School, testified as to the condition of the life preservers Joseph Gaffney the Chief Engineer of North Brother Island, saw a woman in the water with a life preserver which split in half and floated away from her.

Jacob S. Jacobs was a survivor; he had a hard time getting life preservers down from the Wire cages; he never saw the crew during the disaster.

Mrs. H. W. Turner testified that the life belts powdered when removed from the rack and the cork dust blinded her. She tried to find a good one but, gave up after three tries and jumped to a tug with her child in her arms. Her nephew and her sister died.

Mr. John Kircher testified that his youngest child, Elsie (age 7) could not swim, so he put a belt on her. She sank like a stone. She was the only child in family to die.

Annie Kip (age 11) a resident of 3rd Ave.; told how her cousin drowned wearing a life Preserver.

Thomas Ryan worked the chowder stand on the Promenade Deck. He pulled down life preservers and put one on the boat's steward who jumped overboard and sank with the ship's money bag; Ryan then helped a boy to get ashore.

William J. O'Gorman, Jr., Coroner, Testified that he felt the life preservers were life destroyers that had acted as weights to drag victims down. He also stated that the hose would have needed to run with water for 10 minutes in order to be effective.

William A. Ortman was in charge of the Ice Cream Booth; he saw the crew fail to connect the hose to the standpipe and he noted that the hose was thin linen and leaked like a sieve.

O.M. Habuer He was a salesman for the New York Belting and Packing Company, Ltd., the supplier of the fire hose. He stated that the hose should have passed a test of 350 pounds hydraulic, or a working pressure of 150 pounds. The *General Slocum's* pumps could not generate more than 100 pounds of pressure, which he noted was equal to most of the city's tall buildings.

George Owen noted that once the hose failed, the crew disappeared and was not seen again onboard the boat.

Thomas Henry Barrett Inspector of Boilers, Port of New York. Since the *General Slocum* carried no cargo, she had no hold and did not require steam pipes for fire safety. Also, a hold was secured with hatches. Therefore he did not apply the safety law concerning steam piping into all compartments to the *General Slocum*[4]. When asked if he could show anything in the law which specifically covered this type of steamer having no hold, he said he could not.

At this point, Representative Goulden, a juror, asked Barrett what would have been the objection to putting a pipe in that storeroom. "None at all" was the reply and he admitted that it would have been a good thing. Robert Jacob, a shipbuilder on the jury, questioned Barrett and maneuvered him into admitting that there is no such thing as a ship without a hold

James A. Dumont, Inspector General, 2nd district, New York, of the Steamboat Inspection Service explained that the ruling of the U.S.S.I.S. was that river vessels unless cargo carrying, had no hold. Therefore the *General Slocum* had no hold. It was not part of Lundberg's duty to test the life preservers other than for large and obvious holes or tears and that the needed straps were attached.

Inspector Rodie of the Steamboat Inspection Service, testified that the

General Slocum's life preservers had been inspected at the time the last three hundred were purchased in 1895; two hundred had been repaired. Only a random sample of life preservers were inspected at the manufacturer, and the inspectors took their word that all in a pile were the same. As the preservers fell apart, they were tossed into a forward cabin, but no new ones were bought for the *General Slocum* after 1895.

John W. Fleming, Inspector, Port of New York; the *General Slocum* had no hold, so he did not check forward compartments for safety equipment. He and Lundberg had worked separately. He had tested the boilers and engines and found them in good condition. He had not put pressure on the fire hose with the donkey engine, nor did he go to the forward compartment. He also stated that the *Slocum* did not need piping to compartments other than the fire room [boiler room].

Captain Dean, Police Harbor Squad, It was his opinion that storerooms, boiler rooms, and galleys should all be lined with fireproof material.

Captain William Henry Van Schaick had been the captain of the *General Slocum* since 1891. He Was wheeled into the inquest, the first time, from the hospital on a stretcher. After the coroner and the lawyers conferred, it was decided that he should return to the hospital. He had sustained burns to his hands and head and eventually lost the sight in one eye as a result of the fire. His spine was injured when he jumped overboard at the island and landed on some rocks; he also sustained a broken heel. It was noted that he would be crippled for life and that his nerves were shattered. The coroner told reporters that if the captain could not testify on Monday (this being Friday), the Inquest would go forward without him, since he was not necessary. The following Monday, he declined to answer if he had anything to do with the fire apparatus. It was noted that when launched, the *General Slocum* was licensed to carry 2,500 passengers but, in 1895 that had been raised to 2,750. He did note that over the years no new life preservers had been bought but, in that time the inspectors had only condemned four or five and Capt. Van Schaick had done the same with another fifteen to twenty. Also three hundred had had new straps attached. He never discussed any of this with Barnaby. Captain Van Schaick also observed that Mr. Barnaby had been aboard prior to this excursion and thought that the ship looked fine. No one inquired about the life preservers. Inspector Lundberg had rejected only one as dirty and did not test the hose. He insisted that there had been three or four fire drills this year.

When he first heard fire, he was in the pilothouse and three or four lengths north of Sunken Meadow[5]; he went to see how bad it was and was driven back to the Hurricane Deck [with pilot house] by the fire. He called to beach at North Brother Island as the ship was already lost. He told the pilot to "skin the dock and beach with the starboard side to the island, so people could get off away from the fire. Van Schaick then took up a position about fifteen feet in front of the pilothouse and directed the grounding from this vantage point. After answering the district attorney's questions, the Captain turned to the jury with these remarks;

> "I want you gentlemen to understand that I did all in my power that could be done under the circumstances at the time of the accident. I could not turn the boat and beach her any sooner than I did. I remained standing in front of the pilot house until we struck. Then I started to go downstairs. A volcano of fire had broken out along the decks and was sweeping up the stairways. I could not get down the stairs. Some one yelled, "Jump for your life, captain" and then I

jumped overboard. I struck in shallow water on the rocks and broke my knee. My clothes had been on fire. I was stunned. "A large, fleshy woman and a young nurse girl came to me and dragged me up over the rocks to the shore. Then I managed to crawl up to a small tree and I lay down under it. As I Looked toward the *Slocum* I saw the pilot house and the upper decks fall."

At the conclusion of testimony, the captain was returned to the hospital, where he spiked a high fever. Had the *Slocum* not caught fire, the captain had planned to retire at the end of the season. Interviewed afterward, the captain reported that he knew there was fire at 132nd Street [halfway between Sunken Meadow and North Brother Island]; when told that Weaver had said they knew at 128th Street he amended his statement to match Weaver's and then stated that they could not have beached at 129th, as some suggested. Concerning the statement of the boy who tried to report the fire, the captain remarked; "how could a boy tell me at Blackwell's Island? I was in the pilothouse and the door was locked!"

John Wade, owner and engineer of the tug *Wade* saw that the captain and one pilot remained at their posts in the pilothouse until after the beaching but, Flanagan did nothing to help in the rescue. Van Schaick had beached in the best available place.

Foreman Thorn read the recommendations of the Inquest:
- The catastrophe could have been avoided and criminal prosecutions should follow.
- The crew was not properly equipped or trained.
- The President and members of the Board of Directors were responsible for provisioning the ship. Therefore they were guilty of criminal negligence in the failure to maintain properly the fire and lifesaving equipment aboard.
- Captain Van Schaick was guilty of criminal neglect of duty for permitting unsafe conditions to exist. He was also criminally responsible for the accident.
- Captain Pease was also guilty of criminal neglect of duty for permitting unsafe conditions to exist and He shared in the responsibility for failure to properly equip the ship with fire and life saving appliances.
- The Mate, Flanagan, was a coward and was criminally responsible for failure to perform his duty and for impersonating a mate, since he had no license.
- Lundberg had not performed a proper inspection. He had failed to report the true condition of the vessel to superiors

and was criminally negligent for the incompetent and careless inspection of the ship.

Therefore, the persons listed below were bound over for the Grand Jury on charges of criminal negligence/ Manslaughter in the second degree. Under New York law, when criminal negligence resulted in death, then the indictment must be brought as manslaughter.

William H. Van Schaick,	Captain of the *General Slocum*
Edward Flanagan,	Mate of the *General Slocum*
John A. Pease,	Commodore of the Fleet, Knickerbocker Steamship Co.,
Frank A. Barnaby,	President, Knickerbocker Steamship Co.
James K. Atkinson,	Secretary, Knickerbocker Steamship Co.
Charles E. Hill,	Director, Knickerbocker Steamship Co.
C. DeLacy Evans,	Director, Knickerbocker Steamship Co.
Robert K. Story,	Director, Knickerbocker Steamship Co.
Floyd S. Corbin,	Director, Knickerbocker Steamship Co.
Frank H. Dexter,	Director, Knickerbocker Steamship Co.
Henry Lundberg,	Probationary Inspector[‡], Steamboat Inspection Service

In an interview prior to the start of the hearings, an unnamed counsel was quoted as follows;

> "A man's duty depends upon what his judgment at the time of the danger may [allow]. It would be a harsh law to hold a man criminally guilty when he had done [any] thing to his best judgment even if that [judgment may] prove to have been in error."

The jury disagreed. Pending the hearing before the Grand Jury, Lundberg and Flanagan were held on $1,000.00 bail, the Directors, $5,000.00, and Captain Van Schaick, $5,000.00. The crewmen of the *General Slocum* (Brandow, Coakley, and Trembley) were committed to the House of Detention as witnesses for the Grand Jury.

At the close of proceedings, the defendants were required to pay bail. Captain Pease, accompanied by his sister, Mrs. Caroline Armstrong of Brooklyn, went to the coroner's office to surrender. Mrs. Armstrong signed his bail for $5,000.00 and handed over the cash. Mr. Dexter made a cash bail of $5,000.00. C. DeLacy Evans, director of the company, chose to telegraph from Rye Beach, New Hampshire that he would appear and furnish bail whenever he was wanted. The response was – forthwith! Robert K. Storey was in the North Woods but was expected back in communication soon.

Only Captain Van Schaick[°] was indicted on two counts of involuntary manslaughter and on a third count of Neglect of Duty by failure to maintain proper fire drills aboard the vessel *General Slocum*. All the others indicted were given minor fines and reprimands. The press had a field day which equaled any modern checkout counter tabloid. Undoubtedly, they helped to fuel New York's passions about this case. Here is a sampling:

"Not one of them had ever been instructed in a fire drill, or had ever learned the station he was to take or the duties he was to perform in case of fire or panic. Only a few of them were in any way familiar with the boat. Not one of them saved a human life when the *Slocum* burned."

"One, a land laborer, a few days on the boat, testified that he leaped into the lifeboat when it was lowered and swamped it. The captain, according to the testimony of the second pilot, Edwin Weaver, was not in the pilot house at any time after the fire was reported. The engineer, according to the testimony of the assistant engineer, Brandow, was not in the engine room after the fire was reported."

"The mate, Edward Flanagan, a house smith, acting as mate without a license, in violation of law, did nothing toward marshalling the crew and instructing them as to what they should do after the fake fire hose, at sixteen cents a foot, had burst, and their first futile, senseless effort to use the hose without taking out the false washer had failed of effect. He made no attempt to use the hose attached to the other standpipe forty feet aft of the forward standpipe, and on the starboard side, clear of all flames, according to the testimony of all the witnesses."

"And added to these evidences of criminal economy was the unpleasant spectacle of the United States Steamboat Inspector [Robt. S. Rodie] responsible for the condition of the apparatus on board the *Slocum* deliberately refusing to testify or in any manner aid the Coroner and the Assistant District Attorney in ascertaining the facts connected with the frightful catastrophe."

In an interesting sidebar to the inquest, Police Commissioner McAdoo instructed his precinct captains to report to him on the general provisions for the handling of excursion crowds at the piers in the thirty-six affected precincts. Although the captains reported generally good conditions on the piers, their evaluations of the boats were not so encouraging. Sharp criticism was made of the narrow stairways on the boats and the general class of employees used to crew the vessels. Captain Dean, of the Harbor Police, recommended that the carrying capacity of many of the excursion boats be cut down and that storerooms for oil, paints, and other inflammable materials, as well as the boiler rooms and galleys, be lined with fireproof material.

Thousands of letters poured into the New York and Washington offices of the Steamboat Inspection Service, filled with the comments, criticisms, and inventions each writer felt would prevent future occurrences. General Uhler commented, "...I do not know whether any of them are useful. It is not my province to decide. It seems to me, though, that the only way to prevent such horrors as that in New York Harbor is to require all the boats to be built fireproof. Time and experiment may prove that this is possible." In fact, in 1904, ocean going vessels were already governed by many of the fire and safety regulations and building practices used today. Only inland or river craft were exempted and the *General Slocum's* usual routes along the coast placed her in this category.

Coroner Wm. O'Gorman on scene at North Brother Island

[1] The counsels for Mr. Barnaby and the Knickerbocker Steamboat company and those for the officers and government inspectors argued heatedly but, were allowed to continue only as a courtesy, since only Mr. Garvin and the representatives of the U. S. District Attorney were mandated to participate in the Inquest.

[2] This commonly used technique "erased" ink from paper; however, like the imprint of penciled lettering on paper, which remains after erasure, acid also often left an imprint, which was easily detected

[3] Accounts vary on this point –was this the only lifeboat lowered from the *Slocum*, or another, possibly from the *Wade*, or a privately owned and commandeered rowboat brought to the scene by a good Samaritan, there may never be a definitive answer.

[4] This piping would allow a compartment to be flooded with steam in the event of fire, much like modern sprinklers. The argument presented here and in other inquiries was a poor attempt by the Steamboat Inspection Service to justify its lack of adequate inspection procedures.

[5] Several reporters wrote that the captain and several of the crew had "put their heads together" to come up with a plausible story regarding their actions and the position of the ship when fire was discovered prior to the inquest. This seems highly unlikely because each of the crew and the captain had been taken to several different hospitals and held there under arrest with guards on their doors until the inquest had begun.

Proceedings of the Federal Grand Jury
Indictments handed down
July 30, 1904

Judge e. B. Thomas, Criminal Branch of the U.S. Circuit Court
U. S. Assistant District Attorney Ernest Ellsworth Baldwin [1]
U. S. Assistant District Attorney Henry A. Wise

Former Judge A. J. Dittenhoefer represented; Frank A. Barnaby, James K. Atkinson, F. H. Dexter, chas. E. Hill, c. DeLacy Evans, Robt. Storey, Floyd S. Corbin, Captain Pease

Other Representation;
        Captain William H. Van Schaick
        John J. Fleming
        Henry Lundberg

David D. Wylie Foreman of the Grand Jury

Called to testify:

Benjamin F. Conklin	George Owen	Herman Burger, Eng'r., gas works
Everrett Brandow	Daniel O'Neill	E. 139th St.
Edward Van Wart	Thomas Lyon	
Edward Flanagan	Walter Payne	John A. Woodman
William W. Trembley		James Gaffney
T. Collins, Deck Hand	Rev. Julius Schultz	LuLu McGibbons
General T. H. Barrett	Charles A. Lang	
J. H. Fleming		
Henry Lundberg		

As the proceedings began, Mr. Atkinson was subpoenaed and declined to answer because he was already a defendant. He was also served with a Writ of *Dues Tecum* (to produce the company books), but he could not because Coroner Berry already had them. Twice General Burnett requested bail for Judge Dittenhoefer's clients amounting to $25,000.00 each.

An interesting detail of Inspector General Dumont's testimony related to his earlier years in the Steamboat Inspection Service. He had been a supervisor when the *Seawanhaka*[2] burned off the Sunken Meadows (largest steamboat fire in New York prior to the *General Slocum*) and it was he who had insisted at that time that the inspectors and not the assistant inspectors were responsible for lax inspections. In 1904, the reverse of that situation was the norm; assistant inspectors were being made fully responsible for the inspections, even when they were probationary and not fully trained. (See Steam Boat Inspection Service) As if to emphasize the service's convoluted and bureaucratic thinking, supervising Inspector Rodie's only comment to the press concerning the Grand Jury's deliberations was that the local board had written to Master W. H. Van Schaick immediately after the disaster, calling to his attention his duty to make a report. The pundits vocalized their opinions that the Grand Jury would hand in a presentment condemning the lax methods of the Steamboat Inspection Service along with their indictments; however, an admiralty lawyer, interviewed by the press, had a different perspective. He noted that the inspectors, even if guilty of neglect in their inspection of the *General Slocum*, could not be held responsible or punished as they were not recognized by law. "…Inspections are done by assistants not inspectors. Assistants report to inspecttors who make an oath and issue certificates stating that they inspected the vessels. The law is at fault and simply winks at perjury."

Indictments were handed down for Captain Van Schaick and the inspectors, under Section 5344 –Negligence, which causes loss of life, and is equal to manslaughter. The company officers were indicted, but only for "aiding and abetting".

During the deliberations, the press conducted its own informal survey and found that seventy percent of the life pre-servers on the *General Slocum* were thirteen years old. During that thirteen-year span, the Knickerbocker Steamboat Company had bought only fifteen hundred life preservers when they should have bought five thousand one hundred to be in compliance on both the *Grand Republic* and the *General Slocum*. The manufacturer testified that when stored damp, as aboard the *General Slocum*, a life vest would only last six to seven years. Furthermore, life preservers marked *Edwin Forrest* had been found on some of the victims. Investigation revealed that Frederick Creamer had bought these life preser-

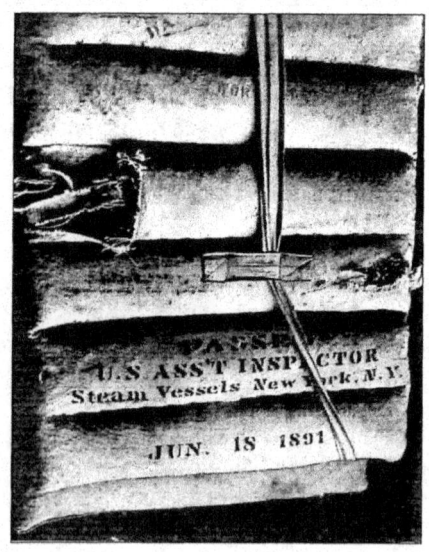

vers in Philadelphia in 1898, when the *Edwin Forrest* was about to be destroyed at the Rickenback Shipyard in Cramer Hill, New Jersey. Mr. Creamer then sold three thousand of those life pre-servers to several New York buyers, including the Knickerbocker Steamboat Company. All of them were filled with granulated cork, which had virtually no buoyancy. The public, first stunned by the disaster, had now become painfully aware of the need for adequate and plentiful life-saving equipment. They thundered out their displeasure in the editorial pages of every New York newspaper, and then began making themselves aware of the state of affairs on the ferries, excursion steamers, and other vessels they rode. Captains and mates were accosted by irate riders almost daily demanding more accessible and dependable safety accommodations. Shock turned to horror. Not only had the *General Slocum* happened to "the poor Germans" it could happen to anyone!

In October, demurrers were filed alleging that a corporation could not be convicted of manslaughter; therefore, the president could not be an accessory to a crime. Further, the indictments did not charge the accused with any crimes and it appeared that the alleged acts of omission on the part of the captain and Mr. Barnaby were the proximate cause of death of the one thousand plus victims. Finally, it was fire and not the acts of the defendants that was responsible for the deaths The prosecution alleged that the life preservers and the fire equipment were the responsibility of the inspectors and not the responsibility of the defendants who were obligated to ask for inspections and were then bound by the inspectors' findings.

In December, demurrers were filed on behalf of the officers and captains of the company and the inspectors in response to the indictments being filed; the court sustained the indictments.

---

1   (1852-1949) After this case he resigned and became a partner of Boothby & Baldwin, which later became Boothby, Baldwin & Hardy.
2   This was the largest steamboat fire in New York prior to the *General Slocum*.
3   Although this fire did not result in the horrendous Number of deaths that the *Slocum* did, it did underscore the first thoughts of both officials and watermen that safety was an issue that would have to be reckoned with.

The Trial of William H. Van Schaick, Captain of the Steamboat General Slocum
January 10 - 28, 1906

Trial Judge:	Judge Thomas	
For the Prosecution:	U.S. District Attorney Henry Burnett	
	Assistant U.S. district Attorney E. E. Baldwin	
For the Defense:	Judge Abram J. Dittenhoefer (Ret.)	
	Dudley Phelps, Jr.	
Officer of the Court:	Marshall Henkel	
Jurors		
Arthur W. Lawrence Foreman	Real Estate	542 5th Ave
Bernard G. Gunther	Furs	184 5th Ave
Ernest E. Malcolm	Agent	31 Nassau St.
Thomas G. Lord	Broker	52 Broadway
George Lush	Real Estate	408 2nd Ave
Charles H. Phelps	Agent	11 Broadway
James T. Latham	44 West 10th S	
A. E. Mc Millan	(excused)	
James Stork	(excused)	
Frank Hanaford	(excused)	
August Linderman[1]	Real Estate	349 3rd Ave
Charles O'Connor	Manager	157 West 68th St
Edward S. Reynolds	Liquors	71 West Broadway
Emil A. Hilsman	Butcher	280 8th Ave
George H. Crome	Insurance	124 West 63rd St
Disney Robinson		{53 East 105th St
		{50 Cotton Exchange
Called to testify:		
Edward Van Wart, Pilot		Thomas F. Conley, Captain of *William L. Strong*
Edwin N Weaver, 2nd Pilot		John H. Monforte, Captain of *Bronx*
Benjamin Conkling, Chief Engineer		James J. Duane, Mate of Massasoit
Michael McGrann, Steward		Oscar M. Hubner, New York Belting & Packing co.; manufacturer of Fire Hose
Flanagan, Mate		Charles P. Castle, Manufacturer of fire hose
Brandow, Assistant Engineer		Oscar Kahnweiler, Manufacturer of General *Slocum*'s life preservers
Walter Payne, Steward		George Kelsch[2], Survivor who saved his wife & one child, but lost another and also his niece
Henry Canfield, Cook		Thomas Fleming, U. S. Assistant Inspector
Jake Van Schaick, Captain & Father of Capt.W. H. Van Schaick		Henry W. Lundberg, Probationary Steamboat Inspector of four months
"Billy" Van Schaick, Capt. Of Iron Steamboat Co.'s *Cephus* & son of Capt W. H. Van Schaick		

Charge: Manslaughter [Sec. 5344 Rev. Statutes]

The trial was brought in the Criminal Branch of the U.S. Circuit Court Federal Building, New York City in the name of; Michael McGrann, Steward (Dead) and "Rachel Roe", passenger (Dead).

At the conclusion of the trial, Captain Van Schaick was remanded to the Tombs, where he remained overnight until his counsel could find a bailsman, Charles H. Ehrenstron, and a judge to accept the bond. Captain Van Schaick was released on $10,000.00 bail pending appeal of his case. In a show of support, the Harbor Master & Pilots Association collected $1,000.00 for the captain's retrial, in superior court but, it was never granted.

In February of 1908, his appeals denied by the state court, Captain Van Schaick surrendered to begin his sentence at Sing Sing Prison in Ossining, New York. During this same year, Fred Dalzell, F.J. Gryner, and Eugene F. Moran went to Washington as a delegation representing the maritime interests of New York. They met with the president and asked for executive clemency noting the captain's age and that he still suffered from injuries incurred during the disaster. They noted that when these factors plus the laxity of the steamboat inspection regulations were considered, his sentence seemed harsh. President Roosevelt assured the delegation that he was more anxious to punish the operators of the *General Slocum*, whom he considered the real culprits, than the captain, who was mainly a scapegoat. He would refer the case to the U. S. Attorney General for investigation But. this seems to have been more placation than intention. Grace Mary Spratt, a long time companion campaigned relentlessly for Van Schaick's release and pardon. But her determination was for naught.

Theodore Roosevelt refused two petitions. Her third petition with 250,000 names was finally granted when President, William Howard Taft, succeeded "TR" in the White House.

 August 11, 1911, after three and one half years in prison, the captain was paroled and on Christmas Day, 1911, President Taft signed the captain's pardon.

The Pilots' Fund gave Captain Van Schaick $45,620.00 when he left Sing Sing, which he used to buy a farm in Perth, Fulton County, New York. Although a surprisingly large sum for the times, it is not surprising that the watermen of New York would have stood behind one of their own for so long and so well. In the press and during the trial, Captain Van Schaick had made mention of the *Seawanhaka*. "I remembered the *Seawanhaka*"... he was quoted as saying as a partial justification for not grounding on Sunken Meadows. Already a Master, he must have shuddered along with his cronies as the magnitude and repercussions from this tragedy spread around the 1880 waterfront. Every pilot and master had the same thought; that what had happened to Captain Smith could happen to him.

In 1880, another wooden paddle-wheeler had burned after grounding on Sunken Meadows in Hell Gate. The repercussions which followed the demise of the *Seawanhaka,* a commuter ferry carrying passengers from the city to Long Island Sound communities, so affected the city' watermen that it led to the formation of the International Organization of Masters, Mates and Pilots.

Seawanhaka had departed 33rd Street, Manhattan on her way to the Sound and was in the same area as the *General Slocum* when her boiler exploded. A fire followed and about one hundred passengers panicked and jumped overboard. All were either drowned or beaten to death by the ship's paddle wheels. The captain, Charles Smith, was badly burned but remained aboard and grounded the vessel saving those who did not jump. The captain was charged with second-degree manslaughter along with the vessel's engineer and eleven other crewmen. Captain Smith died in 1881, two years before the charges against him were dropped.

---

[1] Selected by Marshal Henkel on the street. The marshal had been sent out to find an additional prospective juror when the pool dwindled during jury selection. August Linderman happened to be passing and was a friend of the marshal's.
[2] Separately indicted, charges against him were eventually dropped.
[3] He was tried, convicted, and released on appeals twice before this trial. He could not be found to testify at this trial.

Slocum Inspectors' Trial
January 1905

Judge Thomas	Criminal Branch, United States Circuit Court, Manhattan
Prosecution	General Burnett, United States District Attorney
	Three Assistant District Attorneys inc.Mr. Baldwin
Defense	Mr. Gilbert
Defendants	Henry Lundberg,[1] Probationary Assistant Inspector of Hulls
	John W. Fleming, Inspector of Boilers
Called to testify:	
	Robert S. Rodie, Former Supervising Inspector
	Captain Edward Van Wart, Former Pilot of the *General Slocum*
	Mrs. Julia Turner, Passenger
	Captain John L. Wade, The owner of a tugboat
	Agnes Mundle & William Seager, Passengers

Charges were brought under Section 5344 of the United States Statutes – Neglect in the inspection of steam vessels as the result of which life is destroyed [manslaughter], fraud, misconduct, and violation of the law in conjunction with the inspection of the steamboat *General Slocum*. A key point in the defense was that the prosecution had not proven its case that the Treasury Department rules required the defendants to inspect anything *except* the hull, boilers, and engines of the *General Slocum*. The prosecution based its case on the inspectors' report to Mr. Dunont and Mr. Burnett, in which the local inspectors stated that the ship had life preservers and that they were undamaged.

Since the transcript for this trial was not located during the research for this book, we cannot definitively state what a particular witness did or did not say but, of the few witnesses mentioned in the press, those listed above had testified and their testimony had been reported during the Coroner's Inquest. It is safe to assume that their testimony did not change.

Robert S. Rodie, the now former Supervising Inspector of the Steamboat Inspection Service defended the "business as usual" attitude of the service. Captain Edward Van Wart, the former pilot of the *General Slocum*, undoubtedly reiterated his inquest testimony defending both his actions and the captain's while presenting his recollections of the last excursion. Captain John L. Wade, the owner of the tug *Wade* and a hero of the disaster, identified a life preserver already in evidence and then easily poked a hole in the canvas covering. Captain Wade had been the first boatman to retrieve one of the dangerous appliances on the day of the disaster and hand it over to Coroner O'Gorman. Passengers Julia Turner, Agnes Mundle, and William Seager each testified that their life preservers had fallen apart in their hands. Mrs. Turner described how she had tried to get down at least a dozen from an overhead rack on the promenade deck and how they had split apart in the process.

On February 1, 1905, the case of Henry Lundberg went to the jury. He was tried again in May of 1905. In both instances, the trial resulted in a "hung jury". Probationary Assistant Inspector Lundberg was charged three times by U.S.D.A. Henry Burnett but, was never convicted.

---

[1] Both Lundberg and Fleming were formerly attached to the New York City Office of the Steamboat Inspection Service.

Vendors' Trials
"Particular Species of Infamy" –TR

The Department of Justice in Washington handed down indictments under Sec. 5440 of the Revised Statutes in the District Court of the U. S. District of New Jersey at Trenton for conspiring to defraud the government and prejudice the administration of the Steamboat Inspection laws by putting on the market compressed cork blocks for use in making life preservers each of which contained a bar of iron, six inches long and weighing eight ounces, at its center. The iron was inserted for the purpose of bringing the weight of the blocks up to the legal requirement of "six pounds of good cork" for each life preserver.

Three hundred of the blocks were shipped to David Kahnweiler's Sons, manufacturer of life preservers in New York City. An employee of David Kahnweiler's, described as an expert in handling cork, suspected that the blocks might not be right; broke one open and found the bar. The discovery was reported to the Steamboat Inspection Service in New York and then to the Department of Commerce and Labor. The case was investigated by the Department of Justice with assistance of the Secret Service and an indictment followed. Based on receipts and other documents found in Camden, New Jersey, the Secret Service made seizures at various custommers of Non-Parieil Cork Works. At the time of the trial the Secret Service was confident that all adulterated cork blocks had been seized. On September 29, 1904, J. H. Stone, Manager, H. C. Quintard, Charles W. Russ, and James Russ were indicted. Trials were set to begin after in October.

On October 18th, the defendants retracted their pleas of innocence and filed demurrers. The basis of their motion was that the cork blocks had been made for and sold on the open market and not directly for the government, therefore there was no conspiracy against the government. This was their best defense. The government was not amused.

Before the House of Representatives
House Committee on Merchant Marine and Fisheries
Washington, D.C.

Representative Bassett of Brooklyn delivered a speech before Representative Grosvenor, Chairman of the House Committee on Merchant Marine and Fisheries, concerning the necessity of legislation to prevent another *General Slocum* disaster. ..."While this country is among the most advanced in the world in the construction and equipment of its naval vessels, there is another class of vessels in which it is far behind the other countries of the world. I am speaking of the excursion boats that ply about the harbors of our large cities."

He alerted the congress that the federal law regarding the construction of such vessels had remained unchanged since 1871 in spite of the vast improvements in fireproof construction since that time. Metal sheathing was still only required around stoves in the galleys and around the boiler room. There was no stipulation for safe construction above the waterline. All other areas of the vessel could still be constructed entirely of wood. This was the greatest danger. Owners tended to cut down on the diameter of the wooden stanchions, which held up the upper decks. Thus when they burned, they burned through quickly dumping passengers into a sea of fire. This they justified as needed to keep the sea view unobstructed for the passengers on the lower decks. The law was so arcane that there was no stipulation for electrical inspections of steam vessels in the law. On the seventeenth, Chairman Grosvenor stated that all desired legislation to bring the obsolete steamboat inspection laws up to date would be enacted in the current session. He also stated that the bill embodying Mayor McClellan's suggestions regarding the increased liabilities for owners of excursion boats would also be adopted before the adjournment.

Six bills were adopted in the House, and several amendments were made in committee. The Chairman endorsed them all so that there would be nothing to hold up their passing when they returned to the house. The proposed laws incorporated the following;

1. The once yearly-recommended inspection would become compulsory, whether or not the masters applied.

2. Those vessels, which were out of commission, would be exempt from inspection. (This allowed vessels decommissioned for a year or two to have their safety equipment stripped for service elsewhere without penalty but, the items would have to be replaced prior to applying for a yearly inspection and license.)

3. Inspectors, for the first time, would be allowed to condemn substandard or defective equipment. They would be given the power to enforce the repair, removal, or destruction of defective equipment by stopping the operation of the vessel until such defects were remedied. They would be allowed to revoke a vessel's certificate when a defect was found.

4. The present statutes would be toughened up to require a life preserver or float for each person on board each vessel. (In 1905 certain vessels were exempt from this regulation)

5. Adequate criminal penalties would be included for persons willfully manufacturing or selling defective life saving appliances

6. For those charters, which lasted for more than one day (term charters), the charters of a vessel would be criminally liable for violations of the provisions of the title. (In 1905, only the owners were liable)

7. Officers and directors of a corporation would be considered willfully and knowingly guilty of misconduct in the management of a vessel whereby lives were endangered. (This was to prevent the arguing of the technical "loopholes" such as were presented in the *General Slocum* trials.)

8. Changes would be made in the salary system of inspectors and their duties and those of assistant inspectors would be defined for the first time.

9. The Secretary of the Department of Commerce and Labor would be able to detail assistant inspectors from one port for service in another as the needs of the government might require.

H.R.4154
A Bill for the Relief of the Victims of the General Slocum Disaster
Introduced 23 March 1910
Hearing 20 April 1910

On June 22, 1904, the Legal Aid Society, through Arthur Von Briesen, President, offered to file suit "...to those who may wish to bring legal action for the redress of injuries received in the destruction of the *Slocum* or for the death of relatives." The suit was brought by Theo Sutro and Gustav Voss, volunteer attorneys for the families of the victims and survivors. In it, Charles Dersch, whose wife and daughter had died on the *General Slocum,* was named as plaintiff. A case could not be filed against the Knickerbocker Steamboat Company, because wording in the organization's charter gave the company a loophole which amounted to a disclaimer that passengers rode at their own risk. And on the federal level, Section 4289 of the Revised Statures, stated no personal liability could be charged to owners of vessels used in rivers and inland waters beyond the value of the vessel. The *General Slocum* was severely under insured and a total loss. The wreck was sold by the city for $1,800.00. [This was the equivalent of $1.50 for each *family*.][1] Precedent for the suit was the relief granted San Francisco after the earthquake of 1906 which totaled five million dollars from federal funds. This had been followed in 1908, by an $800,000 relief fund, for earthquake victims in Sicily. These actions gave the families grounds for a lawsuit since they could not sue the government. The Hon. William Sulzer, congressman from St. Mark's district, rose in support of the bill. He contended that the government was primarily responsible, as proven by the findings of the investigating committee;

> "...It was ascertained by governmental investigations and by other inquiries, official and unofficial, that the terrible loss of life incident of this lamentable tragedy could have been prevented if the officials of the government charged with the responsibility had performed their duty in accordance with the law. It was proven conclusively that the life preservers and the fire apparatus on board the *General Slocum* were old and worn out and absolutely useless; and this fire apparatus and these life preservers had only recently been inspected by the government inspectors, who had passed them as being n compliance with the provisions of the Statutes of the U. S. the law, therefore, had been flagrantly violated, and the testimony proves conclusively that if these inspectors of the government had done their duty and enforced the law, the frightful loss of life never would have occurred.

"I bring this matter to the attention of the committee for the purpose of showing that the Government was negligent; that the Government was primarily responsible...."

"All I desire to do is to present the matter before the committee in such a way that the record will show exactly what we are trying to accomplish, and it is this: We know that there is no legal liability on the part of the government, having so far as a right of action is concerned in behalf of these victims. However, ...it is morally incumbent upon the government as a matter of conscience to do something for the equitable relief of these helpless victims."

In August the first of several bills known as Bill H.R.4154 was presented, but was voted down. Representative Goulden continued to introduce similar bills until one was given a hearing before the Claims Committee on April 20, 1910; all claims were denied. If approved, this bill would have taken the case to the Federal Court of Claims for a final disposition. By this time many families had buried their loved ones and taken care of the injured. They had gone through their life's savings in the process, and had been forced to move in with other relatives. Some had died in destitution. Had the bill been passed, it would still have been much too little and far too late. But the seeming indifference of the country, which had nurtured the greatest German community in America, was a heavy blow; the heart of Little Germany was forever broken.

---

[1] Per the United States Revised Statutes, the liability of the owner of a wrecked steamer was limited to the value of the hull, cargo, and engines after the wreck has been raised, if the vessel was seaworthy. If the vessel was not seaworthy its owners were liable for the damages awarded by the courts. The *Slocum* was raised and the hull sold for $1,800.00 Out of this sum the wharfage and other charges had to be deducted first, leaving about $1,200.00 to $1,300.00 to be divided among the claimants. Per the Commissioner, the total amount of claims was $1,475,673.00 The claimants were dived into three groups:

Claims for death	$ 1,361,152.00
Claims for personal injury	80,000.00
The City of New York	34,521.35 [2]

Henry Weidemann, a barber of 79 East Houston St. filed $50,000 claim for the loss of his wife, Caroline. His son also filed a suit for the loss of his wife Helen. Henry's son-in-law, Emil Reichenbach, also filed for $5,000.00 in the loss of his 2 ½ yr old baby. Of all their extended family aboard, only one daughter was saved.

[2] This included raising the hull, searching for bodies, etc., and incidental expenses which totaled $950.00 for draping city hall in mourning.

United States Commission of Investigation Upon the Disaster to the Steamer General Slocum
Offices of the Steamboat Inspection Service, Whitehall Building, 17 Battery Place, New York and Washington, D.C.

The committee was composed of:
Lawrence O. Murray, Commission Chairman — Asst. Secretary of Commerce & Labor
Herbert Knox Smith, Commission Secretary — Deputy Commissioner of Corporations[1]
Commander Cameron McRae Winslow — U.S.N.[2]
George Uhler — Supervising Inspector General, Steamboat Inspection Service
Brig.General John M. Wilson — U.S. Army (Ret.)
C.S. Hoogton [3]

Almost as soon as word of the disaster reached Washington, President Theodore Roosevelt took up the charge to justice. Born and raised in the city of New York, this felt like a personal attack. Furthermore, former Secretary of the Navy Department discerned gross incompetence on all sides of the situation. He thundered about the White House venting his indignation, then settled down to appoint a five man commission to investigate the causes of the disaster under the direction of the Secretary of the Department of Commerce and Labor, George Cortelyou. The "Roosevelt Commission" was to make a totally independent, deep, and complete search of the events. Roosevelt then announced to the Washington press corps that those responsible would be held personally accountable.

Throughout the summer, the commission met in both New York City and Washington, D.C., as circumstances warranted. At the beginning of the hearings, the commission lifted the licenses of Captain Van Schaick, Pilot Edward Van Wart, and Chief Engineer Conklin for failure to do their duty. Van Wart and Conklin were reinstated later. During July, Secretary Cortelyou ordered the re-inspection of two hundred and sixty eight steamships (ferries, steamers, excursion vessels under his jurisdiction) of these, the excursion vessels proved the worst. Mostly older than average, often badly maintained and overloaded, they were disasters waiting to happen.

The commission heard the testimony of many witnesses. At the close of the hearings, the full typed transcript of testimony ran to two thousand pages and included forty exhibits. As the commission noted, this process was wholly dependant on the "public spirit and courtesy" of those called, since the commission had no compulsory powers. Their inability to compel witnesses to attend and testify was at times a serious hindrance to the progress of the investigation. Several of the most pertinent testimonies have been abstracted below.

Peter C. Petrie, an Assistant Inspector of Hulls, had inspected both of the Knickerbocker vessels. On the *Grand Republic*, he found three hundred and twenty five life preservers on deck and twelve to fifteen hundred stored in a cabin between decks. He inspected each one and failed all but one hundred and thirty, which were in the cabin. He alerted Captain Pease to their condition, but had no authority to order the defective items destroyed. He stated that he had asked the Captain why they were there and if he would dispose of them, but received no reply. At that point the boat's engineer was quoted as saying "send them down to me and I'll use them." The Inspector next saw Captain Pease on the date of the *General Slocum's* inspection and he asked Captain Pease if the failed life preservers were still aboard, to which the answer was "yes". (The Inspector had no authority to condemn the life preservers, but he could have withheld the season's license until an affidavit was submitted claiming that the deficiency had been corrected. Receipt of that affidavit would have released the vessel without further inspection.)

Robert S. Rodie, the Supervising Inspector of Steam Vessels, 2nd District, New York, reported that the Board of Supervising Inspectors had sent up a list of recommendations with one hundred items which encompassed much of the Roosevelt Commission's report, but it was not acted on because it arrived just as the department was being transferred from the Treasury to the Department of Commerce and Labor early in 1904. Implementation of those recommendations might have forestalled the calamity.

Daniel O'Neill, a deckhand from the *General Slocum*, testified before the commission that there never were fire drills. O'Neill further stated that the hose couplings did not correspond [would not screw together]. On Flanagan's order he had disposed of many worthless life preservers, but he did not know how to put one on.

The commission hired their own vessel to follow the course of the *General Slocum* on her last trip. They made copious notes on the conditions of the shoreline, water depth, speed, and wind, anything that might have had an impact on the disaster. They then set out to construct a timeline of the June 15th trip similar to the one included in this book. This timeline and maps of the East River transit were included in the report to the President, as was their list of thirteen deductions. These were included as pointers to things that should be avoided in future. Foremost among them was the assertion that the *Slocum* was probably typical, in almost all particulars, of all excursion boats in New York Harbor and, doubtless, elsewhere. The commission made note of the peculiarly helpless condition of the passengers in case of disaster and the several conditions arising from her predominantly wooden structure that contributed to the high loss of life. These they felt were exasperated by the inefficiency of the crew through lack of drills and discipline and the failure of safety equipment from negligence. Next, they turned their attention to the inspectors, captain, pilot, mate, and engineer all of whom, they said were culpable for not fulfilling their responsibilities both before and during the disaster. The master was to know that his vessel was fit for the service she was engaged in; the pilot was responsible for navigation, especially the avoidance of collisions and groundings while adhering to the Rules of the Road. The mate should maintain cleanliness on the ship, the good condition of the equipment, and the discipline of the crew. Likewise, the engineer was responsible for the engines, boilers, and steam fire pumps, etc. Finally it was the responsibility, either direct or through other crewmen, for each of these to care for the passengers in times of crisis or catastrophe.

The commission's conclusions rested even more heavily on the Steamboat Inspection Service. There was an inadequate corps of inspectors in the Port of New York; inadequate supervision by supervising inspectors over inspectors; and inadequate supervision by inspectors over assistant inspectors. It was also noted that there was a reluctance of vessel owners to maintain proper life saving equipment and fire fighting equipment in a proper condition. In conclusion, they noted that

> ..." While it is true that it is the business of the Steamboat Inspection Service to see that proper safety appliances as required by law are provided, this by no means relieves the owner from a similar legal and moral obligation, nor from the liability for the maintenance of proper crew discipline. The commission is of the opinion that the owners of the *General Slocum* are censurable in a high degree for the inadequate and improper conditions prevailing on board this vessel, and that, whatever may be their technical legal liability, they and their executive agents share largely in the moral responsibility for the awful results of this disaster."

On October 2, 1904, the President approved the report and ordered all officers of the Steamboat Inspection Service who were involved with the *General Slocum* dismissed for "Laxity and neglect in the performance of their duties". Only the Supervising Inspector General of the Steamboat Inspection Service was spared.

The commission commended Everett Brandow. He had remained at his post and answered all bells until his service was no longer required without knowing, from his post [below decks], what his chances of saving his own life might be. In contrast, they condemned Captain Van Schaick and Chief Pilot Van Wart. Van Wart was considered next to the captain in responsibility and therefore equally guilty. Many of the harbor's best river men, captains, pilots and mates, came out in support of Captain Van Schaick. Based on their knowledge of the wind, tides, and other conditions on that part of the river, they felt that he had done his best; on treacherous waters he had done exactly as he should have done. No one was interested.

The Whitehall Building

---

[1] His primary focus was to examine harbor conditions
[2] He also made an independent report to the Navy Department in which he suggested that the Steamboat Inspection Service might be transferred to the Navy Department in order to tighten discipline and remove bureaucracy
[3] Hoogton was not an official member of the commission but, was assigned to attend the closed-door hearings by U.S. District Attorney Burnett.

Steamboat Inspection Service

Ch.2: Sec.2 —Steam pipes are to be installed from the boiler to the hold and Compartments for the Purpose of extinguishing fires

Ch.2: Sec.9 —No loose hay, loose cotton, loose hemp, coal oil, crude or refined Petroleum shall be carried as freight or used as stores aboard steamboats.

These two extracted regulations were hotly debated in the inquest and later in the trial of Captain Van Schaick. If properly applied, they were lifesavers, but they had been ignored and the Steamboat Inspection Service found itself tied hand and foot by its own carelessness.

From the moment that the Steamboat Inspection Service was placed under the Department of Commerce and Labor and for nearly a year thereafter, the Board of Supervising Inspectors had worked to accomplish a complete revision of the department. This included a period of almost two months during the summer of 1903 when all the Supervising Inspectors were convened in Washington to work exclusively on this project. The work culminated in substantial recommendations for revisions to the laws, rules, and regulations, which governed the service. These were submitted to Secretary Cortelyou. Having made their recommendations, the board continued working toward the implementation of these revisions in order to have everything in place at such time as the revisions might be made law by the Congress. The law was prepared and introduced as Senate Bill No. 5306 on March 29th of 1904. Action, however, was suspended awaiting Congress' proposed revision of the statute law but, the revision was not enacted.

In the estimated appropriations for the fiscal year ending June 30, 1905, the Secretary of Commerce and Labor had requested $100,000 for agents who would make "investigations regarding the manner of conducting the public business in the various bureaus, offices, and services of the Department of Commerce and Labor, with the object of securing more uniform, economical, and business-like methods of administration". Had this appropriation been granted, the Secretary would have used part of it for employing special agents, directly under his own control, to investigate the actual prevailing conditions in the Steamboat Inspection Service.

In another precognitive move, Secretary Cortelyou issued an order under Section 4496 of the Revised Statutes, concerning overcrowding on passenger steamers. With the onset of summer and the excursion season, there was good reason to have a particular concern regarding the overcrowding on regular steamers and excursion steamers. They often transported numbers of passengers well above the limits of their certificates of inspection or their excursion permits. Therefore, as of May 23, 1904, all chief officers of the customs service and collectors were to enforce the provisions of the statute on all such vessels coming or leaving their ports nationwide. Extra effort was to be made on Sundays and the Fourth of July, and other state or national holidays. The inspectors were to head-count passengers as they embarked to make sure that overages did not occur. Offending vessels were to be reported to the department for prosecution. It was hoped that such obvious vigilance would in and of itself reduce violations. Nothing has been found to date that would evidence the implementation of this order in the Port of New York.

Re-Inspection of Vessels

On July 7, 1904, as directed by the Secretary, re-inspections of all passenger steamers in the Port of New York were begun. Eight hull inspecttors and eight boiler inspectors were reassigned from other ports to help. Per the Roosevelt Commission's report, each inspector made written reports noting the deficiencies and shortages found on each vessel and also any new equipment on board, which seemed to have been acquired as a direct result of the *General Slocum* disaster and not the result of customary precautions. Two hundred and sixty-eight vessels were written up with as much attention given to the lifesaving and fire equipment as to the boilers and hulls. Life preservers and fire hoses were given special attention, for obvious reasons. The inspections revealed that generally, lifeboats and life rafts were in decent condition and adequate in number. On average, the inspectors reported approximately one third of all such gear was new since the issuance of the previous season's certificate. The commission decided that in reality, had the *Slocum* not burned, the percentage would have been less than one third of the numbers noted at the time of the re-inspection. In other words, in any other year, the ship owners would have re-stocked only one ninth of 1904's new equipment purchases. (For example, if on a given ship the requisite number of life preservers had been one hundred, in the year 1904, the owners supplied thirty six new life preservers; in an average year they would have supplied no more than four.) Below are the tables from the inspectors' report showing the results of the re-inspections of life preservers and fire hoses.

Table A
Percentages of Deficiency in Life-Saving Apparatus on Various Classes of Vessels After the Disaster

Deficiency	% Life-preservers	% Fire Hose
Short	2.06	1.22
Repaired	4.34	....
Condemned	7.92	9.11
Total	14.32	10.33
Two-thirds new Equipment	**3.95**	**7.94**
Total % of deficiency	18.27	18.27

The commission's report considered table A as helpful in understanding the mindset of the owners at the time of reinspection, but that it was not as revealing as Table B. Note the great differential between the highlighted column for excursion vessels and all others. The commission stated that Table B indicated the actual inefficiency of the Service with greater accuracy. No other group of vessels surveyed came close to the dangerous levels of unsafe equipment found on the excursion steamers. However, the great differences found among the various classes of vessels were probably more reflective of the attitudes of the ships' owners than of the level of inspections made by the Steamboat Inspection Service. It was also noted that the nature of the excursion trade, where boats were only in use for four or five months out of the year, probably had a greater impact on owners' attitudes than among other classes of owners who recouped their expenses over the entire year.

Table B
Percentages of Deficiency in Life-Saving Apparatus by Type

Deficiency	% Ocean Passgr.	% Inland Passgr.	% Excursion	% Ferry	% Towing Passgr. License
Life-Preservers					
Short	0.41	0.61	5.19	0.30	1.96
Repaired	4.61	3.68	6.51	2.26	2.86
Condemned	3.03	4.91	15.91	2.73	4.91
Total	8.08	9.23	27.61	5.29	9.73
Two-thirds new Equip.	0.11	0.27	5.39	9.04	3.92
Total % deficiency	8.19	9.50	33.00	14.33	13.65
Fire Hose					
Short	....	1.92	3.00	....	1.62
Condemned	14.02	9.61	14.36	4.46	4.85
Total	14.02	11.53	17.36	4.46	6.47
Two-thirds new Equip.	1.55	7.25	8.99	11.07	11.33
Total % of deficiency	15.57	18.78	26.35	15.53	17.80

Although the re-inspections, which included hulls and boilers, showed a fairly satisfactory condition of equipment, the commission felt that the same reasoning as applied to safety equipment would apply to hulls and boilers. If anything, regardless of the class of the vessel, it was of greater interest to the owners to keep these portions of their ships in good repair.

Safety equipment might never be needed, but a breach in the hull or a faulty engine could keep a vessel out of operation for many days, thus reduceing the company's revenues. Thus, the condition of this type of equipment was deemed more likely the result of the owners' pecuniary interest than the efficiency of the inspections.

After the re-inspections, the inspectors testified before the commission. All noted that large proportions of safety equipment were defective. Life preservers were found with covers so rotten that they could not withstand normal handling; and many which were not as bad, still required repairs such as new straps, buttons, etc. Many new life preservers were found in lifeboats without the proper complement of oars, automatic plugs, painters, and lifelines. Cabins were in a highly inflammable condition due to both contents and construction. Fire hoses consistently failed pressure testing and burst when pressures varying from five to one hundred pounds were applied; some leaked so badly that no pressure at all could be obtained at the nozzle. In other cases, hoses blew off couplings or couplings failed to have proper threads. Large shortages of hose were found on some vessels and substantial amounts of new hose were found on others. Hand fire pumps were in no better shape than the hoses. All of these findings only served to underscore conditions found on board the *General Slocum,* which should have been eliminated prior to the start of the 1904 season. The gravest charge of the commission, however, was reserved for the inspection of the *Slocum's* fire hose; no pressure test had been made at all.

At the time of the fire, the hose was pulled down and connected to the standpipe. The steam pump was running and the water was turned on for the hose. The nozzle was brought to within a few feet of the place of the fire and at least four crewmen were ready to operate it. *All this transpired while the fire was still at a point where it might have been contained.* But the hose burst in two places almost as soon as the water was turned on and almost simultaneously pulled off its coupling. Based on the dimensions of the steam pump, the commission calculated the pressure generated by the system and determined that it would have been insufficient to produce water pressure anywhere near one hundred pounds on the hose. Therefore, if the hose had been capable of withstanding 100 psi, or something close to this pressure, the pump could not have burst the hose even if the hose had been badly kinked or its nozzle completely closed. The hose burst because it was defective, *thus nullifying the chief fire protection of the boat.* Therefore, had the hose been of regulation strength, the fire probably would have been extinguished and the loss of life averted. *Therefore the negligent work of the Steamboat Inspection Service was one of the fundamental causes of the disaster.*

Re-Inspection of the Steamboat Inspection Service

The Roosevelt Commission felt that there were several causes for the sorry state of affairs of the Service. First, they noted the inadequate size of the force in New York but, then commented that even if there had been unlimited manpower; the results might have been the same. Nevertheless, the corps of inspectors should be enlarged.

Second, the commission noted that another obstruction to the work of the Service was the agitated opposition of the public; every time inspectors tried to enforce precautionary rules, which resulted in personal inconvenience to the traveling public, they complained loudly. This they noted had even shown up during the New York reinspections when a particular charter had anticipated taking an excursion steamer trip on a given day, but when the charter arrived at the pier, they discovered that the vessel had been tied up for reinspection. Although some in the party had been very outspoken concerning the need for inspections after the *Slocum* disaster one month earlier, they now demanded that their boat be released for the trip without the reinspection. Beyond the public's reluctance to be inconvenienced, was the third reason for the steamers' sorry state; the reluctance of owners to keep their equipment in proper shape, and this applied even more so to excursion steamers. *However, this was not justification for negligent inspections.*

The fourth and most important reason was the Steamboat Inspection Service itself! In a word, supervision, or the lack of it, had directly led to the disaster. Supervising Inspectors did not adequately supervise their Inspectors; neither did the Inspectors properly supervise the Assistant Inspectors. This was especially distasteful since it was the Assistant Inspectors who did the vast majority of the inspections. The superior officers were neither making proper inquiries to ascertain whether such inspections were being properly done, nor checking on or verifying the work of their subordinates; furthermore, there was an almost total failure to instruct or direct their subordinates in the proper performance of their duties. All of this they concluded had led, to the total breakdown of the service due to subordinates not being kept up to the proper efficiency by their supervisors, but left to perform as they saw fit. *Their conclusion was that the Steamboat Inspection Service in the Port of New York was neither adequate in upholding the standards of inspection laid out in the law, nor in furnishing sufficient protection to life and property aboard the harbor vessels under their charge.* The individuals that the commission felt were the most culpable in this regard, were primarily Robert S. Rodie, Supervising Inspector of the Second District and secondarily James A. Dumont and Thomas H. Barrett, his local inspectors. The commission also noted that they had found no evidence that this laxity existed due to corruption or improper motives, but that the lack of intentional malfeasance was no excuse.

Recommendations of the Commission

The commission summarized their thoughts on their investigation;
> "The Commission is earnestly of the belief that by far the most important part of its work in connection with this disaster has to do with the future. During the long investigation which the Commission has carried on it has been impressed with the magnitude and horror of this terrible

catastrophe, and increasingly anxious that the lesson thereof shall not be wasted, but that when the facts are properly understood there will be produced such results in legislation and in departmental action as will materially increase the safety of life and property involved in traffic on the public waters of the United States."

They went on to note, that they were not trying to lay out all the details of a new plan but, rather were outlining in general, changes that needed to be made.

- The Construction of Passenger Steamers
  The *General Slocum* was merely a "shell of highly combustible, frequently painted, extremely dry wood –a tinder box of the greatest possible inflammability." Furthermore, she was *"...not abnormal; but typical"* of the hundreds of vessels in the harbor and elsewhere. They noted that considerable suggestions had been made regarding the use of such materials and that these hazards should be reduced through the use of fireproof bulkheads, etc. Noting that this was not a panel of marine construction experts, the commission suggested that a commission of such experts be convened to discuss and make recommendations along these lines.
- Safety Equipment
  - Life Preservers
    - Under the 1904 statutes, there was no provision for having one life preserver per passenger and per crewman. The statute was very general and a recommendation was made to revise the statute to read that there should be one such appliance for each person aboard.
    - Noting that the regulations concerning the quality of life preservers and the types of materials from which life preservers could be constructed was deficient, the commission stated that there should be specifics regarding the contents, covering, straps, and
    - methods of testing employed at the factory. There should also be regulation of the places and methods of storing and hanging of preservers; and that
    - inspectors should have to report the age of all such life preservers on board the vessels they inspected.
  - Hose
    - The regulations did not require testing of hose or its construction. The commission's recommendation was that hose be lined in rubber to withstand the

water pressure and also be coated in rubber to reduce the rate of deterioration. Therefore, regulations should be amended to stipulate the quality of new hose to be good and such that it would not deteriorate from one inspection to the next.
- All hose couplings on a steamer should be of the same size and of standard thread making them interchangeable, except for reducers, where used.

- Hand Fire Extinguishers
  - These should be required (or hand grenades) on excursion steamers and;
  - should be placed at intervals aboard where they would be available instantaneously.

- Steam Fire Branches
  - Efficient carbonic-acid gas systems should be allowed as an alternative for steam fire branches, at the owner's option and if found equivalent by the Board of Supervising Inspectors.
  - Steam fire branches should be inserted into cargo or freight holds only, as directed by the Department decision forbidding the use of such branches in compartments used as cabins.

- Lifeboats
  - Should be provided and;
  - Be secured in such a way that they might easily and quickly be set adrift.

- Responsibility for Equipment
  - The present law made the determination of responsibility for faulty equipment difficult and intended to leave this responsibility wholly with the captain.
  - Theoretically, no captain should take a boat from the dock if the equipment was defective; however the captain knew that if he refused to sail under such circumstances, he would be promptly dismissed. Thus he often had to bear the burden of this responsibility on his conscience as well as the danger of criminal penalty.
  - The law should be amended to make possible charges of criminal liability to the individual owners and caterers, the officers and agents of corporate owners, in any case where their vessels were navi-

- gated with defective equipment and the owners, etc. had, or should have had knowledge of such conditions.
- To this end, the captain should also be required to periodically report to the Steamboat Inspection Office on the condition of his vessel's equipment. The same types of reports should be made for fire and boat drills. This would absolve the captain of the responsibility for informing his management since the Inspection Service would send out the notification; thus placing the responsibility on the owners.

- Officers and Crew
  - The law was deficient. It did not define the responsibilities of the captain as regarded navigation or equipment of his vessel, *"leaving this question almost entirely to custom."* These duties needed to be outlined if for no other purpose than to facilitate holding the captain responsible in case he failed in his performance of those duties.
  - This should also apply to licensed officers.
  - Since the current statutes *"...do not require, except on ocean-going vessels"* licensed mates, this should be amended to read that at least one such *uniformed* officer should be required on the large steamers, especially the excursion steamers.
  - If possible, provision should be made to improve the personnel of these crews. Although it might be extremely difficult to procure men of better qualification, this should be recognized as fact and the statutes and regulations amended to require *frequent fire and boat drills*, and for other discipline as may be necessary to properly train such crewmen.
  - The commission also recommended that when there were trials for the revocation of licenses, where either the license had been revoked or suspended for more than six months, an appeal should be allowed on behalf of the defendant from the decision of the Supervising Inspector to the Supervising Inspector-General.

- Inspectors' Powers and Duties
  - Under the existing regulations, inspectors had no authority to condemn improper or worthless safety appliances or to see them destroyed. The commission recommended that this be changed. *The inspectors should be able to withhold a license until such time as*

worthless or dangerous equipment was removed and destroyed, or if feasible, repaired to an as new condition. The inspectors should be able to pull a certificate when a violation was discovered or an inspection refused.

- Inspectors should be given the power to require a boat and fire drill on board a steamer at any time.
- Inspectors should be given the power to order the inspecttion of a vessel at any reasonable time, so as not to unnecessarily inconvenience the owners. This was in direct contrast to the requirement that the request for an inspections was left to the owners.
- Inspectors should be given the power to enforce the requirements of the law and a sufficient force of men should be added for this purpose.
- Regulations should be strengthened to require *all life preservers to be taken from their usual location and inspected by handling each one* to determine if its cover, straps, etc. are in good condition, and not just by casual observation.
- The rule requiring the Inspector of Boilers to also inspect the standpipes should be clarified, to remove its ambiguity regarding which department is responsible for the inspection.
- Pressure Tests made on fire hose *should be standardized* at one hundred pounds per square inch, as indicated by an attached gauge and all hose aboard should pass this test.
- Both hand pumps and steam fire pumps should be *tested by actual operation.*
- Hull inspectors *must examine all compartments* to see whether inflammable conditions exist and to have such conditions promptly reported and remedied.

Having picked apart every aspect of the Steamboat Inspection Service's regulations, practices, and stumbling blocks, the commission turned its attention to the internal construction of the Service itself. They first analyzed staffing; noting that even with the increase of inspectors brought into the Port of New York after the *Slocum*, the number of inspectors was still inadequate. These men had of necessity worked late into the night every night in order to complete their task. Manpower needed to be increased dramatically. The burden might also be eased by reapportioning inspections throughout the year and the addition of temporary inspectors taken on for April through August when the highest rates of excursion traffic were expected. Also, there was an unquestioned need to police the steamboat traffic between inspections in order to ascertain whether the proper condition of the steamers was maintained. At the time this fell under the duties of the Collector of the Port but, the commission felt it should also become a part of the Steamboat Inspection Service's mandate.

Salary, while not a major cause of the problems, was noted in the commission's report as a vicious system. An inspector was paid according to the number of vessels he inspected and passed in a given year, thus encouraging lax and hurried inspections. Instead, salaries should be fixed as opposed to predicated on the number of inspections.

At the level of the local Board of Inspectors, it was also noted that assistant inspectors had no predefined responsibilities under existing law. This, the commission noted, should be changed and the local board should stipulate the assignment of the assistant inspectors' responsibilities. The board also came under scrutiny for its system of blindly signing certificates. Since the board did not do the inspections, they did not know the details of certain facts, which they verified by signing the certificate. Instead, the commission felt that the Assistant Inspectors, who actually did the inspections, should sign the certificates, so as to certify the correctness of the actual facts stated therein and then the signatures of the local board should approve those certificates. Thus the board would only be approving the general principles upon which the certificates were based.

Age limits were also advised for those inspectors who actually did the inspections, since much physical exertion was involved. Finally the commission noted that whenever practicable, samples of new inventions of safety and fire equipment should be tested and samples of those items kept as part of the records of the Board of supervising Inspectors. This, it was noted, was exemplified by the discovery of kapok life preservers. These were a newer type preserver than the cork block type, which was supplied on the *General Slocum* but, when tested, kapok became a worse anchor than the disintegrated cork-dust life preservers, which were cited as one cause of death in the *General Slocum* disaster. Finally, the commission turned its attention to the manufacture of the life preservers noting once again, the defective kapok devices and the infamous cork vests augmented with bars of iron. They advised that the making and selling of defective life saving appliances was currently not a crime. If such defective articles were not discovered during routine annual checks by Service inspectors at the factory, or those sold to the ship owners did not conform to the initial samples approved by the local board, no penalty could be attached to them later. Well-defined statutes should impose penalties on the makers of such devises.

Again, stretching the parameters of their duty, the commission sought to change the legal definition of a ferryboat. The common rationale was that while an excursion steamer could be a ferryboat but, a ferryboat could not be an excursion steamer in a regulatory sense. Therefore, excursion companies sometimes tried to make their vessels ferries in order to work under less restrictive regulations. The intention was that since ferries had all passengers on one deck; had wide access and egress; and traveled over short routes, their safety standards did not have to meet the more restrictive demands made upon excursion steamers which could have three decks, limited access and egress, and often traveled long distances non-stop.

Finally, the commission asked that some accommodation be made in the law for charterers. They noted that the charterer, who rents a boat for one day and has no control over her equipment or crew, should not be liable. But some charterers hired steamers for all of a season, or a significant part of the season and went so far as to staff with their own hires. In this case, the commission felt that the charterer should be at least partially liable since during the time of the charter the charterer had control of her equipment, crew, etc. and had taken on the position of the owner.

Final Thoughts

The General Slocum disaster should not have happened. No disaster should happen, but they do. Then as now, the public, who thought little about it, over night, became experts in marine safety and knew better than the veteran watermen how things should have been handled. This is human nature. Greed, passing the blame, pushing men and machines to the ragged edge of their limits are also elements of human nature. But, it is also our nature to strive toward what is right and good. We work to make things better and, when possible, to make amends for our mistakes. There is no compensation for the loss of a loved one. There is only the memory of times spent together, of exploits and adventures, quiet moments of great tenderness, and love. For most of the survivors and their family remnants this was enough. For a few it was not. Some chose to end their suffering by their own hand, some did so unconsciously by drinking themselves into early graves. Some developed other unexplained chronic complaints that eventually did them in. It seemed that no one who lived near the church on 6th Street could stand the many silent reminders of what had been. Some of the residents left the neighborhood immediately, for others it was a long goodbye culminating in an eventual move to the Upper East Side or the Bronx. Older survivors moved away to live with relatives. Even the church moved uptown and eventually merged with Zion Lutheran. But this is not the legacy of the *General Slocum*. It is my personal belief that the long term outcome of that horrible day and its aftermath are our legacy hewn, like the marble of a monument, from the mountain of their sacrifice.

In the years following the disaster, many survivors remarried. Perhaps this was hastened or at least was made a viable alternative when the first searing pain of grief abated and the children's need for a parent became clear. Some of these new unions also produced children, which helped the healing process. Life went on. While this was an immediate result of supreme consequence to each of us who traces our family back to New York's worst disaster before September 11th, 2001, it is far from the only result or the most far reaching.

Steamboats soon made their way to the boat cutters' yards. The public became leery of these graceful wooden relics and their profitability plummeted. The Knickerbocker Steamboat Company could not recoup from the disaster, and went out of business. Iron Steamboat Company, for whom Captain Van Schaick had sailed before the *General Slocum* was commissioned, lingered a bit longer than the competition. The combination of the company's name and their advertising ... "Iron Steamboat can't be burned" promoted the notion of their safety. Captain Van Schaick's son, also a William Henry, [aka "Capt. Billy"] continued to sail Iron Steamboat's *Cygnus* for at least a decade.

Within hours of the disaster, every federal, state, and city politician and regulatory official connected with ships, commerce in the harbor, or safety was scrambling to determine if he was at risk of being buried under the blazing coals of publics outrage. From the president down, each agency, regulation, and custom associated with inland shipping was brought up for examination, and many were found wanting. Had a less vigorous and outspoken man held the presidency, the results might not have been so complete and lasting. But Theodore Roosevelt immediately charged in and declared his personal interest. This was more than politics; he had been raised in New York and held New York offices before his rise to the Navy Department and then the presidency. This was personal! Impaneling a commission to investigate in the Department of Commerce and Labor's Steamboat Inspection Service was a tangible proof of his statements to the press that he would personally follow the proceedings.

As the agency responsible for the Steamboat Inspection Service, the Department of Commerce and Labor was itself put under the lens. There could be no quiet reassignments of negligent officials and inspectors to other parts of the country. There was no where to run! The steamboat inspectors had complained before that their hands were tied and so they were prohibited from doing their duty. But this time neither the government nor the public was interested. There was plenty of blame all around. Lundberg, who was indicted, was only a probationary inspector when he visited the *General Slocum* in May of 1904. If the consequences of that inspection had not been so dire, he might have been forgiven for not withholding the seasonal certificate, which would have kept the *Slocum* tied up in her berth until her violations were corrected. His supervisors, in the port of New York, had done what they had always done. They sent out the underling while they themselves sent non-committal reports with bogus oaths of inspections to Washington. At the inquest, they argued that they were only responsible for the boilers and the hull. Some inspectors specialized in the one, and others in the other division of the inspections. A quirk in the law, which classified the *General Slocum* as an inland waters vessel, because she seldom sailed out of sight of land, and therefore was governed by inland regulations, was strenuously argued. It was a rationale worthy of a child caught in the cookie jar. The regulations were rationalized as that any ship, which carried passengers, was not a cargo vessel and therefore had no hold. They further stated that if a vessel had no hold, there was no need for fire fighting steam lines in compartments other than the boiler and boiler controls room. Also the interpretation of the regulation was explained as concluding that since boilers and hulls were stipulated, any other inspecttions, such as to life saving appliances was merely a courtesy to the owners. These inspections thus carried neither a requirement for compliance nor a penalty for noncompliance. What was not mentioned was that any initiative on the part of an inspector might have jeopardized future favors from the owners. The Chief Inspector for the port, by his own statement, sought exemption from responsibility because he had only recently been reassigned from Albany, and had no steamboat experience. In the end, Teddy's famous stick came down across all their backs and most of the senior inspectors in New York and Washington were gone.

    The shake up was the beginning of the end for the corrupted and bribable service and the beginning of modern legislation. In future inspecttions would include fire fighting, life saving, and safety equipment and infractions of the regulations would carry harsh penalties. Inspections became mandated as well as regulated. Even the greenest crewmen would be required to attend fire fighting and safety drills. They would be instructed in how to deal with the passengers in such emergencies and the necessity of communicating with the pilot and/or captain at the first hint of trouble. Many of the *Slocum* regulations remain in force today; those that have been changed have been strengthened. Paradoxically, had the *Slocum* been considered a sea going vessel, she would have been regulated by laws already in place which covered these circumstances.

    And what of those indicted? Lundberg was indicted four times, but never convicted. There was also a report, which the author has not been able to substantiate, that the probationary inspector high tailed it out of New York as soon as he could and was spotted in South America several years later.

    The pilots and mates of the *General Slocum*, whose licenses were lifted at the time of the inquest, were all reinstated with the exception of Captain Van Schaick and Edward Van Wart. No further actions were taken against Van Wart. Similarly, the officers of the Knickerbocker Steamship Company paid fines amounting to a slap on the wrist and walked away.

    Why was Flanagan, seemingly the guiltiest character among the crew, never brought up on charges? His testimony continually vacillated. But was it due to a fear of the possible punishment or of the prevalent lynch mob mentality? It might also have been "witness-itis", that phenomenon commonly documented by police at the scene of automobile accidents where no two witnesses' accounts match. Since his entire testimony at the Inquest is no longer available, we know that he was repeatedly tripped up by both prosecutors and the more reliable witnesses and that within the space of an hour he changed his first knowledge of the fire several times from about 87th Street to 120th Street, a difference of a crucial mile and a quarter. But we cannot conclusively say why. The Coroner's Inquest panel singled out his conduct and desire to assign blame to others; they recommended his indictment, yet he was not indicted by the Grand Jury. Flanagan lost nothing but a season's pay and resumed a life of drifting from foundry floor to waterfront and back again. By family accounts, we know that he carried the visions of that day to the end of his life, which was hastened by alcohol. Some might see this as a punishment equal to the tortures endured by the surviving passengers but, it also denied them and us of a full examination of his actions.

    Captain Van Schaick, who was tried in the press almost before the ship stopped moving, died a broken old man half blind from a severely burned eye and hobbled by bones broken in the grounding. He was the only one who served a prison sentence. Was this right? There are still many unanswered questions. As master of the vessel, he undoubtedly deserved a portion of the blame. But, how much? We have all heard the old rube that the captain is master of his ship. Under admiralty law and military tradition, this is correct on the open sea. Until this century, it was impossible for a captain to communicate with his superior officers or ship line when not in port. But inland navigation was different. Since the *General Slocum* fell under inland waterways laws, Captain Van Schaick was responsible to the owners of the vessel in a more immediate way. He could stop and call, cable, etc. his needs and desires to his superiors. In this regard, he became more of an office manager than an intrepid marine navigator. What about his "chain of command"? Captain Van Schaick testified and was corroborated by Captain Pease and the bookkeepers of both Knickerbocker Steamboat Company and Kahnweiler, the manufacturer of the life jackets, that he did not have autonomy in ordering fire hose, life preservers, and other equipment, which might have made the crucial difference in the disaster. It was Captain Pease, master of the Steamboat *Grand Republic* and Commodore of the Line, who approved requests and ordered supplies and equipment for both ships. It was he who decided on the allocation of various supplies, and it was he who ordered the yearly inspection and received the report. It was he who year after year chose to outfit his own vessel, and not the *General Slocum*.

So, is Captain Pease, a man of at least as much sailing experience in New York waters as Captain Van Schaick the one to blame? His ship was built on identical plans to the *General Slocum* and he sailed similar routes. Chances are good that if we could inspect his ship on the day before the *Slocum* disaster, we would find a vessel in only slightly better condition. But if he was channeling needed supplies and equipment to his own benefit, wouldn't he have a far superior vessel when inspected for fire and lifesaving equipment? Not by much. When the *Grand Republic* was inspected, shortly before the *General Slocum*, in the spring of 1904, she also had serious delinquencies in her apparatus. Both Captains Pease and Van Schaick had to answer to the front office. Interviewed before the inquest, the officers of the company admitted that they knew little of sailing (one of them had a thriving dental practice at the same time) and were just interested in growing their investment in the Knickerbocker Steamship Company. They were not interested in investing in the ships, only the dividends derived from them. In fact the president of the company, Frank Barnaby, had been aboard the *Slocum* at about the time of the 1904 inspection, and noticed nothing amiss.

So, should Captain Van Schaick have taken the responsibility on himself and approached the head of the company with his concerns? Not in 1904. This was still the era of the Robber Barons. Many businessmen on the rise fancied themselves the next baron. No subordination, no unasked for opinions, and certainly no out of the chain-of-command requests would have been entertained. Van Schaick himself, in the inquest and interviews notes that to bring up an additional expense was to risk retribution from Pease, and dismissal from the front office. For a man in his sixties, that must have been a formidable reason for remaining silent. His world did not include the safety nets we expect. He was too old to find other employment. He could be ruined very swiftly! It is difficult at best to put ourselves in another's shoes, much less their head. We can never know for certain what Van Schaick was thinking. But by the same token, we cannot entirely understand other items that appeared in the press. For instance, while reporters dug out old stories which showed he had had seemingly "bad luck" or was particularly inept with the *General Slocum* in the decade preceding the disaster having run aground or collided with other vessels many times; the harbor men of the Pilots & Masters Association loudly stood by their man; the harbor was full of groundings and collisions. They had all had their share, but actual damages to ships or harm to passengers was rare. Van Schaick was a good captain! When he was convicted they literally rode to the gates of prison with him; worked for his parole; and raised thousands of dollars for him to use to start a new life. These were eloquent testimonials for a fallen comrade and rose beyond the level of just supporting a fellow member of the Association.

No one knows every factor which came into play that day. We see greed in the lack of maintenance to the ship, carelessness in how inspecttions were conducted, and the indifference of every agency and individual that came in contact with the *Slocum* as obvious components of the disaster. We are horrified that a manufacturer of the most crucial of component used in life saving equipment, the life preservers' cork blocks, would be augmented with hidden iron bars. We cannot understand how, when the company which sewed the life preservers discovered this and confronted the cork supplier, that supplier continued in his practice. What is less obvious is how little attention the public paid to their own safety; those antique vessels were considered as safe as one's front parlor. Water skills, so basic to our lives, were not routine; those who saved themselves were among the few who knew how to swim. Finally, even Mother Nature had a hand in the tragedy. Had no breeze been blowing down the river, the fire would not have moved so fast. Had the tide not been running at about ten knots, finding a site to ground would have been somewhat easier and fewer victims would have been forever lost to the sea.

We can not readily understand life in that great metropolis of 1904 when few laws protected the public. What we do see is the traumatic birth of a citizens' based consumer movement. Many of our assumptions concerning public safety public began with the changes in maritime regulations arising directly from this tragedy. What might have been the immediate response of politicians to pacify the public with knee-jerk reactions and short lived outrage became instead the systematic redefining and upgrading of federal and local laws. Inspectors received the authority to withhold a certificate. Their pay became based on the number of inspecttions made. Ship construction, including the use of waterproof compartments, more fire resistant materials, and better systems for the use and storage of flammables were brought into line with standards already considered crucial on sea-going vessels. Crews Began receiving proper training. Fire drills, lifeboat drills, and instruction in the direction of passengers in an emergency, became mandatory. The vessel ratings, [total passengers boarded], were standardized and ratings assigned on the drawing board could no longer be raised for the company's benefit in succeeding years.

Of no less significance and equal importance were the public's reactions. From June 15th forward, passengers no longer boarded a steamer, ferry, bus, train, etc. without looking about for a life preserver or fire hose. Americans began to feel responsible and empowered to act in their own best interests. Private citizens took it upon themselves to face down captains, pilots and crewmen when possible violations were noticed. The papers were full of their indignation for months after the tragedy. But after the shock and hysterics were past, the mindset continued. We still look over our shoulders in hotels, restaurants, and on public transportation for safety necessities.

Of equal importance, Americans began teaching their children to swim. No longer irrelevant or an idol pastime, it became a badge of courage, and then just another normal part of raising a child. In once extreme and now subtle ways those Angels of Innocence lost in The Gate still watch over us. When the *Slocum* Victims' Monument was dedicated in 1905, a banner strung across the adjacent plaza read "Let us not have died in vain." The prayer is still being answered. Our legacy from those who died remains public awareness, the legislation to back it up, and the empowering of the individual to protect himself from greed and indifference so long as we remember the Terrible Day and the Most Awful Grave.

BIBLIOGRAPHY

Books

Allen, John L. M., Chief Examiner <u>Alphabetical List 1331 Names comprising all persons known to have been on the Steamer "General Slocum" at the time of the fire on June 15, 1904</u>, compiled for Commissioner of Accounts, New York, May 17, 1905.

Armstrong, Warren, <u>Fire Down Below</u>, Chapter 5, John Day Company, Pond View Books, New York, 1968.

Bettman, Otto L., <u>The good Old Days – They were Terrible</u>, Random House, New York, 1974, Pg.180.

Bunting, W. H., <u>Steamers, Schooners, Cutters and Sloops</u>, Published for the Society for the Preservation of New England Antiquities, Houghton Mifflin Co., Boston, 1974, Pgs. 78-81, 98-99.

Burgess, Robt. F., <u>Sinkings, Salvages, and Shipwrecks</u>, American Heritage Press, 1970, Pg.157.

Butler, Hal <u>Inferno Fourteen Tragedies of Our Time</u>, Chapter 4, Dorset Press, New York, 1975.

Cudahy, Brian J., <u>Around Manhattan Island and Other Maritime Tales of New York</u>, Fordham University Press, New York, 1997, Pgs. 80-88, 124, 243-251.

Davis, Lee, <u>Man-Made Catastrophes</u>, Facts on File, New York City, 1993, Pgs. 240-1.

Ellis, Edward Robb, <u>The Epic of New York City</u>, Chapter 38, Kodansha International, NY, Tokyo, London, 1997.

Giles, Dorothy, <u>A Candle in Her Hand</u>, G.P. Putnam & Sons, New York, 1949, Pg.179.

Government Printing Office, <u>United States Congress House of Representatives Committee on Claims [from old catalog] "General Slocum" Disaster. Evidence before the Committee on Claims of the House of Representatives on H.R. 4154 for the relief of the victims of the General Slocum Disaster</u>, 1910, LC Control Number 10035814.

Government Printing Office, <u>Evidence before the Committee on Claims of the House of Representatives on H.R.4154 for the relief of the victims of the "General Slocum" Disaster</u>, April 20, 1910, Washington, 1904, LC Control Number VXCp.v.131.

Government Printing Office, <u>United States Commission of Investigation on "General Slocum" Disaster Report of the United States Commission of Investigation upon the Disaster to the Steamer "General Slocum"</u>, October 8, 1904, Washington, 1904, LC Control Number 17004068.

Hanson, John Wesley, Editor, Told by Survivors and Rescuers <u>New York's Awful Excursion Boat Horror, The Ghastly Story of the Heart Rending Tragedy at Hell Gate Vividly Portrayed by Pen and Picture, Scenes on the Flame Swept Decks of the Ill Fated Pleasure Boat, General Slocum, Truthfully Delineated</u> 1904.

Horberger, Eric, <u>The Historical Atlas of New York City</u>, Henry Holt & Co., New York, 1994, Pgs. 98-9, 122-3.

Jackson, Kenneth J., General Editor, <u>The Neighborhoods of New York Brooklyn</u>, Citizens Committee for New York City, Yale University Press, New Haven, Ct., 1998

Kisseloff, Jeff <u>You Must Remember This an Oral History of Manhattan from the 1890's to World War Two</u>, Johns Hopkins University Press, Baltimore, 1989, Pgs. 90, 91, 102-4.

Longstreet, Stephen, <u>City on Two Rivers Profiles of New York Yesterday and Today</u>, Hawthorn Books, Inc., New York 1975, pgs.120-122.

Marcuse, Maxwell F., <u>This Was New York</u>, LIM Press, New York, 1969, Pgs. 14-16.

McAdoo, William, <u>Guarding a Great City</u>, Harper & Brothers Publishers, New York & London, 1906.

Morris, Paul C, & Quinn, Wm. P., <u>Shipwrecks in New York Waters: A Chronology of Ship Disasters from Montauk Point to Barnegat Inlet, from the 1880's to the 1930's</u>, Parnassus Imprints, Orleans, Mass, 1984, Pgs. 77-79

<u>New York Extra: A Newspaper History of the Greatest City in the World from 1671 to the 1939 World's Fair</u>, EricC. Claren Collection, Castle Books, 2000, Pg.233.

Northrop, H.D., New York's Awful Steamboat Horror [Memorial Ed.], Philadelphia, Pa., National Publishing Company,1904, 432 pgs., front., plates, ports. LC Control Number 04026220.

Obenzinger, Hilton, New York on Fire, (The General Slocum June 15, 1904),The Real Comet Press, Seattle, 1989, Pgs. 54-67.

Ogilvie, J. S., Editor & Compiled by, History of the General Slocum Disaster by which nearly 1200 lives were lost by the burning of the steamer General Slocum in Hell Gate, New York Harbor, June 15, 1904 J. S. Ogilvie Publishing Company, New York, 1904, 251 pgs. Incl. Plates, Port, Diagrams, LC Control Number 04016790.

Newton, Edward Douglas, Editor, Disaster, Disaster, Disaster Catastrophes Which Changed Laws, Franklin Watts, Inc., New York, 1961, Pgs. 67-86.

Oppel, Frank Tales of Gaslight New York, Chapter 4- "The Story of the Slocum Disaster" (*Van Pelt, Daniel, Leslie's History of Greater New York, Vol. I, Arkell Publishing Company, 110 5th Ave. New York, 1898, Pgs. 488-89.*), Castle Books, NJ, 2000 Edition, 1985, Pgs.37-57.

Paine, Lincoln P., Ships of the World, An Historical Encyclopedia, Houghton Mifflin Co., New York, 1997, Pg.201.

Rattray, Jeannette Edwards, Perils of the Port of New York: Maritime Disasters from Sandy Hook to Execution Rocks, Dodd, Mead, 7 Co., New York, 1973, Pgs. 124-130.

Rust, Claude, The Burning of the General Slocum, Elsevier/ Nelson Books, 1981, 148 pgs., Incl. Photos, Map, Drawings.

Richie,David, Shipwrecks General Slocum (American Excursion Steamer), Facts on File, New York City, 1996, Pgs.77, 87-88, 116, 142.

Stokes, I. N. Phelps, The Iconography of Manhattan Island 1498-1909, Arno Press, New York reprinted 1967, Vol. 2, Pgs. 94-5.

Werstein, Irving, The General Slocum Incident the Story of an Ill-Fated Ship, 1965 The John Day Company, 159 pages.

Workers of the Writer's Program of the WPA for the City of New York, A Maritime History of New York, Pgs. 204-5.

Books – Descendantcy of Captain Van Schaick

Holgate, Jerome, American Genealogy, George P. Putnam, New York, 1851

Honeyman, A. Van Doren, Johannes Nevius, Schepen and 3rd Secretary of New Amsterdam Under the Dutch, 1st Secretary of New York Under the English; and His Descendants A.D. 1627-1900, Plainfield: Honeyman & Co., 1900, Pgs. 59-141, 157-158, 391-392.

Howe, Shirley Swift, The Greenman Family (Vol. 2)

Peckham, Harriet C. Waite Van Buren, History of Cornelis Maessen Van Buren

Newspapers

In 1904, New York was a city of newspapers delivered morning, noon, and night. While we have centered coverage based on the New York Times, the other papers listed below carried extensive coverage. Many larger libraries either have the microfilms or can obtain them through interlibrary loan for June of 1904.

    Brooklyn Daily Eagle – Available on microfilm in many libraries
    The Globe & Commercial Advertiser
    The New York Times
        See News Index for 1904 "Marine-Gen. Slocum," and also; 1906 "Ships" – Index and Microfilms are available in many libraries
    The New York Daily Tribune
    New York World
    The New York Evening Post
    Long Island Newsday
    The Queens Courier
    The New Yorker Volks Zeitung (German Language Newspaper)
        See Index to, 1878-1920 Vol. 1, Vol. 2, Pub. Heritage Books, Inc., 2000
        Microfilm of this paper is available at the NYPL, Call Number ZY

These New Jersey newspapers also had extensive coverage, and were in areas of the state with large German communities (see notes on microfilm above).
    The Hudson Dispatch
    The Bergen Record

Magazines

Cosmopolitan Magazine, "The Riverside Hospital", Jacob A. Riis, 1892, Pgs.291-298.

Harpers Weekly, Harper & Brothers, New York, June 25, 1904.
Munseys Magazine, Frank A. Munsey Company, (175 5th Avenue New York, New York), June 1904

The Literary Digest, Topics of the Day: "The Slocum Tragedy", Vol. XXVIII. No. 26, Whole Number 740, Funk & Wagnalls Company, New York, June 25,1904.

Randel, William Peirce The Flames of Hellgate, October/November 1979, American Heritage Books, Pgs. 62-75.

The Van Schaick Kinfolks Quarterly, Michigan, Vol.1, No.1, Pg.5-6.

Other Documents

General New York University Alumni Catalog, Medical Alumni 1883-1907, General Alumni Society, New York, 1908

General Slocum Survivors, Organization of the VXCE Annual Memorial Services (Organization of the "General Slocum" Survivors Annual Memorial Services) 1905-1930 (Incomplete)

Gustav Scholer Papers, 1855-1929/ General Slocum Disaster, New York Public Library, 5th Avenue & 42nd Street Rare Books & Manuscripts Division, , Room 328.

Fire Report- Fireboat *Abram S. Hewitt* on the burning of the *General Slocum*, New York City Fire Department

Letters of the Trustees, Jan 2, 1903- Nov 12, 1904, Letter to P. B. Brooks, May 11, 1903.

McClellan, George B., Mayoral Papers, year 1904, files MGB- 1 through MGB-115, Department of Records & Information Services, Municipal Archives Research Room 120, 31 Chambers Street, New York City.

Minutes of the Board of Trustees, Dec 11, 1903- Dec 30, 1904, "Acting Superintendent's Report, Mr. Michael J. Rickard, Acting Superintendent of Bellevue and Allied Hospitals" #238 June 22, 1904.

3rd Annual Report, Jan 1,1904- Dec 31, 1904,Bellevue & Allied Hospitals, City of New York, Martin B. Brown Press, New York, the "General Slocum" Disaster, pg.23-24.

Journal for the Seventeenth Annual Excursion, St. Marks Evangelical Lutheran Church, Wednesday June 15, 1904, Pamphlet viewable as Xerox copy of original, New York Historical Society, New York City.

"Who Was Zophar mills?" Zophar Mills Company web page < http://www.zopharmills.com/pages/hist.html >

OTHER SOURCES FOR RESEARCH

Organizations

    Brooklyn Historical Society
    128 Pierrepont St.
    Brooklyn, New York 11201
    (718) 222-4111
    < http://www.brooklynhistory.org>

    Bronx Historical Society
    3309 Bainbridge Ave
    Bronx, New York 10467
    (718) 881-8900
    < http://www.bronxhistoricalsociety.org>

    New York Family History Center
    125 Columbus Avenue
    New York, New York 10023
    (212) 873-1690

    New York Historical Society
    2 West 77th St. (Central Park West)
    New York, New York 10024
    (212) 873-3400
    < http://www.nyhistory.org>

    New York City Fire Museum
    278 spring Street
    New York, New York 10013
    (212) 691-1303
    <nycfiremuseum@nyexcel.com>

New York City Municipal Archives / Research Room
31 Chambers Street
New York, New York 10007
(212) 788-8590
<http://www.nyc.gov/html/doris/html/1vital.html>

New York City Police Museum
100 Old Slip
New York, New York 10005
(212) 480-3100
<nycpolicemuseum.org>

The New York Genealogical & Biographical Society
122 East 58th St. (between Park & Lexington Aves.)
New York, New York 10022-1939
(212) 755-8532
< http://www.nygbs.org>

Library of Congress
101 Independence Ave. S.E.
Washington, DC 20540
(202) 707-5000
< http://www loc.gov >

The Steamship Historical Society of America, Inc.
300 Ray Drive Suite #4
Providence, Rhode Island 02906
(401) 274-0805
< http://www.sshsa.net>

Libraries

New York Public Library
5th Avenue at 42nd Street
New York, New York 1008-2788
(212) 930-0830

Jersey City Library, New Jersey Room
472 Jersey Avenue
Jersey City, New Jersey 07302
(201) 547-4503

Hoboken Public Library
500 Park Avenue
Hoboken, New Jersey 07030
(201) 420-2347

George F. Mand Library
New York City Fire Academy
Randall's Island, New York 10035
(212) 860-9200

Web Sites

http://www.geocities.com/general_slocum/

http://www.dizzycity.com/nbor.asp?busLID=lav-83st&nbor=Yorkville "Yorkville"

http://www.ezl.com/~fireball/Disaster12.htm "1904 Paddle Boat Fire"

http://www.queenscourier.com/spclissue/slocum/slocum1.htm

http://www.queenscourier.com/spclissue/slocum/slocum2.htm

http://www.lihistory.com/7/hs743a..htm

http://www.nydailynews.com/manual/news/bigtown/chap17.htm

http://www.web.simmons.edu/~wrightj/nyc/disaster.html

http://www.numa.net/exped/slocum/slocum.htm
(This site deals with the re-discovery of the remains of the General Slocum, which was salvaged and rebuilt as a barge only to be lost again off the southern New Jersey coast.)

http://www.catalog.loc.gov/
(This page, which is part of the Library of congress' web site, is the portal to researching the collections)

Cemeteries

Evergreen Cemetery
1629 Bushwick Ave.
Brooklyn, New York 11207-1849

Flower Hill Cemetery
North Country Club Dr.
Port Washington, New York 11050

Linden Hill Cemetery
323 Woodward Avenue
Ridgewood, Queens, New York 11385-1162

Lutheran All Faiths Cemetery
69-29 Metropolitan Avenue
Ridgewood, Queens, New York 11379-1683

Mount Olivet Cemetery
65-40 Grand Avenue
Masbeth, Queens, New York113-78-1683

The Greenwood Cemetery
255 5th Ave. & 25th St.
Brooklyn, New York 11215

Woodlawn Cemetery (YMCA Plot)
Webster Ave. & E. 233rd St.
Queens, New York 10470

German Lutheran Churches

Zion- St. Marks Evangelical Lutheran Church*
339-341 East 84th St.
New York, New York 10028

Trinity Lower East Side Lutheran parish
602 East 9th St.
New York, New York 10009

St. Johns Lutheran Church
81 Christopher St.
New York, New York 10014

St. Mathews Evangelical Lutheran Church
8th & Hudson St.
Hoboken, New Jersey 07030

St. Johns Lutheran Church
300 Bloomfield St.
Hoboken, New Jersey 07030

St. Mathews Evangelical Lutheran Church
83-85 Wayne St.
Jersey City, New Jersey 07302

* This is the name & address for the descendant congregation of St. Mark's on 6th St. When the community dissolved, the church followed its former members uptown and combined with Zion Lutheran.

Vital Records

New York City - Department of Records & Information Services
Municipal Archives
31 Chambers Street
New York, New York 10007
(212) 788-8590

New York City Health Department, Vital Records
125 Worth Street, Box 4
New York, New York 10013
(212) 788-4520

Hudson County - Hudson County Clerk
595 Newark Avenue Room 101
Jersey City, New Jersey 07306-2301
(201) 795-6112

Chief Personnel- Staff Services Section
1 Police Plaza
New York, New York 10038
(Information on work history of personnel by mail only)

After the disaster, the New York newspapers vied with each other to be first to publish pictures of the victims. These were republished in Ogilvie's, Hnson's and Northrup's books during 1904-5.

MINNIE CHRIST
Minnie Christ was with the Muth family and perished in the wreck of the burning steamer.

MARY DUCKHOFF
The body of fifteen-year-old Mary Duckhoff was one of the first to be identified.

HENRIETTA AND HEDWIG TIMM

MISS LOUISE HEINZ

GRACE GADE

EVA SCHNEIDER

JULIA WORTMANN

ANNIE BLUMENKRANZ

MARGARET SMITH

WINIFRED CRAGER

OTTO BOENHARDT

1. ANTONIO SCHWARTZ
   SAVED 22 LIVES
2. CHARLES CORDES
   SAVED
3. LUCY BORST
   SAVED
4. TILLIE HANFT
   MISSING
5. ROSIE ASCHE
   MISSING
6. CHARLES KUNSTNER
   SAVED

ELLA BOENHARDT

MARGARET GIBBONS

MISS M. BRIDA

MRS. ANNIE BACKMAN AND HER BABY
The bodies of Mrs. Backman and her child were found among the drowned victims on the beach.

THIS GROUP SHOWS THE PHOTOGRAPHS OF TWENTY-EIGHT WOMEN AND CHILDREN WHO WERE ON THE ILL-FATED STEAMER "GENERAL SLOCUM."

Lists of Persons aboard the General Slocum June 15, 1904

This list is a composite of the various lists printed in 1904, and later as noted below. To the best of our knowledge, this is the most complete listing of who was aboard the General Slocum on June 15, 1904 ever compiled. Every effort has been made to locate and include all available period references to passengers, crew, and concessionaires aboard. These references include several editions of the New York newspapers, including obituaries, during June and later during 1904, the Journal of St. Mark's Lutheran Church for the 1904 outing, the Annual Report of the Department of Public Charities of the City of New York for 1904 (Alphabetical List 1331 names .....,"General Slocum" compiled for the Commissioner of Accounts, published May 1905) and later information taken from the trial accounts in January, 1906, the Congressional Committee on Claims (H.R. 4154, 20 April 1910), published obituaries, and correspondence with descendants themselves.

The list is arranged as family groups wherever possible, and has been cross-checked in the following manner. A database was compiled of all listings of individuals, the family or group they were with (extended family, friends, co-workers, etc.), and their addresses. These were drawn from the newspaper accounts of the days following the disaster, the Coroner's Inquest, and trials. These were then annoted with any personal information or condition (dead, injured, uninjured, not aboard, etc.) as noted in the papers. Because there were many variations in the spellings of German surnames, the data was arranged by the most reliable spelling encountered. When one considers the fact that many of the newspaper reporters and typesetters working the story in June of 1904 were either unfamiliar with German spelling conventions or in such a rush to deadline, that many mistakes were made, then the reason for so many mistakes becomes obvious. Therefore, it also became obvious that the list would require reliable cross-checking. Of the documents available 98 years later, two emerged as the logical ones to use. These were the Department of Charities Report list, which was compiled by the coroners and their clerks on scene at the disaster and at the morgues. Since accuracy was at the core of their function, it is reasonable to assume that even with the clamber, hysterical relatives, and the need to keep the process moving, that greatest care would be taken by these individuals to obtain correct information. Therefore the list deferred to the Department's listings of addresses and names of individuals in each family group. Of the several spellings for each surname already collected, the list also deferred to the Department's list at this point. Once this was complete, and extraneous individuals moved to a separate list, the Congressional Committee's affidavits were consulted. Since this list was compiled several years after the fact, it provided two otherwise elusive elements. First, the family surname was given by a member of the household at a time of minimum stress, and therefore can be considered most reliable in terms of accuracy in spelling. Second, each affidavit recorded what the family members were doing at the time of the disaster (job title, school, etc.), if the affidavit was filed outside of New York County (Manhattan), New York, and what their incomes were, if any. In most cases this still information would have been unavailable by traditional research methods. Finally, each affidavit states what the person's condition is at the time of the filing (dead, capacitated, etc.) and sometimes, what if any income they were generating in the spring of 1910. The greatest drawback to this list is that only about 200 families participated in the claims and so they are the only families subject to this final scrutiny.

Each family or individual below may be listed in one of several ways. Those arranged as individuals, may be listed alone or as "id'd by" (identified by) a friend or relative. Where a relationship was listed in the source material, it has been listed here. Those persons arranged as family groups, may be under the father, or elder relatives' surname, or by the name of the person identifying or "looking for" relatives at the morgue. This structure was followed to allow for more interconnections between family groups than the traditional surname groupings. Pointers to extended family relationships and / or friends of the family have been included as (see also…). The result of these efforts is listed below as Validated Names.

If the family or surname being researched does not appear on this list, two other resources should be consulted in this appendix. First, the Surname Variations Table which shows all forms of the name encountered across all the data collected; and second, the Non-Validated Names listings, which appear after this list. The Non-Validated Names are individuals/ groups / families which were listed in initial accounts, but which do not appear in later documents and may be erroneous reports on the part of reporters or outright fictions to enhance their reports. They have been listed with addresses which appeared in those accounts and which may link them back to validated names, when evidence of their association with a validated family is found. While every effort has been made to corroborate and validate family / group associations, without the input of living descendants, no guarantee of accuracy can be made. If in reviewing this list, the reader finds misinformation concerning his family, please contact the author so that corrections can be made in a later edition.

Key to Abbreviations:

B   Bellevue Hospital   Lb   Lebanon Hospital   R   Rikers Island
F   Flower Hospital   Ln   Lincoln Hospital   Rv   Riverside HospitaL
H   Harlem Hospital   Pbn Presbyterian Hospital

Disinterred – Body was buried with the unidentified; then removed for burial with correct family
(n) Not onboard   (m) Missing*
(NYT, BDE) –Notations such as this at end of not refers to the source for the note; see bibliography
(D) Died*   (s) Survived
(I) Injured   (u) Uninjured
b. born in
( ) age of individual
1-3 digits refers to the body number assigned by the coroner so that individuals and their belongings could be matched for identification purposes when the body was claimed.
4 digits refers to the death certificate number for this individual, because of the location of the disaster, most were given Bronx death certificates, those listed as other boroughs have been identified as to the borough of origin, where known. All are now resident in the Municipal Archives of the City of New York.

* Initially many victims were listed as missing (m). Later, when a body was recovered and identified, the designation was changed do (D) for died and a death Certificate was issued. In the case of those individuals who were recovered and never identified and those whose bodies were never found, either designator may appear. Sometime the family assumed that the person was dead and referred to them as such in later documents, sometimes the most recent document still carried the city's original designation of (m) missing. The author has updated the lists from (m) to (D) as further and more recent documentation allowed.

Validated List of Persons aboard the General Slocum

A

Abendschein, Mrs. F. (Margaret) –325 E. 18th St. #3316
    Mary (34) (D) –Chairman of Sunday Excursion Committee, Sunday school teacher, Assistant Sunday school superintendent –She died trying to save some of her charges (BDE) –Id'd by her brother, George

Abesser, Henry (m) 128 E. 4th St. –Electrician / electrified bell hanger, he had a shop on the street floor of the family's building. His daughters were sent to stay with friends in Brooklyn. (BDE)

Mrs. Emilia / Amilia (36) #3386
    Kate (8) (I) –Lb
    Henry, Jr. (6) #3388
    Emma (10) (I) –Lb (still in Dr.Care, 1910)
    Emil (11) (m)

Abrahams, Isaac (26)–166 Ave C #3766 –Musician

Acherman, Mr. Xavier (n) –406 E. 5th St. –Woodworker   Id'd
    Barbara (29 / 30) #3047 b. Germany
    Caroline / Lena (1) #3050 – Her father found her in a coffin on the pier with a stranger. He snatched her up and would not relinquish the body to the attendants for several minutes.

Addicks, Ernst (n) –49 Ave A –Fine Confectionery & Ice Cream Saloon, 53 Ave. A   (see Stüve)
    John (15) –Clerk $6.00 #3647
    Martha (11) (m)
    Mary (9) (m)
    Annie (8) (I) –F
    Ernst (6)(I) –Ln Head, leg & hand burned
    Margaret/ Maggie Hedenkamp (2), adopt.daughter #2836   (see Hedenkamp)
    Margareth Stüve (65) (I) –Ln Lame left leg grandmother

Alber, Jacob G. (n) –628 E. 138th St. –Pressman
　　Eva (17) (D)
　　Sophie E. (16) –Operator #2748

Albrecht, Joseph (50) (I)–201 E. 10th St. –Delicatessen Id'd
　　Selma (44) (D) b. Germany
　　Martha (9) (u)

Alt, Mr. (m)–14 Lafayette Place (see Salvation Army)
　　Ella Hoffman, niece (14) #3473
　　Henry (13) (u) –Astor Library

Alfeld, Carl –339 6th St.　Id'd
　　Anna (45) #3033 b. Germany
　　Tillie Brum, step-daughter of Mr. Alfeld (16) #3427

Amann, Henry (13) (u) –77 3rd Ave.　(see Elwanger)
　　Lulu Elwanger (11) (I) lived with Mrs. Amann

Anger, Charles A. (52) (D)　–357 E. 62nd St.
　　Minnie, Mrs. (24) (29) (D)
　　Carl -Asst. Supt. Of the Sunday School
　　Charles F. (29) (I)

Anger, Mrs. Charles (m) –523 E. 83rd St.

Anger, Rosa (19)–1365 3rd Ave. #2907
　　Katie (18) (I) –Lb

Ansel, Eugene –103 E.4th St. – Eugene Ansel Fine Delicatessen, 103 E. 4th St. - Earlier on the 15th Mr. Ansel had had a message from Germany that his father had died. Perhaps already distraught from that notice, he identified five different people as his wife at the Alexander St. Station. He also gave the police the initials in her wedding ring, by which she was identified. She was not one of the five.He Id'd
    Louise (28) #2974 b. Germany
    Alfred (4) #2973
    Eugene (6) (D)

Armand, Mrs. Anna / Annie (27) (I) –Ln –shock Nervous Prostration -334 6th St.
    Stella (8) (I) -burns of face & body –Ln
    Lillian (1) #3037

Ambrust, Ferdinand –116 E. 4th St. –glass worker
    Barbara (45) # 3330 b. Austria
    Edna (6) #3804
    Mamie / Mary (13) (I) –Lb burned on legs, } Both were saved by a boat 1910 card cutter supported by father
    Florence / Florie (10) (I) –Burned on legs, }   1910 passementerie, self supporting

Arnoldi, Mary –73 1st Ave. –Cook
    Eleanor / Ella (11) (D)

Augur, George –1365 3rd Ave. Id'd   (see Bernius)
    Rose (19) (D)
    Carl A. G. (D)
    Elizabeth, wife (not known if she & the children were aboard)

B

Bader, Eliza (70) (I) -127 New York Ave., Jersey City, NJ –Ln

Bahr, Lillie (7) –424 E. 9th St. #3584
    Louisa / Lizzie (32)(m)
    Ida A. (12) #3344

268

Baist, Mamie (u) –23 Ave B.
　　Lillian (13)　#3745

Balmer, Joseph, a waiter, -123 1st Ave.　Id'd　　(see Leitz, Wollmer)
　　Mary, wife (35)(D)
　　Joseph W. (16) (D)
　　Augusta (8) (D)
　　Minnie (8) (m)

Balser, Gustavus (57) (I) Id'd. –137 Ave. B –Balser's Pharmacy, Balser's Wild Cherry Cough Balm
　　Mary
　　Catherine A. (32) #3068

Balzer, Nicholas (57) (I) –422 E. 8th St. –Driver –Tried to cut lifeboat free with knife, but could not because it was wired down. He swam ashore.
　　Amilia (46) (D)　　　　　　　　　　　　　　　　　　　　　　　　　　　　(see Braun, Belzer)

Balzer, August (s)　　　　　　(see Fickbaum)
　　Catherine (D) Wife
　　Fickbaum, Peter J., (n)　Balzer family friend, saloon keeper
　　　　Freddie　(D) son

Bandelow, William (n) -84 7th St.
　　Louise A. / Eliza (29)　–(Step-daughter of Mr. & Mrs. William Schwickert, 420 E. New York Ave)　# 3439 b. Austria –Lutheran Cemetery
　　George (3) (m) – (was reported seen alive after the disaster by Capt. Burke of the Life Saving Corp., but he was never found and may
　　　　　　　　　　have been adopted by a well meaning woman he was possibly seen with at the Pavillion in Carnarsie)
　　LuLu /Louisa (5) #3321 –Lutheran Cemetery

Barth, Mary (6) (D) –87 Ave A. (see Mattes, Singer)

Bauer, Caroline (42) (D) –31 Beekman Place　(see Vetter)
　　Harry (27) (D)

Baumann, Margarite (6) −526 6th St. #3581
    Otto (5) #3410
    Magdalena (31) #3138

Baumler, Frederick (n) −433 6th St. −Woodworker (see Zausch)
    Margeretta / Marguerite (35) #2740 b. Germany
    Frederick (11) (u)
    Annie (12) #2879
    Charles (10) #3680
    Amelia (15) #3741 −Saleslady $4.50/ wk.
    Margaret (85) (D)
    Annie (43) (D)

Beck, John −313 E. 9th St. −Barber (see Horway)
    Christina / Christian (56) #2788
    Horway, William
    Horway, Adelia (6)

Becker, Clara (20) (u) −1157 Lexington Ave.
    Mary (60) (m)
    Lilly (27) (m)
    Amelia (25) (l) −Lb
    Barbara − 1157 3rd Ave. rescued by Off. Scheuing

Becker, Frank (n) - 1010 E. 178th St. Id'd
    May (Mary), wife (50) (m)
    Theodore Frank (3) (D)
    Mary (25) (l)

Beckman, Henry C. W. (n) −1894 3rd Ave −Grocer
    Arthur Id'd;

Annie (23) #3995
Anna Margaretha (7 mo.) #3996
Margaret (23) #3996

Behnen, Herman (n) –E. 48th St. –Janitor
Annie (13) #344 #2808

Behrendt, Hermann (n) -88 E. 3rd St.
Marie (40) (I) –Hands burned looking for; – Mrs. Behrendt wandered away from the police station possibly delirious and definitely injured. Testified, still bandaged from burns, at Inquest Trial that a life belt saved "her Annie"
Lizzie (10) #2969
Louise (11) (D)
Clara (8) #2970
Annie (13) (I) –Nervous Prostration, 1910 a dressmaker, no means of support

Behrens, Gerhard (n) –127 Garden St, Hoboken, NJ –Woodworker
Augusta (51) #2721 b. Grmany
Alice (16) # 382 #2962 was in dressmaking $7.00

Behrens, Henry (6) (m) -134 E. 28th St.
Fred (9) (I) –Ln

Bell, Agnes (19) Adopted Daughter #2796 -242 E. 5th St.

Bell, Angus (16) –242 E. 5th St. (see Harris)

Beneke, William C. –420 E. 17th St. –Highway Inspector   (see Gruben)
Mrs. William C. / Mary (30) (D)
Emma Gruben (40) (D)
Carrie (13) (D)

Bensch, Jacob – 401 5th St.
　　Mary (41) #2875
　　LuLu (11) (I) –Lb

Bentz, Julia –333 5th St.
　　Otto (14) (m)
　　Arthur (14) #3133

Berdolt, Augusta (30) #3146 -1050 Prospect Ave., Bronx

Berg, Theodore -158 Goerck St. Id'd
　　Lena, (45) #2922 b. Germany

Bergman, Edward (30) (I) -156 3rd Ave. –Peddler on the *Slocum*

Berhold, Augusta (30) (D) –1050 Prospect Ave. Brooklyn

Bernhardi, William -614 E. 9th St –Coachman Id'd
　　Annie (5) #2717–also Id'd by Frances Bernhardi, an aunt –403 E. 90th St. (also listed at 43 E. 2nd St.)

Bernius, George J. –Driver　　　　(see Auger)
　　Minnie Auger (29) (D) daughter

Bertrand, Henry (46) (m) –730 E. 6th St.
　　Lizzie (45) #3438
　　Arthur (13) (u)

Bindewald Id'd
　　Pullman, William H. (D) brother in law & Secy. Of Sunday School carrying $300.00 check for the charter

272

Birmingham, Catherine (72) (D) –79 Mangin St. -   Id'd by Michael Dillon, her brother (see Dillon)   (see Diamond)
   Katie (55) (D)

Blohm, John -18 Jackson St., Williamsburg, Brooklyn –Butcher Id'd   –Crowd at the Blohm residence was so thick, police Capt. Becker of Hamburg
        Ave. Station had to turn out the reserves.
   Dorothea / Dora (15) #2766
   Margaretha / Margaret (17) #2765 –Bookbinder $5.00/week
   Anna, Mrs. (50) (D)*
   *[BDE] reported that Dora, Margaret, and Mrs. Annie Smith (D) were all sisters of John.

Blohm, John William -573 Central Ave., Brooklyn –Photoengraver
   Margaret (53) (I) –Ln
   Mrs Anna Smith (28) #2763 –sister of John W.

Blumenkranz, Annie (20) –9 E. 106 St. #3466 b. Austria

Blüsch, Katherine (Kate) (25) (D) –41 Ave A   (see Hauff)
   Lizzie (14) (D) –Housekeeper $25.00/mo.

Bock, Charles –69 Marcy Ave., Brooklyn
   Louisa W. (32) (I) –Ln –Saw her daughter drown, family thought she might go insane; she was a former St. Marks Sunday School pupil &
   still member. She had accompanied Mrs. Anna Burkhardt (D), a former playmate of 141 E. 3rd St., Manhattan. When fire started she took
   May, and Anna took Grace. Louisa slid down a pole to hawser to a fender with baby and then slid into water. She fainted from
   exhaustion and lost hold on baby. She awoke as she was being pulled aboard a boat, but they could not reach the child who slid under the
   water as her mother watched. (BDE)
   May Louise (7) #3739
   Grace Edna (5) #3747
   Louisa (69) (D)

Boden, John H. (n) –101 Clymer St., Brooklyn –Laborer (see Lutjen)

Ella (46) #3593 – Was with the Lutjen family, who were friends from their apartment building. She had gone against her husband's wishes. She was Id'd by her wedding ring and jewelry; a gold watch and chain were not recovered and may have been lost in the panic or ripped off by grappling hooks afterwards. Funeral at Hillary Funeral Home, Division Ave. near Clymer St.

Boeger, Franz (n)–910 Putnam, Brooklyn looking for see -Bookkeeper for firm on Monroe St.
    Mrs. Susan L. (32) #3291 –Daughter of Mrs. Caroline Hanneman –NBI burns & injuries   (Hanneman)
    Florence / "Dolly" (3) #2724 –Was called Dollie for her dark eyes and black curls which made her look like a wax doll
    Wilber A. (5) #3992 –Blue eyes and golden hair*
    Philip (9) (D)

*A few days before the excursion, he and his sister had walked into the Ralph Ave Police Station, insisting that they be presented to Captain Miles O'Reilly. When introduced they told of being pelted with rocks by ruffians. The captain assured them they would be left alone.

Boenhardt, Otto (14) #3362 –322 E. 13th St.
    Ella (12) (D)
    Albin (50) (I) –Ln –Waiter 415.00/ week, 1910 unable to work

Boesch, Walburge (see Schelke)

Boesewill, Henry –Stableman
    Meta (46) (D)
    Mathilda (26) (I) –Hands burned –Flowermaker $10.00, 1910 housekeeping

Bohmer, Anna (59) (I) –Ln –306 E. 93rd St.
    Emile (18) (I) – Ln / Pbn –He got his mother to the upper deck where he shielded her with his body; when hope was gone, he held onto her & jumped. They were picked up and taken home, where his burns were dressed, but the pain became so intense that Policeman Slator, of the same address had to call an ambulance for him.

Bollmer, Adeline (7) (D) –123 1st Ave.   (see Balmer)

Bopp, Dora (66) (I) – 74 1st Ave.

Borger, Philip (n) –104 1st Ave

    Pauline (37) #2860
    Philip, Jr. (9) #3565
    Pauline (5) #2893

Borst, Lucy (14) (u) –15 2nd Ave.

Borsum, Henry (D)

Bose, John (n) –135 Ave. A –Packer (1910 affidavit filed in Kings County)
    Anna (54) #2923 b. Germany
    Meta (19) #3481
    Emily (18) (D)
    Henry (14) (u) –shock -1910 clerk $10 own support
    Anna (25) (I) –burned at knee & ankle, shock –Operator of flowers $12.00, 1910 supported by father

Bozenhardt, Emily (38) #2805 b. Germany –110 1st Ave. –Id'd by brother Bernard / Herman Raag   (see Kaaz)
                                             Services at Evangelical Ref. Church, 5thSt. & Ave. B by the Rev. Jacob Schlegel
    Lucille (11) #2933

Brandow, Everett (34) (s) –Catskill, New York -2nd Engineer on the *Slocum*

Brandt, Carl –410 E. 9th St. –Janitor
    Ida (34) (D) #3406 –Forelady $19.00 /week

Brandt, Robert of Brooklyn looking for   (see Zerdes)
    Zerdes, Mrs. Margaret (D)

Brauer, Jeanette R. (5) (D) #3702 –107 E. 84th St.
    Margaret L. (33) (D) #3524
    Seiler, Catherine (72) (D) #3220 b.Germany, mother of Margaret

Braun, Annie (see Balzer)
   Emilia Balzer (46) (D)

Braun, Alfons (see Roth)

Braun, Minnie (D) (see Ehrhardt)
   Minnie Ehrhardt (13) (D)

Braun, Valentine J. – 233 5th St. –Hotelkeeper
   Peter G. (12) (I) –Lb released self from hospital and showed up at St. Mark's to announce he was not dead.
   Mollie (32) (D) #2865
   Elsie (10) (D) #3559
   Walter (6) (D) #3558
   Valentine (6) (D)

Breda, Gottlieb (n) –90 Ave A. –Gottlieb Breda Jeweler, Ave. A & 6th St
   Mamie (23) (D) #3112 –Saleslady

Breda, Minnie (29) (D). #311 -150 N. 9th St., Brooklyn
   Minnie (1) (m)

Breden, Ellen (16) (I) –Ln –383 11th St., Brooklyn –Went with friend, Matilda Mercelis of 370 6th Ave., Bklyn. and Gustav Lutz of Manhattan. She was on the Hurricane Deck. Instantly fire was all around and people on lower decks screaming and jumping. The friends pushed through the crowd and jumped. Ellen told Matilda to jump for the tug, but Matilda thought they would land between and be crushed. Matilda and Gustave tried to restrain Ellen. She came to on way to hospital in grocery wagon. Two strangers brought her home the same day whe released from hospital. Gustave was picked up suffering from submersion. (see Lutz, Merseles)

Brennan, John (48) (u) -354 W. 25th St. –Deckhand on the *Slocum*

Brennan, John (50) (u) -354 W. 25th St. –In charge of wine room on the *Slocum*

Bretz, Mary (28) (D) # 3013 b. Hungry –304 W. 28th St.
  Elsie (7 Mo) (D) #3014
  Edith (3) (D) #3547

Brooks, Margaret (13) #3110 – Ave A.

Brum, Tillie   (see Alfeld)

Bruning, John L. (44)– #2913   –215 E. 12th St. –Services performed by Rev. Junius B. Remensnyder, English Lutheran Church of St. James, 73rd & Madison
  Annie E.– #2914 Id'd by brother John –Burial at Lutheran Cemetery
  Magdalene (12) – #3369
  Grace M. (14) (I) – Ln

Buchmiller, Albert (n) – Painter, Greenpoint Ave. –Res. 79 Colyer St. Greenpoint, Brooklyn –Left work and searched through the night for his family
  Mrs Anna (40) (D)
  Annie, Mrs. (27) #3557
  Arthur Hernberg (9) #2784 Stepson of Mr. Buchmiller
  George (7) # 2783 Stepson of Mr. Buchmiller

Burfield (Burfiend), Mr. J. H. K. –100 W. 106th St. Id'd
  Kate (21) #2811 b. Germany
  John J. (10 Mo) #2761

Burfiend, Margaret (2) #3555 –245 W. 27th St.
  Dora (7Mo) #3518
  Dora (22) #2949 b. Germany

Burfiend, Margaret (60) (I) –Ln -275 7th Ave.

Burkhardt, Anna (66) (D) -141 3rd St.

277

Burkhardt, Peter –43 Tompkins St.
  Albertina (39)– #3471
  May (4) – #3472
  Edward (13) (u) – Also listed at 270 E. 10th St.
  Arthur (13) (s)
  Charles (n)
  George (n)

Burkhardt, – 141 3rd St Id'd
  Schwartz, Louisa, Grandmother (D) –Id'd also by son-in-law Chas. (Jacob) Schwartz   (see Schwartz, Bock)
  Elizabeth, Mother (66) – #3123

Bush, Hilda (10) # 3287 –82 W. 90th St.

Bushong, Henrietta D. (25) (u) –1028 Hudson St., Hoboken, NJ

C
Cahill, Mrs. Anna N. (22) #3028 b. Germany –316 6th St.

Canfield, Henry (53) (I) –421 10th Ave. –described as a "strapping southern negro", had been a cook aboard the *Slocum* for several seasons

Christ, Fred (n) –144 E. 7th St. –Fish Market
  Minnie (13) #2934

Cibilsky, John (n) –91 Ave. B (1910- Affidavit filed Suffolk Co., NY) –Tailor   (see Fickbauhm)
  Kate (19) (s) Housework $16.00/ mo. & board + lodging

Clow, Alfred (n) -54 E. 7th St. Id'd
  Margaret /Mary, (41) #2734

Coakley, John (30) (u) –419 W. 31st St. –Deckhand on the *Slocum*

Cohrs, Freda (26) #3180 –106 (103) Ave A -Id'd by brother Henry
  Michael (m)
  Minnie (28) (m)

Cohrs, Henry -70 1st Ave. –Milkman
  Kate L./ H., Katie (27) #3184
  Frieda E. (6) #3181
  Henry D. (1) #3182
  Freda (26) (D) sister of Henry –Dressmaker $10.00/ week

Collins, Thomas (25) (u) –303 Van Brunt St., Brooklyn –Deckhand on the *Slocum*

Conklin, Benjamin F./ Frank (39) (u) –Catskill, N Y – Chief Engineer on the *Slocum*
  William A. –came to the Alexander Ave. Station House looking for his father

Corcoran, James (20) -680 Washington Ave., Bklyn -2nd Mate on the *Slocum*

Cordes, Hinrich/ Diedrich -417 E.. 16th St. This party was with Mrs. Mary Rosenberger & Mrs. Mary Wolff (see Rosenberger, Wolff)
  Charles F. (17) (I) – Was aboard with his friends Charles Kuentsner & Anthony Schwartz
  Metta M. [nee Hopps] (52) – #2731 b. Germany } both Id'd by Metta's son John
  Fred W. (15) brother of Henry J. – #3194 }
  Henrietta A. (21) – #3193 –Operator $15.00
  Henry J.(20) (I) –Foreman

Crager, Winifred (12) #3375 b. England –222 E. 12th St.
  Julia –Dressmaker

Cragh, Martin –(s) -29 President St., Brooklyn –Deckhand on the General *Slocum*

D

Dauernheim, Anna (58) # 515 #3147 b. Germany –41 3rd Ave.

279

Wilhelmina (26) #3185

Deck, Frank (11) #3002 -395 # 4th St.
Nicholas (2) (m)
Rose (12) (l) –Ln
Lena (31) (u)

DeLuccia, Luigi – 54 E. 7th St.
Meta (45) (l) –Ln badly burned–She persuaded her neighbor, Sophie Siegel to join her family. When fire was noticed, she, Sophie, and the children were near the first deck rail, and were pushed overboard. She could not find any of her children; clung to a paddle until rescued.
Frank (9) #3380
Agnes (6) #3018
Nicholas (3) (m)
Rosa (12) (u) –shock – Lb

Delventhal, Mr. Henry (n) –381 Madison St.
Sophie (19) (l) –Ln
Hattie (28) (l) -Lb

Dengler, Hattie (28) (l) –Nervous Wreck, eyes affected -123 7th St. Id'd
Adolph, Jr. (4) #2727
Harriet (6) (l) –Ln Hands burned

Deppert, August (n) –328 E.6th St. Looking for; -A. Deppert & Co. Mfg. of Worsted & Saxony hosiery & dealers in yarns, hosiery, & Notions
Agnes (62) #3332 b. Germany –Id'd by neighbor, also by son, Charles
Charles
Mary (28) #3378
Charles (10 mo.) Grandson (s) –Lb (was tagged body #144, then revived at hospital)

Dersch, Charles (n) –76 1st Ave –Salesman – Would become the President of the General Slocum Survivors Association
Helene (38) #2998

280

Elsie (16) #3511

Diamond, John F. – 79 Mangin St. –Truck Driver
Catherine (s) & brothers ID'd – In her grief, she tried to jump off the pier, but was restrained by W.D. Howard, a Bellevue nurse
Birmingham, Katherine (55) #2902 b. Ireland –Catherine's mother   (see Birmingham, Dillon)
Frank (4) #3486
May (7) #3548

Dieckhoff, Frederick (n) –121 4th Ave. Brooklyn Id'd.; -Driver for Consumers Pie Baking company. He had tried to get the day off to go, but company was short-handed and he had to work.
Catherine (43) #3578 b. Germany
William H. (4) #137 #3016
Anna C. (17) #315 #3017
Mary (15) #3015
Catherine (13) #3579

Diehl, Catherine (58) #2722 b. Germany –886 Cortlandt Ave

Diehl, Fred (n) -209 5th St. looking for;
Elizabeth / Lulu (34) (m) –She was first identified by another family, the correctly identified by her own family by her dress.
Catherine (57) (D)
Katy (3) (m)
Elsie (7) #3585
Josephine (11) (I) –Ln –She was caught in a crowd and carried under the boat. She received a broken arm and other injuries and was being cared for by relatives in the Bronx.

Dietz, Rose, (14) (I) -438 6th St.

Dittrich, George –96 Greenwich St   –Grocer
Adelinde (35) # 473 #3217 .
Herman (5) #3391

Alfred (9) (D) #464 #3218
George (11) #3754 Disinterred
Emma (3) #3216

Doering, Mr. (Rev.) George E. (n) –12 State St. –German Lutheran Mission for Immigrants, Ellis Is. & Supt. of Lutheran Emigrant Home, 12 State St Ida D.(34) wife #22804, Man. d. 6/22/04 Submersion, pneumonia & burns –She was buried from ST. Peters German Lutheran Church on W. 22nd St.
Ida (11) #3513 –Service conducted by Rev. Leo Koenig and Pastor Richter of St. Matthews German Lutheran Church, Hoboken, burial
Gustav (9) #3080 was at Lutheran Cemetery with her two children (BDE)
Edna (6) (I) –Ln taken for dead & then revived (You Must Rem.)
Ida (84) (D)

Doerrhoefer, Barbara (42) (I) –Ln –Burns of head, back, arms -1910 can do only light housework –121 Ave. A. –saved by Off. Scheuing
Fred (11) #3345
Frieda (13) #3670
Mamie (6) (I) –Burns over body –Ln 1910 unable to work
Catherine (48) (I) –Ln

Dreher, John -310 E. 25th St. Id'd.;
Catherine) (10) #2794
Conrad (4) #3448
Angelica (34) #2795 b. Germany

Drues, Mr. Hermann –54 E. 4th St. Id'd;
Catherine, (65) #2921 b.Germany –Janitress $8.00/ Mo.
Freda (28) #2920 b. Germany –Sister of Louisa
Henry H. (7) #3750
Lillian /Tillie/ Millie (2) #3431
Distelhorst, Louisa –Office cleaning

Duls, Julia (m) –103 Ave A –Milliner

Pauline (54) #3186 mother

Dunn, Harry -2112 3rd Ave. (also listed as Sherwood Park, New York) –Clerk This was a party of 11, 3 survived (BDE)
Julia (27) #3056 –Julia & Fanny were associated with the Harlem Branch of the Salvation Army. Col. Miilce & Capt. Green conducted the services (BDE)
    Arthur (4y 4mo.) #3060
    Irwin, Fanny (33) #3058 –Lived with Mrs. Dunn

E

Ebeling, Alfonse -77 1st Ave. Id'd ;   –After Identifying his wife, he fell faint for an hour (BDE)
    Emma, wife (34) #3383
    George H. P. (5) #3490

Eckerstorfer, Charlotte (21) #3396) –313 E. 26th St.

Ehrhardt, Minnie (s) –69 1st Ave. Id'd   (see Braun)
    Minnie (13) #3136

Ehrhardt, George –151 E. 4th St. – Undertaker
    Pauline (33) #3650 b. Germany
    Elizabeth (2) #3467
    Clara (6 mo) (m)

Ehrlich, William –Cabinetmaker   (see Herbolt)
    Margaret Herbolt (75) (I) –Face & hands burned, nervous, died since then of Injuries –Sewing $8.00/ week

Eichoff, William (24) #2844 -230 Ave. A.

Eimer, Philip -84 Stockholm St., Brooklyn –Painter
    Charles (13) #3355
    George (11) #3347

Kate (45) #2782 b. Germany

Eisler, Jennie looking for
    [no name] widowed mother
    Burkhart [no name] wife (D) supported 6 children, their father was home bedridden with rheumatism and minding four other children

Elk, Adelaide (23) #3209 –306 E. 6th St.
    Francis (2) #3499

Ell, John (37) (u) –99 1st St. –Stationary Engineer
    Matilda (34) #3441
    John L. (13) (14) (s) (I) – All four had gone to the engine room to watch the machinery just before fire broke out, rescued by gasoline launch.
    Paul (11) (I) –Ln * died later
    William (9) (I) –Ln in hospital for 4 mo. Permanently injured & disfigured
    Connie (Zimmerman) Ell   (see Zimmerman, Gray, Greenwal, Hans)
    Fishman, John (friend of John's) –410 5th St. –Jumped overboard and swam ashore
    Gray, John (friend of John's) –309 E 14th ST. –Jumped overboard and swam ashore
    Greenwall, Albert (friend of John's) –326 E 14th St.
    Hans, Otto (friend of John's) –310 E 14th St.

*Paul & Willie were the last victims released from a hospital, 4 months after the disaster. Paul was asked to write his story for a magazine. When fire broke out and the panic began, Willie was in the bow, kneeling in prayer. Paul found him and grabbing his hand jumped. Both boys were burned on the head, neck, and back. After multiple grafts and new ears, they were released looking like patchworkds. They were only told of their mother's death when they arrived home. Willie would not believe until he had searched the house for her.

Eller, Joseph (65) (I) –Ln –219 E. 13th St. –He clung to one of the paddle wheels with two small children until rescued.
    Matilda (46) #3320 b. Germany
    Ella / Elsie (16) #3319 –Was aboard with friends including Albert Frese (who was a "little soft on" her) & Charles Kuentsner

Eller, James – no address (musician at Metropolitan Opera House) –Ln

Elwanger, LuLu (12) (I) –77 3rd Ave. – Lived with Mrs. Amann   (see Amann)

Engel, LuLu (21) (I) –Ln –359 W. 47th St. -1910 LuLu (Engel) Mann, deponent, Saleslady $5.00/ week, now home injuries to spine, hip bones, nerve center

Engelman, John –425 E. 12th St –Tugman, for NY, N.H. & H. Line / Bookeeper –testified –Saved himself and swam ashore Id'd;
   Louise (29) #2929 – possibly a sister of Mrs. Lottie Sierich, BDE refers to them as the Lander Sisters   (see Sierich)
   William (6) #3434
   Edna (5) (u)
   Anna, sister of John (D)

Enz, Christina (62) #3517 b. Germany -184 W. Broadway

Erdmann, Anton -346 E. 9th St. –Agent
   Margaret (38) #3404 b. Germany
   Alma (11) #3429

Erklin, Otto E. (R.) (30) (u) –1028 Husdon St., Hoboken, NJ –Coal Merchant at 1 Broadway Came looking for other members of his party, they were former members of St. Marks. He was accompanied by Mr. Fulling, a paper Mfgr, 53 Crosby St., Manhattan   (see Klenen)
   Gertrude (1) (u)
   Anna (32) (I)
   Theodore (6) #3393
   Stephen (7wks) (I) –(youngest child to survive) – Rescued by his nursemaid, Louise Garling (Gailing) (12) (s)– Nutley, NJ; She managed to get a life preserver, put it on and jumped overboard (20') with the baby. An "officer" of the boat, already in the water held the children up until they were rescued by a rowboat. Another account says that when Louise jumped, she sank twice with the baby, but came up second time by a board which supported them until they were rescued. (see Heroes)

Esche, Rose (16) #3416 b. Germany – Millinery $5.00/ week -88 Ave A
   Clara (47) (I) –Still suffering Nervous Shock

Eysel, Jennie (9) #3105 –203 Ave. A Id'd by Nicholas Stocker, same address   (see Stocker)

285

F

Feldhaus, George (11) (D) –526 6th St.
  Elsie –Washing & Scrubbing
  Joseph (12) (m)

Feldhausen, Margaret (47) #3083 b. Germany -50 W. 8th St.   Id'd by Cortland Meyburg, a friend, - 35 W. 8th St.
  Nicholas (12) #3082

Ferneisen, Henry (s) –40 7th St. –shock – Ln
  Emma (33) (I) –Ln
  Henry G. (10) (I) –Ln
  William F. (8) (I) –Ln
  Marie (7) (I) –Ln

Fettig, Peter (47) #3374 –120 2nd Ave.   –Peter Fettig Sole Agent Sheboygan Mineral Water BS-99 2nd Ave.   (see Ulrich)
  Mrs. Christine (32) #3615 b. Germany Disinterred
  Elsa (2) #3633 Disinterred

Fetzke, Frank –211 E. 5th St. –Fur Dresser *
  Augusta / Gusta (38) #2876 –Janitress $18.00/mo.
  Elizabeth E. (15) #2877 –Clerk $5.00/week
  Herman C. W. (8 Mo) #2878
  Hattie J. (12) (u) Nervous ever since
*In another account, Mr. T. Felzke is quoted that "If not for her (his daughter) I would be in the river now. We struggled for years and got things piece by piece, we moved here one month ago, I guess for the funerals." (North)

Fickbaum, Peter J. (n) – 91 Ave. D – P.J. Fickbohm Choice Wines, Liquors, & Cigars same address Id'd; This family belonged to many organizations inc. United Brothers Lodge 356 F. & A. M. –Rev. John C. Palmer performed services at Emanuel Chapel, 735 5th St.
  Maria (38) #2861 b. Germany   (see Balzer, Paulsen, Cibilsky)
  Ernest (12) # 3677

Fred (10) (l) –Lb burned hand & Leg –Clerk
Maria (Mary) (14) #2862 –Id'd by her uncle, Henry Paulsen, 66 Cumberland St., Brooklyn
Kate Cibilsky (18) #2947 –Servant

Filskow, Karl -170 E. 4th St. –Tailor
Meta (20) (l) –Lb Pneumonia & Brain Fever –Lace worker $10.00/week – 1910 not working
Antonia (44) (l) –Lb Hemorage of Stomach, face & hands burned –Nurse $15.00/ week –1910 not working

Finkenagel, John 439 6th St. –Coal Vendor; This family also listed at St. Mark's Place, might have moved recently thus two addresses (NYT)
Cahtharina (34) (m)
Cattharina (12) (l) –Lb
Mary (10) #3608 Disinterred
Clara (8) (m)
William (6) (m)

Firmeisen, Elizabeth (34) #2885 –80 1st Ave.
Fred (40) (u) –Salesman

Fischer, Henry (n) –108 E. 1st. Ave –Baker
Emma [nee- Daerzbacher] (30) #3743
Edna (6) #3480
Lilly (4) #3420
Barbara Hegyi (20) #3365 b. Hungry –Servant

Fischler, Edmund –314 E. 9th St –Lunchman   (see Junck)
Antonia / Anna (6) #3455
Erma (8) (D)
Erna (6) (D)
Hertha (7) # 3666
Bertha Junck (57) (D) Mother-in-Law –Servant $10.00 / week

Fitch, Edwin S. (55) (n) – Decorator wallpaper and paint, $30.00 / week; Emig. from England 1902; 1910 Nervous Shock, insane, no support – Affidavit filed in Columbia Co., NY.
    Clara (D)
    Minnie (18) (D)
    Catherine (16) (D)
    Emma (14) (D)

Flanagan, Edward (25) (u) – 445 W. 28th St. -Mate aboard the *Slocum*

Flegenheimer, Lena (19) #2804 –684 Greene Ave., Brooklyn

Fleischer, Henry (16) #3631 Disinterred –322 E. 13th St.   (see Schaefer)

Foelsing, Henry –1914 3rd Ave. –Painter & Decorator Id'd;
    Amelia (nee-Moeschen), (35) #3207
    Ferdinand (9) #3071
    Elizabeth (16) #3069 –Typewriter & Stenographer $12.00 / week
    George (10) #3070

Folke, Anna (50) #2859 b. Germany) –257 Ave B
    Dora (75) #3205 b. Germany
    Ludwig (41) (I) –Lb

Foster, Maggie (59) #3219 -79 Ave. A

Freck, Charles (14) #3336 –409 E. 5th St.
    Edward (13) (I) –Ln
    William (11) (I)

Freeman, Jenny E. (33) (u) 433 W. 52nd St. –Negro Stewardess on *Slocum*

Frese, Ferdinand (Fred) (50) (l) –Lb Right hand burned — Ferry Hotel, 509 E. Houston St., Ferdinand Frese, Proprieter –Affidavit filed in Kings Co., NY. Also a director of the Consumer's Brewing Co., Brooklyn
   Meta (46) (I) –Lb –Mr . & Mrs. Freseand Anna, were viewing engine room when fire started, and so escaped. Meta died in 1947 at 92
   Anna (20) #2842 -426 E. 15th St.
   Albert F. (16) (u) – sorted mail at Funk & Wagnall's publishing house; jumped from bow, swam to safety & lived to become firm's treasurer. (Ellis.)
   Mrs. Lillie Pfeifer -1910 her name appears on affidavit signed by Chas. Pfeifer 6/15/04 was their first anniversary   (see Pfeifer)
   Anna (15) (I) –Lb – Her hand had seared to the rail, & she had to pull loose. She jumped and hit a rock, broke her teeth. 3 friends killed.

Frey, Elizabeth (34) #3688 –84 7th St.
   Frederick (1) (m)

Fritz, George (collector for Lipmann's Brewery) -1235 Park Ave. Id'd (see Gieser)
   Alma (47) #32756 b. Germany
   Catherine Gieser (25) #3461 daughter

Froelich, Charles (n) – 301 W. 96th St.
   Lena (23) #2814 Id'd by cards

Fulling, Mr. (n) –a paper manufacturer of 53 Crosby St, New York   (see Erklin, Klenen)

Fulling, Matilda (29) (u) –110 W. 129th St.   (see Olsen)
   Edmund (1) (u)
   Clara Olsen (16) (I) –Servant

Funk, Michael (12) #2772 –33 Ave. A

G

Gade, Henry –405 5th St. Id'd
   Grace (16) #3331 –Id'd by brother Henry

Gaffga, Elbert J. (20) (u) –72 Howard Ave., Brooklyn –Oiler on *Slocum*

Gailing, Louise (16) (u) –30 Hudson St., Hoboken; home –Kearney, New Jersey -Daughter of Christopher Gailing (see Erklin)

Galewiski, Samuel (m) -54 7th St.
    Flora (36) (m)
    Helen (6) #3465
    Morris (3) #3329

Gallagher, John (n) –424 E. 15th St. –Clothing Store
    Veronica (32) #3348 b. Germany –her first marriage was to an Uhlmeyer
    Catherine (11) (I) –Lb
    Walter (9) #3052
    Regina Agnes (9 mo) #2980–Id'd by her uncle, Jacob Ottinger –280 Ave B (see Ottinger)

Gambichler, Harry (13) (u) –404 E. 5th St.

Gardiner, Freda (8) (s) – Disobeyed parents to go (Inferno); Picked up by a rowboat off E. 138th St. she was with her Aunt Louise, on the main deck.

Gartner, Louisa (23) #3627 b. Germany – 748 Westchester Ave.

Gassmann, Mrs. Michael / Henrietta (41) (I) –Lb -128 E. 4th St.
    Minnie I. (7) #3447
    Frank J. (11) #3495
    Michael J. (5) #2849

Geiser, Henry (n) -1225 Park Ave – Collector for Lippman's brewery   (see Fitz)
    Kate (25) #3461

Geisler, William (16) (u)–201 Ave. A Id'd
    Lena (43) (I) –Ln

Edith, (19) #2868

Geissler, Mrs. Minnie (35) (m) –439 6th St.
　Ella (5) (m)
　Louis (10) #3190
　Lillie (7) (m)

Gemeimer, William J. (n) – Freight Agent –895 Jefferson Place, Bronx
　Grace T. (3) #2978

　Anna S. (30) (u)
　William Henry (5) #2958

Geminn, Elizabeth (nee Mauer) [no address]　　　　(see Mauer)
　George Mauer (52) (D) father –Musician $30.00 week
　Margaret (47) (I) mother –later died of injuries
　Matilda (14) (D) sister
　Clara (12) (D)

Geudert, Emilie –Janitor　　　　(see Stenger)
　Francis Stenger (widow) (32) (D) daughter –Dressmaker $14.00 / week
　Rose L. Stenger (10) (D) granddaughter

Gerdes, George (n) –341 Rivington St. –George Gerdes choice Groceries, 339 Rivington ST. cor. Mangin
　Christina (28) #351 #2952 b.Germany
　Henrietta (27) #267 #2951 b. Germany sister of George –Servant $18.00 / mo.

Gerdes, Henry (70) #3516 b. Germany –430 Kosciusko St., Brooklyn – Former members of St. Marks (oldest male lost at age 80)
　Margaret (66) #2813 –Family claimed that she wore $300.00 in jewelry, but the coroner's office could only account for a wedding ring

German, Frederica (46) #3588 –315 E. 18th St.

Catherine F. (17) #2791
Frieda F. (15) #3067

Gerstenberger, Richard (39) #3676 –the Hall Building. –"All the family lost", Their funeral was held in Central Hall and had representatives from the unions & lodges who met in the hall. They had no children. Buried Lutheran Cemetery.
Anna (32) #3403

Gettler, Caroline / Lena (68) #2866 –231 (233) 5th St. –Tailoring

Gibbons, Thomas (42) (u) –225 E. 5th St.
Margaret (14) #2736
Ella (12) (D)

Giessman, Moritz M., –114 E. 4th St. -Janitor
Lena (16) #3307 –Passimentari $9.00 / week
Christina (47) #3856 b. Germany

Gillis, Charles, Jr. (16) #3129 –512 5th St. –Office Helper $7.00
George (13) #3128

Goetz, Edward –80 1st Av. –Butcher
Catherine (28) #2847 b. Germany
Albert (2) #3772
Eddie (5) (D)

Goetz, Leona (13) #2995 –337 5th St. – was on the *Slocum* with her friend, Wanda Rokanski (10) (D) of the same address.

Goss, Adolph (no validated address, reported as 27 Stanton St. or 97 7th St.) –Repairing
Marie (48) (D)
Gertrude (28) (D) –Bookkeeper $8-12.00/ wk.
Anna (27) (D) –Bookkeeper $8-12.00/ wk.

Goss, Gertrude (27) #3661 –97 E. 7th St.

Mary (59) #3505 b. Germany

Graf, Lottie (18) #2746 -56 W. 54th St.

Grafing, Mr. –908 Ave A
    Lillie, (29) #2858

Graner, Rudolph –100 University Pl. –Tailor
    Louis O. (22) #3771 –Goldbeater $18.00/wk
    Carrie Lullmann (daughter) (24) #3509 b. Germany –Dressmaker $9.00/wk. (see Lullmann)

Greenhagen, Ernest (14) (I) –176 Lewis St.

Grelka, Julius –845 (345) E. 15th St. – Cigarmaker
    Emelia (55) #3101
    Olga (14) #3102 – Flower Maker $4.00/ wk.
    Agnes (10) #3100

Gress, Barbara (37) #3773 b. Germany –526 6th St.
    Lillian (10 mo) #3208
    George (8) #3405

Gress, Otto (43) #3469 b. Germany –134 7th St.
    Elisa, (40) #3118 b. Germany –Id'd by her brother Frank Frederich
    Walter (13) #3119
    Clara (12) #3505

Grewe, Henry (17) (16) (D) –54 E. 7th St. (see Stegman)
    Frederick W.J.A. (14) (D)
    Annie M. (45) (I) –Ln

William H. (16) #3398

Grey, George (13) (u) –309 E. 14th St. (see Ell) –Jumped onto the tug *Director*, the tug had so many jump on her stern, that some slid back into the Water George held onto a little girl by sitting on her so she would not slide back off the slippery deck

Griesel, Emma (15) #3457 –117 E. 2nd St.

Grimm, Frederick –314 E. 9th St. Id'd
    Selma (33) (I) #2720

Gringel, Kate (32) (I) –439 5th St.

Growald, Emil, –56 7th St. –Peddler
    Elsie (10) #247 J #3074 b. Austria
    Fred (12) # 3758 b. Germany

Gruning, Helena / Lena (29) #35736 b. Germany –45 E. 7th St.   (see Pienning)
    Henry C. (6) #2840
    Charles (4) #3614 disinterred
    Helene (8 mo.) #3637 disinterred

Grunwald, Albert (13) (u) –326 E. 14th St.

Grunwald, Amelia A. (35)(m) –257 Ave. B
    Richard (5) #674 #3634 buried as unknown

Gross, Gustav –90 1st Ave –Baker, G. Gross Vienna Bakery, Coffee & Lunch Room
    Emma (43) #3038
    Bruno (5) #3032
    Freda (21) (u)
    George (13) (u)

Curt (11) (u)

Gruben, Emma (40) #2759 –420 E. 17th St. –Id'd by her nephew John L. Distler (see Beneke)
   Carrie (13) #3004
   Mary Beneke (30) #3435

H

Haag, Ella (13) #3409 –158 1st Ave
   Susie (48) #2816 b. Hungry
   Aranka (20) (I) –Ln

Haag, Louis – 210 E. 14th St., elderly, came with wife to find; –Shoemaker
   William (14) #3299
   Wilhelmine (12) #3508
   Emma (9) #3300

Haas, Rev. George C. F. (50) (I) – 64 E. 7th St.(Rectory) Held onto underside of paddlebox until rescued –shock – Ln   (see Schultz)
   He was burned about the head & legs (Dr. Geo. Sankin / Semkin attended)   (see Tetamore)
   Anna S. (46) #2930 –Pastor's wife –Id'd by Rev. Julius G. Schultz , Luth. Church, Erie Pa. –buried beside her sister, Mrs. Wm. A.
      Tetamore of Bushwick Ave., in Lutheran Cemetery
   Hanson, Mrs E. J. (70) (m) –Rev. Geo. Mother-in-Law   (see Hanson)
   Gertrude (13) #3617 –disinterred from unidentified plot when identified by her clothes and moved to family plot in Lutheran Cemetery
   Emma L.. (31) (I) – sister of Rev. Geo. & the church organist, was missing for 2 days before found –Ln –burned so badly that she passed
      out during the funeral for her sister and was assisted by her physician and a trained nurse.
   George (n) –son of Pastor; had to confirm reports of disaster to congregants on the 15th . Tutor in Latin at college of city of New York.

Hagenbuchter, Mrs. Kate (m) –2112 3rd Ave
   Mary (32) (D)

Hager, Johanna (no address) –Housework   (see Moeller)
   Bernard Moeller (brother-in-law) (38) (I) –Dept. Manager $63.00/ wk. d. 6/23/04 of injuries (submersion)

      Velesca (nee Hager) (29) (D)
      Edgar (4) (D)
      Arthur (6) (I) –burned head and body

Hanneman, Caroline (56) (I) –Ln shock (m) –415 E. 5th St. – Housekeeper* –Andrew Hannemann Lisenced Plumber & Gas Fitter
                                        –NBI burned on arm, hands, ears, dislocated rt. Foot.
      Mrs. William A. Boeger, daughter   (see Boeger)

* Little Wilber Boeger wanted to go from the lower deck to the upper, so Grandmother took him. When fire started, someone grabbed the boy away. Caroline tried to reach rest of family below, but could not, but she was convinced they had jumped ans so did she. She landed in three feet of water and was pulled out by rescuers. Little "Dollie" was found nearby.

Hanson, Margaret (70) (m) –64 E. 7th St. (see Rev. Haas)
      Elizabeth (67) (m)

Hardekopf, Edward – 343 Rivington St.   –Edward Hardekopf Dealer in fine Dairy Products
      Meta (40) #2863 b. Germany
      Henry (15) (u)

Harms, Ida – 312 E. 14th St. – Storekeeper
      Mrs. Jennie (of Dreamland show) came with brother Herman looking for
      Hermann (21) (D) –electritian $26.00/ wk
      Rosa Wallace, friend (11) (D) –214 E. 11th St.
      Otto (14) (I) –said he saw Rosa & Herman sitting together at NBI, but family could not locate them – burned over entire body

Harris, David (58) –242 E. 5th St.-Flower Factory owner
      Sylvia (10) – #3765 Adopted daughter
      *Agnes Bell, adopted daughter (19) – #2796   (see Bell)
      Lena

*Descendants of this family question this person as family member; however, coroner's records list her in this family.

Hartmann (Hardtman), Mary (45) # 3098 –309 E.10th St.

Clara (11) (l) –Ln –She was tagged body #24, and taken to the 35th Precinct, Alexander Ave. where she suddenly revived after several hours of being presumed dead; per affidavit sher returned to school, but still suffered nervousness
Marjorie ,/ Margaret, (15) #3099 –had graduated public school & was confirmed
Jacob W. – Tutor in German at the College of the City of New York
Marie, (21) S. I. Teacher $600.00/ yr. – Insane Asylum
Reitz, Willie (13) (s) – found his cousin Clara Hartmann in hospital
Jordan, Mrs. (D) – 37 3rd Ave (see Jordan)
    Jordan, Pauline, friend of Clara.
    Jordan, Catherine (s) friend of Clara

Hartung, Magnus, (n) – 342 E. 21st St. Id'd -Tailor
    Louisa (47) #3433 –Tailoress $7.00/ wk.
    Elsie (6) #2909 –Id'd by her brother Frank
    Clara (11) #3646
    Carmelia (13) #2908
    Harry (16) (u)
    Francis (18) #3325 –Piano teacher $12.00/wk. St. Marks Sunday School teacher & sang in choir
    Minnie (24) crippled for life (l) –Ln St. Marks choir – Housework 420.00/ Mo. Burned on body –not working

Hauff, Fred, – 142 E. 3rd St. –Butcher
    Mathilda (14) #3306 – Was in High School
    Agnes (11) (l) –Ln
    Blusch, Katie (25) #3476 – Housekeeper

Hauser, William (49) (u) –317 Bowery

Havemeyer, Emma S (36) #3143 –1499 1st Ave.
    Ernest F. W. (7) #3144

Hayden, Wilhelmina L. (23) (u) –138 2nd Ave.

Heckert, Eva (32) (I) –Ln –88 Ave A –Peter Heckert Dealer in Choice Beef, Veal, Mutton, Lamb, Poultry & Game, 88 Ave.A   (see Link)
   Julia (8 Mo) (I) –Ln #2935
   Annie K. (11) #2725
   Maggie (9) (I) –Ln
   Cecelia (6) (u)

Heckman, Lillian (5) #3643 Disinterred –525 E. 12th St.
   William (3) #3582
   Catherine (24) #3744

Hedencamp, John (55) #2800 –805 6th St. Id'd. –J. Hedencamp Carpenter & Builder 805 6th St.   (see Addicks)
   Margaret (50) (I) –Ln
   Margaret L. (11) #2801
   Frank (8) #3487

Heerz, George –412 6th St. –Varnisher Id'd
   Herminie M. (32) # 2773

Heilshorn, George (3) #3512 –181 Waverley Place
   Margaret (33) #3679

Heins, Fred –397 E. 4th St.
   John C. –344 E. 4th St.
   Mrs. John C. / Bertha (46) (u) –She clung to a paddle, supporting another until she was rescued by a Negro crewman. She rejoined the family at home.
   George } (15)(u) twins, both good swimmers; George picked up by a tug; Theodore swam to Randall's Island. They met at
   Theodore}   home.
   Frank (11) #2811 – had broken his leg and was still in a plaster cast.

Heins, Henry (n) - 300 Front St. – Bartender Id'd
   Annie (40) #3832

Henrietta (10) #3833
Margaret (7) #3834 – floated 8 miles & was found at Ft. of Clinton St., 2 blocks from home (Inferno)
Ida (14) #3485

Heins, Martin (n) –240 9th Ave. – Bartender
Annie (29) #2831

Heinz Henry, –97 Ave A – Henry Heinz Oyster House 97 Ave. A
Louise (20) #3566 – Domestic $20.00/ Mo.
Dina (10) (m)
George (17) (I) –was on the upper deck and last saw his two sisters kneeling in prayer with flames all around; they all fell into the water*
Henry (12) (I) –Ln went dumb from the shock of the event**
Johanna (44) #2892
* George testified that the deckhands deserted the passengers. He had gtone to the top deck and Captain asked him to calm the passengers behind the wheelhouse. He climbed on the rail to get the crowd's attention, but was pushed over by them. His head hit the mud bank bottom where he could see others stuck in the mud. He was rescued by a tug.
**Described as somewhat reserved and shy. He lost speech for 3 days. He climbed a pole to the top deck, then fell when the deck fell into the 2nd deck, then fell into the water. Henry held on to the wheel paddle until rescued by a man from the tug.

Helmke, Clara (15) (I) –Before and after the disaster a stenographer $8.00; Internal Injuries –Was aboard with friends including Albert Frese & Charles Kuenstner

Hencken, Mrs. Lucy G., (41) #3402 –169 South 2nd St., Brooklyn
Charles D. (18) #2809 –Died saving Anna Frese of Houston St..
Lucy G. (15) (I) –Ln –left her mother on top deck to find brother, below. While searching, she found three babies, which she brought to her mother. She found but could not reach the brother. Returned to upper deck, but mother and babies gone. She was rescued by William Major of the tug Theo, after she jumped. Listed as suffering immersion and half crazy from loss of mother. She realized her loss before leaving the Slocum and jumped to die, because she could not swim and did not wish to live without mother and brother. Another account says she was held up in the water by a August Lutjens, Jr., who had forced her to jump. A third account says she did not know who had helped her in the water.

Henes, Annie (66) #2895 –507 5th St. (see Ehrlich)

Henrichs, Theresa (Tessie) (16) #3493 –416 Brook Ave, Bronx
    Amelia (18) #3492

Henry, Sarah (12) #3078 –225 5th St.

Henzler, Jacob –154 1st Ave –Laborer
    Augusta (11) #3311
    Amelia (18) #3629 Disinterred –Milliner $5.00
    Jacob (8) (I) –Ln – Now at school injured eyes and heart trouble

Herboldt, Annie (75) (u) –107 W. 103rd St.

Herbolt, Margaret (76) –416 E. 5th St. (was the oldest female survivor)

Hergenberger, Henrietta (48) #3166 b. Germany –22 St. Marks Place

Herman, Kate (52) #3044 b. Germany –168 1st Ave.– Id'd by her son Henry of 410 E. 5th St.
    Emily (25) #3049
    Elsa (3) #3569
    George (1) #3338
    Lucy (21) #3048

Hermberg, George (7) (D) –79 Coyler St., Greenpoint, Brooklyn (see Buchmiller)
    Arthur (9) (D)

Hessel, Caspar–801 147th St., Bronx –Unable to work (see Schnitzler)
    Wilhelmina (64) #2984 b. Germany –Vest sewer $12.00/ wk Id'd by Schnitzler, Edward, Patrolman, son-in-law –10 Gouverneur Pl., Bronx

Hettinger, Mrs. A. J./ Elizabeth (26) #3551 –127 1st Ave.

Hetterich, Elizabeth (30) #3436   –420 E. 15th St.
 Adolph (8 Mo) #3640 Disinterred
 Emil (3) #3356
 Robert (6) #3003

Heuer, Herman –129 Division St –Tinsmith   (see Hauer)
 Herman (11) #2956
 Dora (7) #190   #2954
 Mary (17) #3542
 Dorothy (39) #2956 b. Germany
 Adolph (13) (u) –Nervous ever since; Tinsmith –support trifling
 Lena #460

Heyrish, Joseph –423 E. 16th St. –Engineer
 Katherina (50) #3006
 Bennie (12) (l) –Ln

Hiller, William -404 6th St. Id'd
 Christina (66) #3022 b. Germany
 Gottfried (64) #3564   –Sexton of St. Marks

Hinkel, Gustav –227 E. 7th St. – Baker
 Lillie (9) (D)
 Gustav, Jr. (12) (u)

Hirt, Mrs. Mary (65) #2968 b. Germany) – 611 Columbus Ave.   (see Smith at same address)
 Id'd by Mrs. Louise Buffo, daughter –82 W. 90th St.

Hlavacek, Annie (22) (m) – 428 E. 71st St.

Hoelder, Mary (79) #3072 b. Germany –169 Ave A   Possibly identified by daughter-in-law Kate
          Mary (43) (I) –Corset Trimmer $6.00/wk –Nervous prostration, but working

Hoffman, Erna (15) (u) -321 E. 9th St.
          Muller, Florintine (59) #3141 b. Germany –Grandmother of Erna

Hoffman, Fred C., (24) (u) – 73 2nd St. – Fireman   (see Workman)
          Sophie, (53) #2931 b. Germany
          Workman, Jennie (23) (D) –116 Lake St., Jersey City, NJ; Friend of Fred's sister-in-law
          Frank, Fred's brother (n) –336 New York Ave., Jersey City; Shipping clerk
          Cecelia, (28) #2837
                    Edna (3) #3308
                    Raymond (4) #3556

Hoffman, Henry –40 Lafayette Pl. (Astor Library) –Machinist   (see Alt)
          Ella (15) (D) –Cook $15.00/ mo.

Hoffman, Henry A. –170th St. & Washington Ave., Bronx –Sheetmetal Worker
          Elizabeth (56) #1394 –Id'd by letter she was carrying
          Mary (29) #2990 –Dressmaking $8.00/ wk.

Hoffmann, Therese –Passementerie
          Muller, Florentina (59) (D) –Passementerie $10.00/ wk   (see Muller)

Holler, Barbara (56) #2750 b. Germany –334 6th St. –Bur. Lutheran Cemetery Rev. W. W. Gillies Rector of Epiphany Chapel on Stanton St. led home & graveside services

Holthusen, John (58) (u) –138 2nd Ave
          Clara (25) (u)

Horway, Charles -313 E. 9th St. –Butcher   (see Beck, Norway)

302

Johanna (31) #2785 –Id'd by William Beck, brother – 310E . 8th St.
Carl (1) #2787
Della (5) #2786

Hotz, Fred W. (37) (u)   –319 5th St. –Bookkeeper
Anna B. (27) (D)
Martha W. Bertha (11) #3384
J. William (8) (I) –Lb –Nervous ever since

Hubschman, Morris (29) (I) –Ln -640 6th St. Waiter on the General *Slocum*

I

Iden, Henry A. (20) (u) - 100 E. 4th St. Id'd
Grace (5) #2918
Henrietta H. (10) #3479
Minnie 18) #2919
Anna (12) (I) –Ln
Zabilansky, Mary –Servent #36,838 b. Bohemia d. 9 Nov. 1904 pneumonia
Schwartz, Miss (s) – a friend saved by Henry
Emma (13) (D)
Julia (24) (D)

Illig, Mary (40) (I) –Ln -Nervous palpitation of the heart –Washing, etc. –433 5th St.
Margaret (5) #3454
Conrad (8) (I) –Lb
Elizabeth (6) (D)

Irwin, Fannie (30) (D) – 2112 3rd Ave –Salvation Army Lassie   (see Dunn)

J

Jacobs, Jacob S., (28) (u) – 1760 Madison Ave. –Candy stand on General *Slocum*

Jentz, Lizzie (27) #3389 b. Germany –277 Greenwich St.

Johnston, Margaret (s / n)
    Tyson, John (39) (D) Brother-in-law –Fireman $75.00/ mo

Jonck, Bertha (57) #2852 b. Germany –314 E. 9th St.

Jordan, Henry J. – 37 3rd Ave –Henry J. Jordan Mfg. of fine cigars (see Hartmann)
    Pauline (16) (I) –Ln friend of Clara Hartmann
    Catherine (20) (I) friend of Clara Hartmann

Joseph, Rudolph –45 3rd Ave –Hotel Keeper
    Margaret (31) (I) –Lb –Right leg & arm broken
    Frank (8) #3189 –Lb

Junck, Bertha (57) (D)   (see Fischler)

Justh, Joseph – 105 E. 8th St. Id'd – cabinetmaker
    Emilia (34) (D)
    Leontine S. (11) #441 #2953
    Adelheid (12) #3401 (formerly Id'd. as Yetta Youst [NYT]
    Emelia (14) #3399 –Ass't. Bookkeeper $3.00 /wk
    Margaret A. (1) #3756 Disinterred
    Joseph F. J. (7) #3623 Disinterred
    Etelka (12) (D)

K

Kaaz, Herman – (Poss.) 110 1st Ave. –Waiter   (see Bozenhardt)
    Bozenhard, Emilia R. (39) (D) sister –same address
        Lucille (11) (D) niece

Kaffenberger, Anna C. (14) (I) –Ln –436 6th St. –An only child, she took the 3rd Ave. train home after her rescue. She had burned hands & a huge bruise on her Forehead gotten when she had jumped from the top deck to a tug and people landed on her. (see Motzer)

Kalb, Augusta F. (21) #3127 –84 7th St. (Possibly wife of F. Kalb; Wholesale & Retail dealer of milk & cream – 517 E. 5th St.)

Karl, Barbara (62) #3544 b. Germany –314 6th St. (see Klein)
    Bertha (25) (D) – Corsetier

Karle, Amelia (12) (I) –Ln – 56 7th St.

Kassebaum, Henry (53) (n) – 28 Enfield St., Brooklyn; Found party of 12   (Affidavit filed from Bucks Co., Pa.)
    Catherine (52) (I) –shock, submersion, mental anguish – 196 Guernsey St., Green Pt, Brooklyn*
    Henrietta (30) (I) – H – Double fracture above ankle, broke femure –Neckwear business $10.00/ wk -now permanently lamed, occasional
        pain & suffering, now a clerk, – Jumped from the second deck onto a tug   (one account says the Wade)
    Schnude, Henry C. (32) (D) –Paying teller at bank $45.00/ wk also head deacon at St. Marks   (see Schnude)
    Anna C. (32) (D) wife, daughter of Henry Kassebaum
        Grace (4) (D)
        Mildred A. (1½) (D)
    Toniport, Mrs. Frida, (28) (D) daughter of Henry Kassebaum
        Francis (4) (D)
        Charlotte A. (2) (D)

* Catherine jumped and sank twice before she managed to grab onto a paddle box, where she was rescued by a man in a sailboat. When she was landed in Manhattan, a woman with a carriage took her to her home on 138th St., gave her dry clothes and sent her home in the carriage. Later at the temporary morgue, she became so distraught that Bellevue nurses restrained her from jumping and took her to the hospital.

Kastner, Paul (n) –110 1st Ave. – Upholsterer
    Mary (15) #3545 – Cash girl $4.00/ wk
    Paul (3) (u) –picked up by a tug

Kaufman, Maildred (18 mo) (I) –Ln – 121 1st Ave.
    Julia (27) (I) –Lb –Internal injuries which caused kidney trouble; affidavit includes doctor's note that last year she underwent operation for

dislocated internal viscera, also still in treatment now at Babies Hospital for mental infirmity

Kawczynski, Teofil –196 3rd Ave. –Barber   (see Stoss)
    Teofil (15) #2830

Keeler, William, Sr. –112 E. 4th St. -Laborer
    Catherine (13) #3504

Keisel, P. Edward (27) (l) –H –submersion - 266 Ave. A –Upolsterer
    Annie (25) (l)   –H – submersion, shock
    Edward J. (3) (l) – H – submersion, expected to die
    Lillian (5) #3109

Kelch, Margaret (32) (l)-Ln –800 E. 14th St
    Katie (6) #3139
    Elizabeth (8) (l) –Ln
    George (38) (l) –Ln

Kelk, Charles (41) -14 Arthur St. –Policeman on the General *Slocum*

Keller, Fred (14) (u) –115 Ave A looking for his Aunt

Kessler, Barbara (40) #2730 b. Germany –205 E. 7th St.
    Augusta(18) #2802

Kiefer, Louis J. (10) (u) –1592 2nd Ave

King, Katherine (26) #3324 b. Ireland –311 E. 37th St.

Kip, Anna (17) (l) –Ln –1894 3rd Ave. – Rescued by Off. Scheuing

Kircher, John (66) (s) –Testified at the Inquest -185 Russell St, Williamsburg, Brooklyn – Three household under one roof, grandparents & families
of two sons. Sons in business together.
Catherine (62) #3120 married
John H. Jr. (n) , son; searching for;
Elizabeth / Lizzie (37) (l) –Ln –swam ashore –She had persuaded the elder Kirchers to go even though Catherine was sickly.
George N. (3) (l)
Elsie C. (7) #3589 – only child who could not swim, so gave her life belt. "She sank like a rock"
Fred D. (9) (l) –Ln
George (s) –Jumped overboard and swam ashore. He said that they had been on the top deck at the back rail. He searched the island for hours without results. The two brothers continued searching the morgues together for their families.
Margaret (34) #3121
Stacey (7) (l) –Ln
Harold (4) #3591

Klatthaar, Katharine (56) #2751 –506 E. 5th St.
George (14) #2873 –college student

Klein, Annie (22) (l) –Ln –331 E. 16th St.

Klein, Edward (n) –31 Ave. A –Wholesale & Retail Liquor dealer at 34 Ave. A. (BDE, he was founder of the White Mice)
Dinah (73) #3669 b.Germany – service conducted by Rev. Junger, St. Matthews Lutheran Church, Bronx; in service room behind the store
Dina (40) #2723 –died at Ln
Julius (6) #3667
Matilda (10) #3668
Edward H., Jr. (14) (l) –Broken kneecap
Lucy (4) (l)
Elsie (26) (l)

Klein, Gottlieb C. -314 6th St. Id'd –Mechanician   Affidavit filed in District of Columbia   (see Karl)
Emma (25) (D)

Amelia (10) #3630 Disinterred

Klein, Harry J. (17) (I) –Ln –399 Miller Ave., Brooklyn
Nancy (13) #3009

Klein, John (17) (D) 191 E. 3rd St.
Salome (11) (I) –Ln

Klein, Kate (21) #3677 –436 E. 15th St.

Kleinhenz, Philip –196 Ave A –had been sent to Blackwell's Is. For non-support preferred by Mrs. Kleinherz, but was released in time for funeral
Barbara (44) #3514 b. Germany
Carolina #3135

Klem, Jacob –Cabinet Maker
Kate (21) (D) –Dressmaker $12.00 –Was aboard with a group of friends including Albert Frese & Charles Kuentsner

Klemme, August E. (69) (m) –433 E. 9th St.

Klenck, Bertha (40) #339   #3087 –113 St. Marks Pl. –She had tried to put life preservers on the children, but the straps broke, so she went without one.
Charles (8) #3088
William F. (20) #3449 –Clerk $12.00/ wk –half brother (see Klenck, Charles)
John W. (13) (I) –Face Burned –Bank Clerk (from affidavit F. W. Bauer, Guardian for John W.)

Klenck, Charles – 441 E. 6th St. Id'd –Cigarmaker
Wilhelmine (20), cousin #2936 –Saleslady $7.00/ wk
William F. (21) (D) (see Klenck, Bertha)

Klenen, Frederick (n) 1391 Washington Ave., Bronx –with party from Hoboken, former members of St. Marks, Rev.Henry Stoup, St. Johns Lutheran, 119th St.,

Neta (56) #3019 May's mother –Merritt Undertaking Parlor, Man. performed funeral & graveside services at Lutheran Cemetery
Ethel (1) #3020 –taken to Merritt Undertaking Parlor 19th St. & 8th Ave. (see Erklin, Fulling)
May (24) –Lb –Badly burned

Klingert, Theresa (6) (I) –Lb –444 E. 15th St.

Knoffler, John (30) #2981 b. Germany – 89 St. Marks Pl.

Koeppler, Henry G. -192 1st Ave. – Butcher
    Lillie (17) #3687
    Irene (12) #3197

Kohler, Henry A. (40) #2924 – 313 or 315 13th St. –H. A. & G. L..Kohler Brothers, Real Estate & Insurance, 123 4th Ave. btwn. 12th & 13th Sts.
    (Party of 29, primarily wife's side)
    Mary Wife (38) #3553
    Henry A., Jr. (12) #2925
    George L., Brother –143 4th Ave. (in business with brother)

Kolb, Albert (22) (u) –743 202nd St., Bedford Park, Bronx Id'd
    Valentine (68) #3260 b. Germany
    Magdalina (72) #3084 b. Germany

Kopf, John (15) (I) –Ln 586 Grand St.

Kopf, Marcus– 337 E. 9th St., – Barber
    Lizzie (32) #32106 b. Germany
    Emil (10) #3500
    Frances (9) #3211
    Theodora (5) #3498
    Ella (1) #3212

Ellis (18 Mo) (D)
Theodore (6) (D)

Korman, Minnie (24) #3115 b. Germany –402 3rd Ave
Koster, Henry–343 Rivington St. –Driver Henry Koster Dealer in Orange County Milk –family members of St. Marks.
Margaret (46) #33415 b. Germany – Her sister Christine Meyer sent letter to editor ob BDE trying to locate Margaret & children before they were found.
Meta (6) #3752 Disinterred
Anna (9) (m)
Charles (16) (u) –jumped to a tug

Krafft, Lena – 140 (145) E. 4th St.
Louisa (30) (D) –Fitter for Costumer $6.00/ wk

Kraljich, Martin (19) (u) -20 President St., Brooklyn – Deckhand on General *Slocum*

Kramer, Barbara (56) #2891 b. Germany – 70 1st Ave –Janitress

Krause, Sadie (17) (I) –Ln –201 W. 111 St.

Krautwurst, Elsie (s) –114 E. 4th St.
Anna C. (14) #2855 b. Germany – Schoolgirl

Kregler, Charles (n) –257 Ave. B looking for
Margaret (34) #3541 b. Germany
Lizzie (3) (u) –Was reported as the "little girl in red" who was placed in a chair at the Alexander Ave. Station, amongst the dead for several hours before her father found her. She may also be the child from the paddlebox, who reported "Mother is all burned up".
Fred ((10) #3749
Dora (11) (I) –Ln

Kreuder, Mrs. Annie E. (18) (u) –62 W. 97th St.

Mrs Lena P. (28) (I) –H

Kreuder, Marie (29) (single lady) (I) –H –burns – 452 West End Ave – she slid down a pole and held onto a rope in the water until rescued by a coal barge

Kruger, Eva (57) #3196 b. Germany -106 7th St., Brooklyn

Krutsch, Martha (14) (I) –Lb – 513 6th St.

Kubera, William (15) (I) –Ln –357 4th St.
    August (14) #2770

Kuenstner, Matilda (41) (I) –Ln -29 Suffolk St.

Kuhn, Carl (n) -64 Driggs Ave., Brooklyn Elizabeth's parents and brother witnessed the disaster from the shore.
    Anna (n)
    Carl (n)
    Elizabeth (7) –onboard with an unidentified aunt

Kunath, Margaret (15) (u) –406 5th St.

Kunstner, William (12) (I) –Ln –65 St. Marks Place
    Charles (17) (I) –Ln
    Mary (46) (I) –Ln

Kunze, Paul –889 Broadway, Brooklyn -873 Broadway, Brooklyn [1910] -Instrument Maker
    Augusta (20) (D) –Operator $12.00

L

Lahem, Philip –1000 Union Ave., Bronx Id'd
    Clara (20) #3250
    Dora (25) #2964

Lamb, Lillian (7) (D) –645 E. 17th St.
   Amelia (40) (D)
   George (11) (u)
   Frank (15) (D)

Lambeck, Henry W. (n) –427 E. 9th St. – Painter
   Albertina (33) (I) –Ln
   Dora (11) (I)
   Ernestina (9 mo) #3671
   Henry (7) #3342
   Albert (4) #3583
   Herman (14) (I)

Lane, Gustav H. (17) #3377 –227 E. 11th St.
   George (14) #3029

Lang, Emilie (15) #3519 –154 E. Broadway cousin of Charles

Lang, Ferdinand – Cutler & Grinder –Afidavit filed from Pittsburgh, Pa.
   Emilia (15) (D)

Languth, Martha (11) (u) –29 Cooper Square
   Louisa (9) (u)

Lee, Michael (46) (u) -17 Catherine Slip –Fireman on the *General Slocum* who testified as to the condition of the life preservers

Leinberger, Wilhelm –51 St. Marks Place –Ink Factory
   Carrie (30) (I)
   Caroline (2) (u)
   Weber, Frederika (36) (D) (sister-in-law) –Furnished Rooms $15.00/ wk   (see Weber)
      Christina (12) (D)

    Mamie (7) (D)
    Helen (5) (D) } Twins
    Estha (5) (D) }

Lemm, Amelia (40) #2728 -645 17th St.
  George (11) (u)
  Frank (13) #3686
  Lillian (7) #3417

Lemp, Charles T. (12) (I) –108 2nd Ave.
  August W. (10) (I) –Lb

Leonard, Nora   (see McLaughlin)
  McLaughlin, Michael son (12) (D)

Leonard, Paul C. (24) (u) -179 Devoe St., Brooklyn –Waiter on *General Slocum*

Licome, F. (n) –83 7th St. -Engineer
  Minnie (18) #3400 –Neckwear $6.00/ wk
  Frederika (53) (I) –Ln Still suffering

Liebenow, Paul (38) (I) –Lb –133 E. 125th St. ID'd -Bartender 480.00/ mo, Died since   (see Weber)
  Anna Christina (32) (I) –Lb -she died 7/31/1947
  Helen (6) (D) –Not found, perhaps buried with the unknowns in Lutheran Cemetery
  Anna C. (3) #3323
  Adella Martha (6 mo) (I) –Lb (Youngest female survivor)

Liebenow, Martha (29) #3746 -404 5th St.

Lindemann, William H. (10) (u) –110 Lynch St., Brooklyn
Lindenbaum, Annie (58) #3117 b. Hungary -440 E. 12th St.

Linnertz, John (18) (I) –Lb -82 Ave. A

Link, Lena –Widow -76 Ave A
  Eddie (12) #2890. –Id'd by his father He had been on board with Mrs. Heckert and her 4 children (North) (see Heckert)
  Arthur (14) (u)*
  Lottie (8) #2889

* Arthur saw Mrs. Heckert about to jump with baby Julia in her arms. "If you can't swim, give me the child" he said, which she did and jumped. He braced the
baby in a camp chair against a stanchion until he felt the deck giving way. Then he tucked the baby under his arm and made for shore. He was about to go
under when a man in a small boat took the baby. "Take care of the baby, I can keep up alright". He was quoted as telling the boatman. Another account says
he jumped to a tug with the baby. So many of the crowd landed on top of him, that he came to some time later in the pilot house and found out the crush had
injured the baby who died soon after. (North)

Loeffler, Katrine (41) #3312 b. Germany -9 E. 3rd St. Id'd
  Louise (10) #3041

Loesell, John (n) –94 Ave A –Millright
  Margareth (29) #516 #3424 –buried as an unknown
  Katherine L. (6 mo) (D)

Lubbert, August –412 6th St. –Id'd –Cigar Maker
  Caroline (46) (l)
  Charles (12) #2850

Lucas, Robert (32) #3672 –540 (546) W. 39th (49th) St. –Negro coffeeman on *General Slocum*

Luderer, Herman (19) #3478 b. Germany 312 E. 14th St.
  Otto (13) (u)

Luderman, Johanna (45) #2896 b. Germany —4 Smith St., White Plains   (see Sullivan, Boden)
Fred (17) (u)
Hannah (19) (I) –She found a good preserver which saved her when the deck collapsed throwing her into river (Inferno)
She was selected to present a medal to Captain Parkinson of the ferry Massasoit, who had rescued her.
John (15) (I) –Lb
Lutjens, August, Sr. (46) (I) – 101 Clymer St., Brooklyn Id'd – He and son August had burns & shock Mr. Lutjens had charge of the lunch counter on the *Slocum*
   Kathrine (46) #2710 b. Germany –Id'd by husband, John

Lutjens, Margaret / Marguerite (18) #3445 –daughter of Mrs. Lunderman
Lutjens, August C., Jr. (16) (I) –Ln –son of the bar cashier on *Slocum*, badly burned – Testified that men in small boats were robbing bodies & taking payment for rescue from victims. He grabbed a sleeve, thinking it was his sister, but she was torn away. When rescued, he still clutched the swatch of material.
Mrs. Ella Bolton, of the same address as the Lutjens, was also with the group.
Her husband (n) and daughter (n) were searching for her. She had gone contrary to her husband's wishes.

Ludwig, George W. (14) #3073 –413 E. 17th St. Id'd
Fred (10) (u)

Lullmann, Carrie (24) (D) –100 University Place (see Graner)

Luria, Harris -111 E. 4th St. –Cabinetmaker
Lena (17) #2777 b. So. Russia –Cashier $12.00

Lutjen (see Luderman)

Lutz, Gustav Adolf (17) (I) –Ln 148 2nd Ave   (see Vollmer)
Minnie
Vollmer, Joseph (37) (brother of Minnie) –Waiter $25.00/ wk, commit suicide over loss of family
Mary (36) (D) (sister-in-law)
Joseph (17) (D) –Office Boy $7.00

    Augusta (9) (D)
    Magdalene (7) (D)

M

Mahlstedt, Henry (n) –629 E. 146th St., Bronx –Carpenter
    Annie R. (31) (I) –Lb –Hysterical ever since
    Henry
    Louis H. (23) #3267 –Clerk $15.00

Manheimer, Mamie (36) #3317 –86 7th St. –Id'd by Louis Lander, a friend   – Jacob Manheimer was appointed guardian and filed the Affidavit
    Walter (11) #3443
    Lillian (9) (u)

Mannheimer, Sophie #3318 -44 St. Mark's Pl. –Teacher –picked up by steamer *Fidelity*   (see Manheimer, Stuer)
    Lillie (24) (D) Sister of Sophie – her father took her to live with relatives (North)

Mardof, Annie (21) (u) –43 E. 2nd St.

Marschall, Daniel (13) #3742 –127 1st Ave.
    Henry B. (11) #3483
    George (16) (u)

Marthern, Augusta –Furnished Rooms   (see Moller)
    Muller, Martha (36) (D) –Furnished Rooms $100.00/ Mo.
    Henry (13) (D)
    Edwin (5) (D)

Masterson, William S.(16) (I) –62 3rd Ave – picked up by a tug

Matzerath, Edward (13) (u) was landed and walked home –330 E. 6th St.

Mattes, Lizzie (20) #2993 –87 Ave A. (see Barth, Singer)
  Mary (58) #3170 b. Germany Disinterred
  Margarette (57) (D)
  Barth, Mary (6) (m) –Niece of Mrs. Mattes

Maurer, George J. (53) – #3131 b. Germany – 421 9th St.*   –Band Leader –he jumped holding 2 girls by the hand; a man landed on him fracturing his skull
  Margaret (48) #22976, Man. b. Germany– Ln –she slid down a rope into the water, but died 25 June 1904, of pneumonia
  Mrs. Charles (Catherine) Geminn, daughter (s) –body #152   –1551 Ave A.**   (see Geminn)
    Kate (13) #3462
    Emma (55)(m)
  Clara (12) # 3130
  Matilda M. / Tillie (14) #374 #3132

* George's father was also a bandleader who had worked the *Grand Republic*, sister ship of the *General Slocum*. George had been ill recently and only agreed to go on the excursion with his family in hopes the trip would benefit him. The family was sitting near the bandstand and George signaled them to follow him, they ran to where the life preservers were, but each one they tried was rotten. He got the family over the rail, and handed his wife a rope to slide down to the water and hold onto until rescued. He took a daughter in each hand and jumped (Tilly & Clara). Margaret saw a man land on George's head. When his body was found, it had a heel mark on the forehead, which was probably the cause of his death.
**She saw several good life preservers, but they were firmly wired down and she could not get one.

Mauer, Mary (55) (I) –Ln -127 1st Ave.
  George (15) (u)

Mauer, Minnie (3) (I) –Ln   –626 E. 12th St.
  Weireter, Mary (83) #2821 b. Germany (Minnie's Grandmother)

Mauer, Martin –Storekeeper
  Emma (56) (D)

Maver, Mary (55) (I) –Ln -127 1st Ave.   (see Meyer)

May, Charlotte (50) (D) b. Germany -275 Ave.B (see O'Leary)

Mayer, Carl –Carpenter
   Albert (19) (D) –Varnisher $10.00/ wk (see Meyer)

McCarthy, Jerimiah* (9) #3468 -134 Hobart Ave., Bayonne, New Jersey (see Meyer)
   John* (11) (u) –Stockboy –Found in hospital, his hair had gone gray. Their life preservers had fallen apart.
   Meyer, Frances (41) (D)
*per Affidavit, stepchildren of Mr. Meyer

McGrann, Michael (48) #2778 b.Ireland -2161 8th Ave.–ID d by C.G. Peker, a friend –302 116th St. –Steward aboard the *General Slocum* $75.00/ Mo.
   Annie, widow with 5 children –no support
      Joseph
      Agnes
      Florence
      John
      Frances

McLaughlin, Michael (12) #2735 –69 1st Ave. (see Leonard)

Meinhardt, Wallberga (38) #3352 b. Germany –146 E. 4th St.
   Walter G. (15) #3353
   Rudolph H. (14) #3554
   Otto (8) (I) –Ln –came home by himself
   Ruda (19) (D)
   Walde (15) (m)

Meininger, Eliza (29) #2996 -631 Bergen Ave., Bronx
   Henry (19) #3526

Merseles, Matilda (n) (15) #277 * -394 6th Ave., Brooklyn (Dau. Of Peter & Jennie Id'd by John D. Lutz of Demorest, NJ. (see Lutz & Non-Validated Lutz)

* The child was taken to the receiving vault at Greenwood Cemetery after a funeral at home. No final resting place had been decided on as of the date of the funeral. It was reported that the family received many floral tokens and notes of condolence from her many friends.

Meseke, Charles (n) - 508 Robbins Ave. –Harness Maker
    Meta G. / Betty (50) #2747 b. Germany
    Annie M. (16) #2745 (Kings) –Id'd by brother –Shopwork $5.00

Mettler, Robert –338 E. 5th St –Lace Worker
    Kate (32) (I) –Lb –Bruised left side of body (s).- She jumped onto a tug holding the baby. Her other children did not jump with her.
    Elsie (15) #3645
    Albert (11) #3359 –Id'd by his Uncle August
    Robert (10) #3304
    Frederick (8) #3308
    William (4) (I) –Both legs burned –Lb
    George (2) (I) -Lb

Meyer, Albert, Jr. (19) #2864 –434 (454) E. 15th St.

Meyer, August -Butcher
    Katie (7) (D)

Meyer, Frances E. (41) #3351 –134 Hobart Ave, Bayonne, New Jersey   (see McCarthy)

Meyer, George (11) (I) -Lb –430 E. 17th St.
    Louisa (39) #2708 b. Germany
    Eise (7) #2751

Meyer, J. L., Jr. -381 Madison St. –Cashier's Assistant
    Elizabeth C. (22) #2959

William, Meta (60) #3434 b. Germany

Meyer, Louis –Agent
    Meta (59) (D)

Meyer, William -88 Ave. A Id'd   -Packer
    Lizzie (35) #2985 widow, b. Germany
    Edward (10) (D)
    Kate (7) (D) – #2937

Michel William H. (n) –171 Ave A (poss. Near 12th St.) –Horseshoer
    George (10) (s) –Ln –Unable to work
    Maggie (42) #3301 –Stewardess, now housework
    William (14) #3302
    Louise, Mother
    Caroline (12) #3183

Miller, Mary, (28) (I) –Lb –92 2nd Ave.

Miller, Flora (28) #2911 –28 W. 97th St. –ID'd by brother
    Fred, Jr. (26) (u) would later become a president of the Survivors Association.

Molitor, (Molstre) Adolph L. –Central Ave. & Bronx River Rd., East Yonkers, NY (see Salvation Army, Hagenbucher)
    Margaret (34) #3059
    Carl (4) #3675

    Eva (9) #3459
    Joseph (6 Mo) (D)

Molke, Lizzie (10) #3497 –125 1st Ave

Moller, Catherine (39) #3642 b. Germany Disinterred –45 2nd Ave.
    Annie M. (13) #3568

Moller, Fred (2) (m) –998 Ave A
    Henry (7 Mo) #3206
    Fred (39) (u)
    Annie (29) #22308 –died in hospital 15 June 1904

Moller, Martha A. (36) #3086 b. Germany -20 St. Marks Place * (see Marthern)
    Henry (13) #3085
    Edwin H. (5) #3683

*One of first wills filed due to *Slocum* disaster (6/21) –Dr. Robt. Kunitzer (188 Lenox Ave) Trustee of her accounts for the benefit of her children (0ver $100,000) her husband, Emil, became sole benefitiary (she owned several boarding houses). Services were performed by Rev. Samft, in German the interments were at Lutheran Cemetery. A contingent of six patrolmen and a sergeant were sent to control the crowd of about 1,000 outside the family home during the funerals.

Morio, George–121 Pitt St. –Sheetmetal worker
    Mary (14) #3496

Morris, Anna –69 1st Ave –Featherworker
    Katie (15) #2762 –Featherworker $6.00/ wk .

Motzer, Charles -405 6th St
    Anna (38) #3021 b. Germany . –Id'd by sister, Mrs. Elizabeth Schwartz or Schneider –167 Ave A
    Lena (4) (m) - Another account shows Lena as injured –Ln
    Louisa Ann (9) (u) – After her rescue, she went home by the 3rd Ave. train. When she got off, a big man recognized her, picked her up and ran with her to the saloon her father ran. One account says she was rescued by Henry Trowbridge.   (see Trowbridge)

Mueller, Louis C.A. (11) (I) –Ln –100 St Marks Place
    Herman C.A. (9) (I) –Ln
    Minnie A. (7) (I) –Ln

Mullen, Patrick (24) (u) –501 W. 14th St. –Fireman on *General Slocum*

Muller, Annie (25) #2943 b. Germany -406 6th St.
    Elizabeth (6 Mo) (m)

Muller, Annie (37) #3576 b. Germany -41 1st Ave.
    Ernest (8) (I) –Lb
    George (40) (u)
    Annie (13) #3392
    Henry (3) #3422
    Mary (7Mo) #3767

Muller, Bernard (38) #22843 b. Germany d. 23 June 1904 –95 2nd Ave
    Velasca (29) #2715 –Id'd by Otto Grossberg (Rosenberger), a friend –same address   (see Rosenberger, Hager)
    Grover (13) (I) –Ln
    Edgar (4) #2737
    Walter (10) (I) –Ln
    Arthur (6) (I) –Ln Burned on head

Muller, Bruno (n) –Musical Instruments
    Anna (25) (D)
    Elizabeth (6 Mo) (D)

Muller, Edward -368 Bowery Id'd –Musical Instrument manufacturer & dealer
    Hermine (37) #3191 b. Germany
    Helen (8) #3199
    Rosie (13) #3198

    Edward (11) #3488
    Irene (4) #485 #3192

Muller, Florentine (59) #3141 b. Germany –321 E. 9th St.   (see Hoffman)

Muller, Jacob, Sr. (n) –123 7th St. –Clerk in 1910, 1763 Madison St.
    Jacob (4) #3606 Disinterred
    Anna (31) #2726 b. Germany

Muller, Samuel 972) (l) –Ln -111 Norfolk St.

Mundle, Lillian (33) (m) –Dressmaker $10.00/ wk – 11 E. 7th St.
    Arthur (9) # 3343
    Agnes (64) (l) –Ln
    Leonard (n)
    Lilly (11) (m)
    Harold Cloeren (11) (n)
    Caroline (38)

Muth, Anna E. (Eliza) (62) #3094 b. Germany –1254 Lexington Ave –Interment on 19th Lutheran Cemetery (see Muth 785 E. 146th St., Bronx)
    Conrad, Jr. (10) (l) –Ln

Muth, John (36) (l) –Ln -785 E. 146th St., Bronx –Interment on 19th Lutheran Cemetery (see Muth, 1254 Lexington)
    Kate (36) #3092
    Christina (8) #3463
    Katie (5) #3093
    Lizzie (11) #3464
    John, Jr. (3) (l) –Ln

N

Necke, Otto –504 E. 16th St. I'd –Box Maker
    Theresa /Daisy (10) #3195
Neubecker, Lina   (see Volkhardt)
    Volkhardt, Lizzie (daughter)

Neubhi, Hermann –23 Ave. B -Confectioner
  Meta(38) (D)
  Lizzie (9) #3079
  Wilhelmina (7) (m)

Noll, Kate (40) #3440 b. Germany – 400 E. 5th St.
  Theodore (5) #3360
  Katie (11) (m)

Novotna, Meri –190 E. 3rd St. ID'd
  Louis (18) #3012   –Electrician $15.00 /wk

O

Ochse, Frederick (n)) –50 St. Marks Place   –Livery Stable
  Caroline K. (15) #2716 –Lb d. 15 June 1904
  Edward F. (11) #2988

Oehler, Henry (n) –510 6th St. –Machinist, Affidavit filed Kings Co., NY
  Fredericke L. (14) #2798
  Frieda L. (13) (D)
  Anna Maria (53) #3426 b. Germany

Oellrich, William (n) –611 Marcy Ave., Brooklyn – Grocer at Willoughby & Marcy Aves. (Previously in business at High & Pearl Sts.) (also listed at 519 Willoughby St., Brooklyn)*   –Later Poultry Business, Affidavit filed from Sullivan County, NY William's brother Herman and his family were also supposed to go but his two sons not ready in time and the family stayed home. Anna's sister, Mrs Sachmann had invited the family as she was a member of St.Marks.
  Anna (34) #3426 b. Germany
  Helen (19 Mo) #3148
  Wilhelmine (4) (m)
  Fred (5) (m)
  Elizabeth (3) (m)

324

Henry N. (12) (I) –Lb –He was playing with friends on the lower deck. He tried to get back to his mother, on upper deck, but could not. A man in a blue uniform rushed by shouting "Everybody that can swim, jump; it's your last chance." The man then jumped and Henry followed. He was taken ashore by a rowboat.

* Mr. Oellrich intended to attend the excursion, but was called for jury duty. Mrs. Oellrich was then going to keep the family home, but William encouraged her to go. Mr. Oellrich's brother Herman (n) had a business at the corner of High & Goldman Sts.   (see Sackmann)

Oeltinger, Andrew (7) #3764 -91 7th St.
   Arthur (5) (m)
   William (13) (I) –H
   Charles (15) #2823
   Emma (15) #2824
   Kate (38) #3390

Ohl, Charles (n) –340 E. 9th St.   –Charles Ohl, Sanitary Plumbing & Gas Fitting, Tin & Slate Roofing   314-316 E. 9th St.
   Carl (9) #3031
   Charles
   Emily (11) #3407
   Elizabeth (36) (u)

O'Leary, Eugenia   (see May)
   May, Charlotte (51) (D), Mother –Sewing $8.00/ wk

Olsen, Clara (17) (u), servant of the Fulling Family 110 W. 129 Th St.   (see Fulling)

O'Neil, Daniel (24) (u)   –146 Cherry St.   –Deckhand on *General Slocum*

Osborne, Mamie (13) #3341 -303 W. 12th St.

Osmers, Frank C. (n) -49 E. 88th St.
   Mildred D. (6) #3458
   Marie (55) (I) –Ln –burned head to feet -402 E. 83rd St –Housework, now unable to work

Otto (17) (I) –Office work $10.00/wk, now only works once in awhile, brain affected

Owens, George (39) (u) -2 Horatio St.   –Chowderman on *General Slocum*

P

Pauli, Katie (21) (D) –Typewriter $12.00 –26 Ave A   (see Phillipp)
    Elsie W. A. (13) #2888
    Tessie (23) (I) –died of injuries since
    Wilhelmine L.C. (20) #2887

Paulsen, Henry –witness to Affidavit of Fred Fickbohm   (see Fickbauhm)

Payne, Walter (22) (u) -758 Greenwich St. –Negro Porter on *General Slocum*

Perdelwitz, Alvina (43) (I) –89 E. 10th St.
    Curt (15) (I) –Lb

Peters, Helen (28) #2982 b. Germany -121 E. 126th St.
    Lillian (1) #3603 Disinterred

Pfeifer, Police Patrolman Charles E. (n) married 1 year (assigned 58th Precinct, Vernon Ave. Station) –937 Bedford Ave., Brooklyn –Newlyweds, he waved to his wife from the Williamsburg Bridge (Inferno) their first anniversary was that week.
Lillian Meta (nee Frese)* (18) #2880
* She was aboard with her sister. Lillie stepped away while the rest of the Frese family went to look at the engine room and was lost. The Freses were saved. Her funeral was conducted by the Rev. J.J. Heischmann, St. Peter's Brooklyn Church at Bedford & DeKalb Aves., Interment was at Brooklyn Cemetery.   (see Frese)

Phillipp, Anna   (see Pauli)

Pienning, Oscar – 45 7th St. looking for;   (see Gruening)
    Dorothea, mother (58) #401 #3030 b. Germany

Gruening, Mrs. Helen, sister (29)
　　　　Henry Gruening (5) (m)

Plintin, James (22) (u) –111 W. 26th St. –Negro Captain's witer on *General Slocum*

Plunkett, Gerold (12) #3428 –74 E. 7th St.

Podzuweit, Gus (27) (l) –Ln –682 Carroll St., Brooklyn

Polnisca, Paul –320 5th St. –Waiter
　　　　Susie (43) (D)
　　　　Olga (14) #2838 –Millinery $4.00
　　　　Paul (12) (l) –Sick since with lung trouble and unable to work

Pollonde, James (10) (l) –Ln -115 26th St.

Port, Paul C. – 86 E. 4th St. –Reliable German-American Laundry, 303 6th St. nr. 2nd Ave., Paul C. Port Prop.
　　　　Henry L. (15) #3090
　　　　Pauline (13) (12) (m)
　　　　Paul C., Jr. (10) (12) #3357
　　　　Wilhelmina (47) (l) –Ln

Potar, Joe (17) (l) –Ln –17 Humboldt St., Brooklyn

Potar, Lewis (28) (l) –Ln –75 4th St.

Pottebaum,Herman H. (52) #2820 * -61 St. Mark's Place -Importer
　　　　William J. (9) #361 #2819
　　　　Eliza (47) #417 #2818 b. Germany
　*Surviving children from this family went to board in Brooklyn

Prawdzicki, Paul –85 E. 3rd St –Painter – Members of St. Marks, Services by Rev. Zicker, Elizabeth St. Lutheran Church; Interment was at Lutheran Cemetery.*
  Maria (36) (I)–Ln –Burned Doctor's note with Affidavit: "In more or less constant care since for general nervousness, nervous prostration, neurasthenia, which now is becoming more prominent because of change of life. " – Louis F. Bischof
  Gertrude (2) #320 #3125
  Anna F. (15) #380 #3126 –Dressmaker $5.00/wk
  Johanna (1) #3751
  Frank H.(12) (I) –H (1910 Clerk) –May be the boy noted in Munsey's account as "Frank Perditski" who was the first to tell Capt. Van Schaick about fire.
  Henrietta H. (13) #3124 –ID'd by uncle of same address
*The crowd in the street was estimated at 1,500 for the funeral, and a dozen police were sent to maintain order.

Probst, Elizabeth (n) –515 E. 12th St. -Laundress
  Catherine (24) #3137 –Mirror Polisher $8.00/ wk

Pullman, William H. (49) #2807 –Treasurer of St. Marks Sunday School –337 E. 18th St. –BDE he died saving children; Service by Rev. Otto Hoffman of Lutheran Church of St. Paul, So. 5th & Rodney St., Bklyn, his lifelong friend.
  Elizabeth (46) (I) –Ln
  Elsie (16) (n)
  William H. (18) (n)

Q

R

Rammalkamp, Rudolph –130 E. 4th St   – Upholsterer
  Estella (12) 33305
  Elizabeth (44) #3682 –Tailoress $10.00
  August (6) (I) –Ln –severe shock from injuries
Ramus, Phillipina (s) –420 E. 17th St.
  Frederick (60) #3651 –Grocer $15.00/ wk
  Irving (10) #3232
  Erwin (12) (D)

Rau, Wilhelmina (52) (I) –Ln –Nervous shock –52 7th St.   –Janitress $7.00/ Mo; now only housework & supported by daughter

Rebenklau, Elfreda (18) (I) –Lb –23 Eldert St. Brooklyn

Rehm, Katharina (47) (I) –Ln –ruptured Left Leg lame & arm –121 Ave A –General light work $12.00

Reichenbach, Herman G. W. – 241 Stockholm St., Brooklyn – Machinist
    Eleanor (23) (I) –Internal & Nervous Prostration; Still invalid
    Herman H. (2) #2869

Reiss,   Kate (33) #3674 -70 1st Ave.
       Annie (2) (m)
       Rosie (16) #3673
       Lizzie (6) #3370
       Susie (4) #3590

Reuffer, Arthur F. (14) (u) –109 1st Ave

Reuzing, Gertrude (22) (D)* –424 6th St
       Emma (24) (D)*
*number illegible

Rheinfrank, Frederick } brothers, looking for 2 families inc. 14 people - Goerck & E. 3rd St. –Messrs. J. Rheinfrank & Co. Gustave gave
    Gustave   } $6.00 to old man crying for lack of burial money at the temporary morgue
    Rheinfrank, John, Sr. (75) (D) –343 W. 71st St. – #2910 – Death Notice NYT June 18th
    Mrs. Catherina (64) (D) – #3552

    Mrs. J. – burned hands & face, bruised – Ln
    Schwartz, Charles - 141 3rd St. (s)   (see Schwartz)

Richter, Emelia (47) #3025* –404 6th St. –Mr. Richter was William A. Richer's brother –all 5 children buried at Lutheran Cemetery   (see Richter)

Ernest (12) #3215
Emelia (21) #3023
Lizzie (19) #3027
Annie J. (8) #3026
August (15) #3024
Frances E. (10) (I) –Ln

* The father of this family had died several years earlier. Mrs. Henning, Mrs. Richter's mother, took in the surviving children. One son (15) who worked at the Commission House downtown and remained at home. Frances (10) grabbed an overturned boat in the river and stayed afloat until Henry Trowbridge rescued her. They lived on the same street. He also saved Louisa Motzer who lived across the street from his home. He then saved two more people before he gave out to his own injuries, burns on his hands. (see Motzer, Trowbridge)

Richter, William A. (n) –104 1st Ave. –Cashier of the tax office in Queens On the 15th William was at Rockaway Beach with his bowling club, and only discovered what had happened when he returned hom at the end of the afternoon. (BDE) (see Richter)
    Lena (35) #3231 b. Germany
    Fred A. (11) #3230
    Christina / Tinnie (9) #3550 } (see Hill)
    Lydia L. (12) #3549 }

Ringer, Clara (37) #3503 -170 Ave A
    Alfred (11) #3763 b. Germany, disinterred

Reitz Theresa (13) #3039 –90 1st Ave

Roberts, Blanche A. (13) #3737 –198 Guernsey St., Williamsburg, Brooklyn
    Clara A. (38) #2774

Robinson, Edwin (19) (u) -414 W. 39th St. –Negro 2nd cook on the *General Slocum*

Roes, Adele A. (36) #2812 b. Germany –222 McDonough St., Brooklyn
    John J. (9) (I) –Ln

Roeth, Helen (20) #3715 –310 E. Broadway (ID'd by Michael Cphen, a friend, -188 E. Houston St.)

Roscoe, George (40) #3626 b. England Disinterred   –362 Front St.

Rosenberger, John (n) -417 E. 15th St. –Baker   (see Cordes, Muller)
   Mary (44) #2946
   Eliszabeth (7 ½) (D)

Rosenholtz (see Harris)

Rosenstein, Sophia (21) #3432 -127 1st Ave

Rosenagel, Charles – 129 E. 4th St. ID'd
   Annie (38) #2848 – She had just finished sewing a confirmation dress for the elder daughter.
   Lucy (13) (14) (I –Lb) –Claimed her mother's jewelry & was supeonaed to testify at Inquest that crew set up hoses which burst & that they did not try to free boats or help passengers
   Grace (8) (I) –Lb
Another account mentions a Mrs. Rosenagel, Sr., grandmother of 13, all of whom died.

Roth, James –203 5th St. Id'd   – Joseph Roth Artistic Signs Factory: 203-205-207 E. 5th St.   (see Alonzo, Brown)
   Josephine (42) #2932 b. Roumania
   Catherine A. (17) #3413
   Alfons Braun (13) #3612 b. Roumania –Disinterred (nephew of Josephine)

Roth, Louisa (16) (I) –516 E. 88th St.

Rotenberger, Anna (19) #2822 –Servant $13.00 –368 Bowery

Rothmann, William C. –48 ½ E. 7th St. –Wm. C. Rothmann Ales, Wines, Liquors & Cigars, 1st Ave at 6th St.
   Emily (34) #2845
   Thomas F. (8) #3385

William C., Jr. (5) #3494

Rumpf, Annie (34) (I) –Lb –342 E. 9th St.

Ruthinger, Ernest (Edward) –47 St. Marks Pl. ld'd – E. G. Ruthinger, Machinist
Meta (39) #2853 b. Germany
Ernest L. (17) #2854 –Bookbinder $8.00/ wk
Fred (10) (u)
Elsie (15) (u)

Ryan, Mamie (5) #3397 –345 E. 15th St.

Ryan Thomas (28) (u) –S.E. Corner 8th Ave & 29th St. – Waiter on *General Slocum*

S

Sackmann, Diedrich (n) –341 Rvington St. – Grocer (see Oellrich.)
Margaret (38) #2810 b. Germany –Had invited her sister's family to go with her, but earlier in week, Mr. Oellrich was pulled for jury duty. He left his cousin Henry Jaeger in charge of the store and learned the news from him on his return. His wife was intending to stay home, but he convinced her to go.
Charles R. (n) –1963, Former president of survivors' Association; he was playing in street on June 14th, hit by a rock & was in hospital instead of on boat (Perils)
Anna (13) (I) –1963 Secretatry of the Survivors' Association
Margaret (9) (D)
Herman (7) #3502

Sanders, Helen (13) #3091 –416 E. 16th St.-Services were at the Grace Chapel, 14th St. by the Rev. G. Bottome, curate of the chapel (Episcopal)

Sauer, William (14) (I) –Ln –142 E. 2nd St.

Schaefer, Joseph (36) (u) –322 E. 13th St. –Bookkeeper (see Fleischer)
Augusta (35), sister-in-law (I) –Burned on back

Katie (6) (D)
Fleischer, Henry (16) (D) –Lithographer $6.00/ wk

Schaefer, William H. (34) #2825 -50 E. 132nd St.

Schafer, John –77 E. Houston St. –Cigar Making
Fannie (40) #3108 b. Austria

Schaier, Margaret (24) #2882 b. Germany -237 E.10th St.
Julia (8 Mo) #3611

Scharf, Richard -419 E. 9th St. –Printer
Mary (58) Mother #3142 b. Austria –Janitress $30.00/ Mo

Scharz, Eva   (see Schneider)

Scheele, Wilhelmine (s) –14 St. Marks Pl. –Father did not go, bedridden & dying   (see Hagenbacker)
Anna D. (15) #3364 –Millinery $4.00
Clara A. W. (7) #3575
Lawina F. (10) #3363

Schelke, Agatha (s) –331 5th St.
Elsie (8) #3546
Bosch, Walburge (15) (u)   – Elsie's cousin

Schepp, Mary (15) (I) –433 E. 17th St.

Scheuermann, Albert ((6) #3607 Disinterred   –100 E. 8th St.
Minnie (40) #2994 b. Germany

Schick, Minnie (21) #945 –430 E. 15th St.

Henry (2) (I) –Ln

Schiettinger, Louis (n) –754 E. 149th St. – Carpenter
Dora (18) (D) –Office Work $6.00/ wk
Frieda (16) #2997 –Passementerie $6.00/ wk

Schirmer, William (44) (I) -140 1st Ave. –Restaurant
Wilhelme (45) (u)
Minnie (18) (u)
Bertha (16) #3315 –Neckwear $7.00/ wk
Lena (13) #3314 –Neckwear $7.00/ wk
Annie (6) #3684
William (9) #3313

Schmid, Carl –290 E. 2nd St. -Butcher
Sophia (36) #3395 b. Germany
Freda (8) #2957
Charles (6) #2793 b. Germany

Schmid, Catherine (40) #3322 -418 E. 9th St.
Catherine (67) #2835 b. Germany
Otto L. (15) #2834

Schmid, Gottlieb -97 E. 4th St. –Butcher –Affidavit filed from Philadelphia, Pa.
Fannie (34) (D) #3339 b. Germany
Bertha (10) (D) #3540

Schmid, Louis –Bricklayer
Martha (30) (D)

Schmidling, John -119 7th St.–Clerical – one report said that he was found wandering and incoherent on the pier after searching for family all night.

334

Mary R. (59) (D)
　　Emily S. (22) #2905) –Stenographer
　　Anna L. (15) #2904
　　George C. (18) #2903 –Clerk $10.00/ wk

Schmidt, August –Florist
　　Anna K. (31) (D) –Housework & Florist
　　Anna (2) (D)

Schmidt, Annie (3) #2941 -108 Ave B

Schmidt, Clara –Operator
　　Annie (3) (D)

Schmidt, Henry (s) -264 1st Ave –Grocer
Emma (25) #3587 b. Germany – Henry and Emma had gone on the excursion together. When she was not found immediately, he did not report it so she was not on the lists when her body was found at College Point. She was identified by the inscription in her wedding ring "HS to EE 1903. (Emma nee Eckhardt) They were married 15 Mar. 1903 by Rev. J. Geyer. The Rev. Feldman, helping at at St. Marks, applied to Bureau of Vital Statistics for marriage licence and Detective Ross then located Henry.
　　Erna H. (5 mo) #3001

Schmidt, Ernest –149 E. 4th St.–Baker
　　Eva (17) #2912 –Sewing White-goods $6.00 –ID'd by her brother, George
　　Phillipine (27) (I) –Ln –Bruises, fractured nose, nervous heart –Artificial Flowers $12.00

Schmidt, Martha (28) #3103 b. Germany -492 18th St.

Schmidt, Louisa A. (10) (I) –Lb –69 1st Ave
　　Francis (14) (I) –Lb
　　Julia (13) (I) –Lb

Schmidt, William (13) (u) –5 Cooper Square

Schmidt, William –138 E. 7th St. –Milkman
    Emma (54) #3204
    Fred P. (9) (I) –Ln –Fractured skull, d. 6 Apr 1905 of injuries

Schnebbe, Caroline H. (15) #3685

Schneider, August (34) (u) – 322 Stanhope St., Brooklyn* – Coronet player in band, friend of Mary Abendschein
    Dora (32) #3620 b. Germany
    Kate (8) #3616 Disinterred
    Amelia (6) child (D)
    Margaretha (6) (D)

* When August had to go overboard, he grabbed his youngest daughter just as the deck collapsed. His wife and two daughters fell with it. Moments later, the section on gave way and he and the child in his arms, also fell into the water. They were picked up by a tug and taken to the Alexander St. Station, where he searched fo his wife and other children.

Schneider, Fred (12) (u) –512 6th St.

Schneider, John –90 1st Ave.
    Theresa (14) #3040 b. Germany

Schnepf, Fred (15) (u) -645 E. 17th St.

Schnitzerling, Conrad, St. (46) (I) –died of injuries since –123 Ave A – Painter $3.00/ day   Affidavit filed Kings Co., NY
    Elsie (36) #2886 b. Germany
    Fred (5) (I) –Lb –burnt & internal injuries
    Anna (11) (I) –Lb –Burns & contusions, now housekeeper

Schnitzler, Patrolman Edward (104th St. Station) (n) -10 Gouverneur Pl. (btwn. Washington & Park,) Bronx Id'd. (see Hessel)

Christina (28) #2742 –Services for Christina & Catherine were performed by the Rev. Hough Houston of Centenary M. E. Church, Bur. Lutheran Cem.
Catherine (5) #2739 –801 E. 147th St.

Schnude, Emil (n) 196 Gurnsey St., Williamsburg, Brooklyn –Fitter   (see Kassebaum, Torniport) *
Henry C. (33) #2781 (brother of Emil) –Cashier $40.00/ wk
Anna C. (32) #3077 (sister-in-law of Emil)
William (60) #3011 b. Germany (father) –426 E. 76th St.–Grocer $10.00/ wk
Louise (57) #3010 b. Germany (mother)
Grace A. (4) (Neice) #3076
Mildred A. (1) (Neice) #3412
Kassebaum, Annette (D) (s) married -196 Guernsey St., Brooklyn

* Although these two families are not reported as being directly related, their funerals were conducted together out of the Jacob Schaefer Funeral Home at 1023 3rd Ave. Brooklyn. All were buried in Greenwood Cemetery.

Schoeffling, Conrad (n) – 189 3rd Ave. –Barber   – Was so distraught that he identified three different women for his wife.
Maria, (34) #2841
Elsie (3) #3812
Eddie (10) (l) –Ln

Schoenemann, Elizabeth – 746 Home St.
Elsie (15) #3289
John (18) #2898 –Salesman $8.00/ wk

Schoeninger, Gottliebe (60) #3075 –118 E. 3rd St.

Schoett, Christian–98 E. 7th St. –Tailor
Josephine (42) #3064 b. Germany – Id'd by her brother in law, Edward Yost, who was so distraught that friends kept a suicide vigil.
Caroline E./ Carrie (10) #2792
Helen A. M. (4) #3066
Christian J. A. (20) #3327 –Jeweller $12.00 –St. Marks Organist & Sunday school teacher; was with his sweetheart and student to whom

he was secretly engaged, banns not yet published. His cousin Henry Siedewand (D) was also with them.

Schuler, Henry (11) (u) -41 1st Ave.

Schrumpf, Jacob / Jake (n) – 208 Ave B – Retired Patrolman
Lizzie (48) #3350
John Edward (16) #131 #2940
William W. (13) #2939 –Choirboy at St. Thomas'

Schuebbe, John H. –54 Ave. A   –Paper Handler   (see Schumann)
Carrie H. (15) (D) –Laceworker $5.00

Schuessler, Sophie (62) #3543 b. Germany –338 6th St. –Sophie was known as the "Grandmother of St. Marks", her custom was to take the families of her two daughters and three sons on the excursion, this was the first year she went alone. She felt she was too old and stout to carry the responsibility of so many children. She weighed about 400 pounds. Her funeral was held at Beetoven Hall, 431 6th St, which was packed. Because of her size the service could not be held at home, even a specially built coffin was required. Rev. Krueska performed the service.

Schulere, Sophia (u) –15 Stuyvesant St. –Affidavit 2250 Hughes Ave, NYC
Charles H. (18) #3054   } last seen on upper deck just before it gave way, the two were working a bucket brigade
Frederick W.(18) #1053 –Plumber $10.00 }

Schultz, Dorothea (38) #3334 b. Germany –112 E. 4th St.
Rudolph (14) #3333
Henry G. (11) #3046
Dorothea (7) #3624 Disinterred

Schultz, Eugene (s) -130 E. 4th St. –Unable to work
Emma (10) #2846
Pauline (43) (I) Ln –Janitress died of injuries 8 months later

Schultz, Susan (24) (I) –Ln –414 E. 9th St.

Schumacher, Charles (n) –434 6th St.
    Katie L. (14) #3346
    Edward (10) #3768
    Henry (10) (m)

Schumann, Arthur R. (n) –113 St Marks Place –Butcher   (see Schuebbe)
    Alfred R. (7) #3501
    Annie (5) (D)
    Marie E. (30) (I) –Ln –Hands, arms, face burned, unable to work (b. 1874 d. 1956)
    Emma (2) (I) –Ln -Now Mrs. Charles H. Mullmann

Schuessler, August (n) –Baker
    Sophie (62) (D)

Schwartz, Charles, Sr. (n) –141 E. 3rd St. –Stationary engineer
    Charles M., Jr.(18) (I) –Neck & arms burned, apprentice machinist –He and 20 friends made bucket brigade in the bow; ID'd;
        (see Bucheidt, Rhinefrank)
    Louisa (43) #3122 wife of Jacob
    Louisa (45) (D) wife of Charles,Sr.
    Burkart, Elizabeth Mother-in-law (66) (D) –had been at home with the children last 16 years. An interview with Charles, Jr. says that he had carried his Grandmother to the rail, which collapsed dumping them both in the river, he swam around looking for her until picked up by a rowboat, disrobed, dived in again and searched until he found her body and retrieved
    Emily (19) (I) –Face burned, –Shoes
    Louis (10) (I) –Lithographer Apprentice $6.00/ wk
    Anton (16) (I) –Lithographer Apprentice; –Left leg amputated at hip, until today both leg & shoulder afflicted w/ abscesses from friction of strap & artificial limb Interfers with his business very much.

Schweikert, Catherine (64) #3681 b. Germany –216 E. 11th St.

Seelig, Anna (27) #3610 Disinterred –Dundee Lake, Bergen Co., New Jersey –Her husband Frederick had seen her off from their chicken farm ten days earlier, to open a delicatessen in Manhattan. He had been a St. Marks member years earlier and had bought tickets for the excursion. He rushed to the city and identified some of her clothing. She had been buried with the un-identified so he had her exhumed and reburied in his own plot.

Seemann, Meta (26) #3081 b. Germany –single –227 E. 21st St. (see Witte)
ID'd by Henry C. Ahrens –37 W. 132nd St.

Seidenwand, Henry (18) #2775 b. Germany –184 3rd St. (ID'd by his brother Charles) (see Schoett)

Seifert, Henry (29) (D) –215 W. 23rd St. –Social Sec'y for 23rd St.YMCA (Interment at YMCA plot Woodlawn Cemetery)

Seifert, Freda (17) #3394 –541 E. 157th St., Bronx

Siegel, Sophia (22) #3470 b. Roumania –54 7th St. (see DeLuccia)

Seiler, Katherine –107 E. 84th St. (see Brauer)

Siegwart, Caroline (9) #2789 –225 5th St.
Phoebe (6) #2790

Sierichs, William –425 E. 12th St. Id'd – Wm. Sierichs Bottler of Mineral Waters, Genuine Ginger Ale, etc. 421 E. 12th St.
Lottie (38) #2928 – possibly a sister of Mrs. Louisa Engelmann, BDE refers to them as the Lander Sisters (see Engelmann)
Charles W. (13) (u)

Silveria, Frank (32) (u) –Fireman on board the *General Slocum*

Singer, Frederick W. –Salesman (see Bartz, Mattes)
Mattes, Marie (59) (D) mother-in-law
Lizzie (21) (D)sister-in-law –Waistmaker $10.00/ wk
Bartz (6) (D) niece

Smith, Anna (25) #2826 –920 E. 156th St., Bronx

    Beatrice M. (2) #3761
    Mildred (2) #2764

Smith, Bessie (35) (u) -315 W. 41st St. –Negro stewardess on *General Slocum*

Smith, Edward (68) #3326 -228 W. 28th St. –Negro pantryman on *General Slocum*

Smith, [Ex-Alderman, 12th Dt.] –283 Monroe St .–He offered $200.00 reward for his child; Ptlman.Gilderman, Harbor Sqd. & Patrick Nolan (E.138th St.) found her off North Brother Is.
    Margaret (14) #3664

Smith, Flora (39) (I) –1028 Hudson St., Hoboken, New Jersey
    Harry (10) (u)
    Hartley (17) (u)
    Owen (14) (u)

Smith, Mamie (13) #3116 –381 E. 10th St.

Spinz, Augusta (52) #2803 b. Germany –90 1st Ave. –ID'd by son Paul

Stahmann, Charles (49) –55 Ave A
    Anna A. (45) #3297
    Viola (10) # 3298

Stake, Caroline (37) #3372 b. Germany –337 5th St. –Entire family killed
    Minnie (9) (D)

Steeger, Rose –59 1st Ave -Bookkeeper
    Anna (14) #2883

Stegman, Anna (45) (I) –Blood poisoning of hand through bruise caused hand to stiffen –E. 4th St. –Janitress $35.00/ Mo., now housework
    Grewe, Henry (17) (D) son –Clerk $4.00/ wk   (see Grewe)
    Fred (14) (D) son

Steil, Adelaide (15) #3533 –Boston Rd., Bronxdale
    George (12) (u)

Steil, Andrew (41) (n) –55 1st Ave. – Cornoer's note; Has since committed suicide, none of family remains
    Emil (12) #3065
    Frederick (6) #3145
    George (14) #2776
    Lillie (16) #3618 Disinterred

Stein, Pauline (s) –45 1st Ave –Janitress
    Caroline (9) #2999

Steiz, Bessie (18) #3005 –606 E. 15th St.

Stenger, Francis (31) #3437 b. Germany –88 E. 3rd St.   (see Geudert)
    Rose L. (10) #2971   –ID'd by her uncle, William Geudert

Stickel, Margie   (see wolf)

Stoebel, Christina (57) #2719 b. Germany –338 6th St.
    Catherine C. (D)

Stocker, Helen (s)
    Eysel, Jennie (9) (D) daughter   (see Eysel)

Stockermann, Gustave –225 5th St. Id'd –Braider
    Lizetta (50) #3337 b. Germany

Louise (10) #3452 b. Germany
Augusta (15) #3450 b. Germany –Weaver $6.00/ Wk
Hulda (17) #3451 b. Germany –Weaver $8.00/ wk
Anna (22) (D) –Weaver $8.00/ wk
Herman (7) (I) –Ln -Bruised

Stoehr, William -340 6th St. Id'd
Susanna (30) #3214
Henry (5) #3648

Stone, Wilhelmine (49) #2856 -114 E. 4th St.

Stoss, Alwin – 316 2nd Ave –Bookkeeper –Services performed by Dr. Huntington, Grace Church, Broadway (Episcopal), Brooklyn
(see Kawczynski)
Wilhelmina (49) #2856
Edna (10) #2829
Kawczynski, Teofil, nephew
Minnie (43) #2828 b. Germany

Strangfeld, Augusta (12) (I) –Lb –1349 Park Ave.
Christine (46) (I) Ln

Stricker, Clara (23) (I) –Ln –315 E. 9th St
Martha (4) (u) –Ln

Strickrodt, Annie (41) # illegible b. Germany –144 Essex St.
Louis (5) #3592
Charles H. (14) #3738
Elsie (8) # illegible
Henry A. (16) (I) –Ln

Stubenrauch, Anna (21) #3381 –225 5th St.

Stuve, Margaret (65) (l) –49 Ave A   (see Addicks)

Suden, Margaret (30) #2950 b. Germany –61 Jackson St.
Herman (4) #2948

Sudmann, Henrietta (23) #3430 b. Germany –342 W. 122 nd St.

Svoboda, Frances (11) #3571 -170 E. 4th St.
Mary (8) #3572
Francis (11) (D)

T
Tchudy, Annie (37) (u) -41 1st Ave.

Tetamore, Sophie E. (30) #3610 – 1471 Bushwick Ave., Brooklyn –Pastor C.G.F. Haas' sister –in-law , Id'd by Dr. Geo. H. Semken of 858 Lexington Ave.   (see Haas)
Herbert (3) (m)

Tienken, Etta (18) (l) –Spine injured, operated since supported by father

Timm, Mary (36) #3586 b. Austria –211 5th St. –Mr. Timm (n) was so unsettled over the loss of his family that he went temporarily insane.
George (7) #2961
Hedwig (11) #3000
Henrietta (9) #2960

Thoma, Stephen (n) –90 Ave A
Joseph (8) #3113
Christine (34) #3114 b. Germany
Lydia F. (5) #3755 Disinterred

Thormahlen, Elad (n) –100 E. 2nd St. –E. Thormahlen, Prop. German American Laundry

344

Augusta (38) #3423 b. Germany

    Matilda (3) #3310
    Elard H. F. (8) #3309

Tischler, John (14) (u) – 404 (401) 5th St.
    Wousky, Ida (14) –school friend from the same address – When Ida almost fainted, he kicked her shins to wake her, got life preserver on her & told her to jump, when she was too scared, he pushed her over, then jumped. He held onto her hair until they were both rescued by a tug. Went home and told his mother he wasn't drowned. He also got a preserver for a woman and baby. Kept Ida & self from catching fire by dunking   (Am. Herit.)

Toth, Susanna (50) #3043 b. Austria –Janitress at 103 E. 75th St. –ID'd by Det. Sgt. McCafferty

Torniport, Francis A. L. #2779 –198 Guernsey St., Greenpoint, Brooklyn (see Schnude)
    Charlotte A. (2) #2780
    Frederika (26) #3560

Trebing, Fred –223 E. 5th St. –Laborer
    Margaret (42) (l) –Ln -burns on arm, face, hands
    Mary (6) #3453
    William (12) (u)

Trembly, William – Address: Steamer *General Slocum* – deckhand aboard the *Slocum*

Trowbridge, William (18) (u) -422 6th St. (see Richter)

Troell, Albert E. (13) #2851 –405 E. 5th St. – He was identified by his father who then seemed to go insane at the sight of his child in his coffin.

Turner, Julia (26) (l) –2649 8th Ave. –She jumped to a tug with her daughter, Mary, but her nephew and sister died. She testified that life preservers went to powder in her hands and nearly blinded her with their powder.
    Mary (5) (l)

Tyson, John (39) #2897 b. Ireland –Fireman on *General Slocum*

U

Uhlendorff, Louis (n) – 93 3ⁿᵈ Ave. Id'd –Cook
   Louise (9) #3373
   Selma (45) #3106 b. Germany

Uehlien, Otto -416 5ᵗʰ St. –Laborer
   Otto (19) #2942 –Silversmith $8.00/ wk
   Wilhelmine (43) #3349
   Johanetta (17) (I) – Nose & Left leg broken –Lace sewer $6.00/ wk, now housework

Ullmann, Edward–409 5ᵗʰ –Porter   (see Wolf)
   Magdalena (37) #2769 –ID'd by her brother, Ptlman. John Hines 11ᵗʰ Precinct
   Edward, Jr. (14) #2768

Ulrich, Elizabeth (32) #3444   -443 W. 41ˢᵗ St.-Had been a hat model –Bookkeeper $10.00/wk
   Sophie (70) (I) –Ln

Ulrich, Charlotte   –58 Willet St.
   Julia (16) #3638 Disinterred –Bookkeeper & typewriter $7.00/ wk   (see Fettig)
   Fettig (47) brother –Mineral Water $20.00/ wk
      Christina (D) sister-in-law
      Elisa (2) (D) neice

Unger, Francis J. -99 Ave A -Roofer
   Kate (54) #2929 b. Germany

Ulmer, Lizzie (45) (I) –Ln -Hands, limbs injured, nerves affected   232 Ave. A   –Seamstress $12.00/ wk

V

Vaeth, William (55) (I) –Ln –Scalded on head, still suffering -107 E. 4th St. –Gardener $60/ Mo., now supported by wife doing outdoor cooking for restaurants
    William J. (8) #3489
    Emilie (46) (s)

Volkenberg, Elsa (12) #3379 –315 E. 17th St.
    Elizabeth (25) #2874

Van Deuser, Matilda (13) #3813 –171 E. 4th St. (see Schwartz)
    Otillie

Van Schaick, Captain William Henry (60) (I) –Lb –Lived aboard the *Slocum* in season or at apt. on 8th Ave.; off seasn lived at Cohoes, NY. Do not confuse with William "Captain Billy" Van Schaick, this captain's son –also harbor captain for the Iron Steamboat Company Steamboat *Cephus*

Van Tassell, Abel R. (40) (I) –NBI -426 58th St., Brooklyn –Policeman on board the *General Slocum*

Van Wart, Edward (62) (I) –Lb –331 W. 21st St. –Pilot on the *General Slocum*

Vassmer, John H. - 333 5th St. –Grocer & Delicatessen store
    Johanna, Widow (32) (m) Took in sewing
    Johanna (11) #3567
    Wilhelmine (49) –help in business; died of nervous prostration & broken heart 3 years after
    William G.(16) (I) –wounded in face, nervous since –Clerk $15.00/ Mo.

Veit, Otto, –405 E. 5th St. –Policeman from College Point
    Kate (23) (I)
    Emma (2) (I)
    Magdalena (28) #3570 –151 E. 9th St., College Pt., Brooklyn
    Rosa (1) #3574

Vetter, Mrs. Henry / Emma (D) – 730 6th St.
   Fredericka / Freda (17) #3188
   Mamie (18) #3622 Disinterred –Sunday school teacher
   Margaret #8187
   Charles V. (11) #3748

Vetter, Mary (49) #3425 –31 Beekman Pl.
   Fred (16) (l)
   Bauer, Caroline (47) #3653 Mary's sister

Volkhardt, Lizzie (35) #3089 –439 5th St. –Worsted Worker $9.00/ wk (see Neubecker)
   Joseph W. (16) #3104
   Mary (36) #2967
   Magdaline #3411
   Auguste W. #3387

Vollmer, Joseph (37) (n) –123 1st St Id'd –fireman Engine Co. 46   (see Lutz, Wollmer)
   Mary (36) #2967 –bur. Lutheran Cemetery with son
   Magdaline (7) #3411
   Jacob (7) (m)
   Joseph W. (16) #3104 –This was probably the son reported in the Eagle as having been a hero the previous week by remaining at his post as elevator man in an Eastside apartment house when it caught fire, running up and down until all residents were out.
   Auguste W. (9) #3387
   Lutz, Minnie –sister of Mary –128 1st Ave.
   Katherine (57) (m) mother

Volze, George (39) (u) -221 96th St. -2nd Bartender on the *General Slocum*
   Von Krafft, Louise (30) #3140 -140 E. 4th St.
   Von Rekowski, Wanda A. (10) #2867 -337 5th St.

Von Heiden, Alexander*

    Wife
    2 children
*the only known connectin of this family to the *General Slocum* disaster is included in Alexander's obit NYT 2/12/1953

W

Wahl, Herminia (34) (u) –137 2nd Ave.
    Hedwig (11) (u)

Wallace, Elizabeth (I) –214 E. 11th St. –Janitress –Nervous physical wreck since; Note; Dr. Harry R. Purdy; she has been under his care since disaster
    Rose (11) (D)

Walheim, Emily (13) (I) –82 7th St.

Walter, Jacob –Unable to work – 336 6th St
    Elizabeth (63), mother #2817 b. Germany -Drowned; was Laundress 49.00

Warnholz, Henry (21) #2906 b. Germany –Photo Engraver $15.00/ wk
    Louise –housework
    Edward Nicholls

Weaver, Edwin N. (28) (u) – in season lived aboard the *Slocum*; off season 12th St, Troy, New York -2nd Pilot of *General Slocum*

Weaver, Frederick A. (36) #3285 b. Germany –304 E. 9th St. (see Leinberger)
    Christiana (12) #3604 Disinterred
    Carrie (9) #3689
    Mamie (8) #3224
    Helen (5) #3223 } Twins
    Esther (5) #3222 }

Weber, Frank C. (35) (I) –Lb –Burned entire body bedridden 10 months -404 5th St. –Waiter 430.00/ wk   (see Liebenow)

Emma A. (Emily) (11) #3678

Frank C., Jr. (7) #3753 Disinterred
Liebenow, Martha, (29), Annie's sister
Liebenow, Paul (38) (I) (s) –133 E. 125th St. – Lb
    Annie (32) (I) –Lb } Died of injuries 1 year later. Severe burns on back which affected lungs.
    Adella (6 mo.) (s) } these two picked up by tug E.E. –Lb
    Helen (6) (D) – Lb –buried with the unidentified
    Anna C. (3) (m)

Wehmer, Lena (10) #3580 –107 E. 12th St.

Weidemann, Henry –79 (70) E. Houston St. Id'd – Barber BDE article says he, his son, and his son in law, Emil Reichenbach, filed suits against the Steamboat Company for the loss of their wives and the son's 2 ½ yr old child.
    Caroline (50) #2870
    Catherine C. (30) #3442 b. Germany

Weidler, Henry (s) – 411 E. 9th St. –Merchant –Henry signed Affidavit by his mark
    Louise (55) #354   #2857
    Herbert N. (13) #3382 father had recovered son's body in the water, but lost hold of him and the boy was lost (NYT)

Weingart, Ethel C. (6) #2894 –409 E. 5th St.

Weintraub, Carl (19) (I) –Ln -179 Norfolk St.

Weireter, Mary (see Mauer)

Weis, Frederick –532 5th St. – There were 16 family members in this party
    Lillie/ Tilly (43) Widow #3202 –ID'd by her son (John Weis –987 E. 151st St. –Driver)
    Louis (21) #3201   nephew of Mrs. Weis –ID'd by his cousin John
    Amelia (9) #3203

Salome (14) #3561
  Fred (18) #3562 –off work $7.00/ wk
  Jake (12) #3200
  Louis (4) (I) –F –Fractured jaw –now supported by brothers
  Harry /Henry (12) (u) –Now a driver making own living
  Louis (48)

Weis, Katie (24) (I) –Nervousness –167 Ave. A
  John, Jr. (5mo.) #3563
  Emily (10) (D) –ID'd by her brother John

Weiss, Frank W. –1235 3rd Ave.
  Ida May (42) (I) –Ln
  Minnie (14) (I) –B*
  George (15) (I) –Tried to get preservers for all, but they burned, he found another for Minnie, she jumped and was picked up by boat (Inferno)
* She climbed down from the third deck bow to the first, then jumped. She caught hold of a woman with a small boy in her arms and held on. The Massasoit threw a rope which they caught and were pulled aboard. (North)

Weisser, Earnestine (55) (I) –Ln –84 Stockholm St., Brooklyn

Wendelken, Lena (25) (u) –299 E. 10th St.(m)
  Richard (9) (u)

Wenz, Charles –421 5th St. –Tinsmith
  Louise (40) #3061 b. Germany
  Louise (9) #3063
  George F.(11) #3062

Wernz, Elizabeth (30) (I) _Lb –426 6th St. –Clerking
  Annie Wernz, sister (21) #3446 –Clerk 412.00/ wk

Werter, Mary (83) widow –E. 12th St.   (oldest female lost)

Wessler, Elizabeth (30) (I) –Lb -Burns of hands & body –123 W. 106th St. –Janitress 425.00/ Mo Still under treatment

Wicker, Charles (54) #3008 b. Germany -333 E. 3st St –Waiter on *General Slocum*

Widman, Anna M. (63) #2915 b. Germany -127 1st Ave

Wierk, Amelia (15) (I) –Ln   –341 E. 55th St.
        Martha (21) (I)

Wiese, William #439 –216 E. 11th St. –The funerals were conducted by the Rev. Jesse E. Forbes (no denomination or church listed)   (see Weis)
        Caroline (50) (D) #139 #2871 b. Germany
        Emily (12) #2872

Will, Elizabeth -118 7th St.
        August (17) son #3408 –diamond Setter $4.00/ wk
        John G.

Witte, Meta (55) (I) –Lb -Burned on legs – 227 E. 21st St.   (see Seemann)
        John C. (11) #3740
        Seemann, Meta (24) (D), niece –Domestic $14.00/ Mo

Wolbern, Henry N. M. (1) #135 #2815   –1702 Dean St. Brooklyn –Ice Dealer –He was making his rounds and heard of the disaster from a Brooklyn Eagle newsboy. This party included 16 women, all St. Marks members. Hulda was described by her sister-in-law as dark and short.
        Hulda (28) #3474
        Marvin (D)

Wolf, Mrs. Magdalena (65) #2767 –1131 40th St, Brooklyn (see Woll)

352

Wolf, Mamie (27) (u) –221 E. 88th St.

Wolf, Margaret (59) #23301 (Man) b. Germany d. 6/27/1904 –H –307 E. 15th St. –H (see Hines, Ullman)

Wohlfert, Robert –106 7th St.
    Eva L. (35) #3609 Disinterred
    Charlotte M. (8) #3635 Disinterred
    Robert C. (12) #3652

Woll, Julius (30) (I) –Lb –283 Himrod St., Brooklyn
    Frederika (27) #3475
    Frederika J. (2) #3475

Wollmer, Mrs Eliza (59) –243 (246) Woodbine St., Brooklyn (see Vollmer)
    Louisa (22) (m)
    Mrs. Catherine E. (58) (D)

Wood, James (45) (u) -337 9th Ave. –Dishwasher on *General Slocum*

Woods, Katie (28) #3477 –Coytesville, New Jersey

Workman, Jennie (22) #2806 –116 Lake St., Jersey City, New Jersey   (see Hoffman)

Worthmann, Julia (19) #3628 Disinterred –178 Ave A

Wunner, George –524 E. 6th St. –Barber (became Vice President of Survivors Association)
    Lillie (19) #2916 –Bookbinding $8.00/ wk – ID'd by Frank Lander, uncle –130 E. 3rd St.
    Carrie A. (46) #2917

Wurmstich, Albert (39) #3042 –413 5th St. –Musician
    Barbara (37) #3456

Albert (5) #3045
Arthur (13) (I) –Ln –Slight burns, his support, earnings & help from grandmother

Wurtenmberger, Mamie (22) #3107 –55 1st Ave.
   Lillie (2) #3491
   Fred (20)

Wytzka, Ida V. (15) (I) –404 E. 5th St.
   Frank (31) (I)
   Lillie (2) (19 Mo) (D)

X

Y

Z
Zabilansky, Mary (see Iden)

Zahn, Bertha (23) #3460 – 69 1st Ave

Zausch, Mary (28) #493   #3218 –1518 Webster Ave. –Mail order work at Hearn's $15.00 / wk   (see Baumler)
   Doris (61) #71 #3007 b. Germany
   Baumler, Katie (25) (I) –Ln -Exposure & Nervous shock, was Dressmaker $10.00 / wk
      Mary (4) (I) –Ln

Zarges, Peter –132 E. 93rd St. –Janitor
   Mary (48) (D))

Zeidler, Conrad –123 E. 108th St. –Porter
   Ruby (2) #3507
   Anna Clara (25) #472 #2884 b. Germany

Zennegg, Edward -345 5th St. –Porter
  Bertha W. (16) (I) –Ln –Broken leg, jaw; was milliner 46.00/ wk now saleslady

Ziegler, Minnie –370 E. 10th St.
  Emily / Emelia (19) #2837 –White-goods sewer – ID'd by John Schrenck, a friend

Zimmer, Andrew (17) (I) -Lb –13 E 3rd St. – Rescued by Off. Scheuing

Zimmerman, Agnes –229 Bleeker St, Brooklyn looking for   (see Ell)
  Ell, Connie
  William (30) #3421   sister's husband & in band –Musician $35.00

Zimmermann, Charles H. –196 2nd Ave   – Window Cleaning
  Hugo (12) #3665
  Augusta (16) #2843 b. England –Bookkeeping $5.00 / wk

Zingg, Eugene (13) (D) –114 E. 4th St.

Zipse, Fred (n) –335 E. 21st St.
  Louise (11) #433   #2901
  Mary (13) #159   #2966
  Albert (10) #155 #2965
  Sophie (17) (m)
  Sophia (41)) (I) –Ln
  William F. (15) (u)
  Helen (3) #2900

Zundel ,Charles J. P. (7) #3376 –104 1st Ave
  Annie (32) (I) –Ln

Non-Validated List of Persons aboard the General Slocum

A_____, Elizabeth & A_____, Alberta (m) 502 Monroe St.
Addicks, Amelia (75) (m)
Albrecht, Bruno J. – member graduating class College of the City of New York
Allman, Lena , (39) (D) was searched for by her brother-in-law Patrolman John Hines –409 5th St.
Anger, Edwin (m) –243 E. 14th St.
Anger, Gerturde (m) –1365 3rd Ave.
Armbrust, Eda (10) (m) –106 E. 4th St.
 Kate, Mrs. (45) (D)
 Agnes (4) (m)
Ansell, Mrs. Catherine -(D) –H 103 E. 4th St
Asch, Rosie (17) (D) –88 Ave. A
Bachman (Beckman), Mr. H. Z. –1894 3rd Ave., Id'd
 Mrs.[no name] (D)
 Margaret (Anna Margarita) (7 mo.) (D)
Bagley, Mrs. Mary (41) (m) –489 W. 130th St.
 Lizzie (11) (m)
Barcei, Mrs. Lizzie (m) –284 E. 7th St.
Barker, K. (m) –137 Ave B
Batzer, Amelia 946) –422 E. 48th St.
Bauer, Mrs. [no name] (D) –130 6th St.
Beck, Carl (2) –313 E. 9th St.
Beck, Charles -69 Marcy Ave.
 Louisa (s) –Ln a former member of St. Marks, She was trampled and then pushed overboard. Her parents came to her home to console her. She had remained in control of herself on the ride home but, became hysterical at doorstep. Both children were described as very pretty with long golden curls and the Dressed nearly alike.
 Grace Edna (4 ½) (D)
 May Louise (6 ½) (D)

Bedesky, Ettie (m) –85 E. 3rd St.
Behrendt, Fannie (8) (10) (D) -88 E. 3rd St.
Behrens, George looking for
   Fassner, Miss Johanna (20). (m) – 333 5th St
Behrens, Henry (6) (D) –22 St. Marks Place
   Fritz (7) (u)
Beiss, Rosa (5) (m) –70 1st Ave
Belunken (Belmken), Herman –344 48th St.
   Anne (13) (D)
Benning, Magdeline 912) (D) –72 W. 114th St.
   Mary (30) (D)
Bensch, Louise (4) (s) –Lb -401 5th St.
Beonhardt, Ella (12) (m) –322 E. 13th St.
Berdoldt (Bernholdt), Mrs. Fred (30) (D) –41 3rd Ave
Berrens, Annie –127 Goerck St. ld'd
   Augusta (5) (D)
Bertrand, Jacob (31) (m) –730 E. 6th St.
   Richard (16) (m)
   Mamie (14) (m)
   Charles (12) (m)
Betzer, Amelia (46) (D) –422 6th St.
Binn, Mary (36) (D) –311 5th St.
Blausch, Frederick (n) Cabinetmaker, Laurel Ave., Stapleton, S.I. – He dreamed of his daughter's death the night before
   Catherine (23) (25) (D)
   Mrs. Tillie Hauff –143 E. 3rd St.
Blohm, Beatrice (2) (m) -573 Central Ave., Brooklyn –Police Capt. Becker of Hamburg Ave. Station sent corp of police to Blohm home to maintain order on day of funerals. All buried at Lutheran Cemetery
   Blumenkranz, [no name] –9 W. 10th (?) St. looking for;
   Wohmer (Wormer), Carrie (m)
Boden, Edna (43) (D)–101 Clymer St., Brooklyn
   Annie

Boemler, Mrs. Martha (D) –423 E. 86th St.
Bollan, Rebecca (53) (D) –334 6th St.
Bose, Mrs. (34) (m) -135 Ave. A.
Breckerstuchs, Annie (m)-276 1st Ave.
Breda, Thomas (9) (D)
Breum, Emma (29) (D) –411 E. 27th St.
Breun, Mrs Emma (29) (D) –Sullivan County, New York
Briehr (Breher), John -310 E. (25th) (28) 105th St. Id'd
    Catherine (4) (D)
    Kate (11) (D)
Broesa (Brosewald) (Broswald), Mrs Meta (50) (D) –269 Monroe Ave. (St.)
    Matilda (20) (l)
Broun, Peter (12) (u) –233 5th St.
Brower, Margaret L. (33) (D) –107 E. 84th St.
    Jeanette (6) (m)
Brown, Alfonso (Alphonse), (13) (D) -203 (205) (223) 5th St. Id'd by James Roth of same address
    Peter (12) (s) – Lb
    Mollie (32) (D)
    Elsie (10) (D0
    Willie (6) (D)
Brown, Minnie (13) (m) –69 1st Ave.
Bruchard, Mrs. –wife of Physician from Hoboken
Brunick, William (m) –71 E. 5th St.
Brandiski Mrs. Mary, -85 3rd St.
Bruning, entire family (m) -72 W. 114th St.
Buck, George (n) looking for360
Buerskle, Bertha & George -9 1st Ave.
Buffo, Mrs. Louisa (D) –82 W. 90th St.
Burchbaum, Mrs. L. 930) (u) –1028 Hudson St.
Burns, Fred (10) (D) –22 St. Marks place
    Henry (6) (D)

Cabilaskiwiz, Mary (19) (u) –100 e. 4th St.
Case, Mrs. [no name] (m) -2048 1st Ave.
Catlin, Rose (19) (m) –27 Sheriff St.
Christ, Amelia (D)) –144 E. 7th St
Clug, Carolina (54) (m) –468 8th Ave.
Cohrs, Jacob Michael (n) Id'd. Carrie (12) (D) – When he found her body at the pier, he fell to his knees in the melted ice run off, kissing her face until the police pulled him off. (poss. Related to Mrs. Catherine Cohrs)
Collins, Charles (36) (m) –401 W. 23rd St.
Coney, Ellig (7) (s) – 426 5th St.
Cordes, Albertine (22) (D) –417 E. 16th St.
Crofine, Mrs. Lillie (D) –995 Ave A
Dader, Eliza (46) (I) –174 New York Ave., Jersey City, New Jersey
Dananbaum, Minnie (26) (D) –103 Ave A.
Dauernheim, Mrs Henry (29) (m) –41 3rd Ave.
Dauernheim, Minnie (26) (m) –1065 Jackson Ave., Bronx (Brooklyn)
Debricht, Martha (8) (m) –414 E. 9th St., Salvation Army Home
Debrinsky, Mrs. Annia (m) –85 3rd St.
Deering (Dearing), Mr. E. -12 State St.
    Mrs. Jane (m)
    Edna (5) (I)
    Martha (5) (m)
    Gustav (9) (m)
Deiner, Mrs. (m) –895 Jefferson Place.
Deissman, Lena (16) (D) –114 E. 4th St.
Delventhal, Mattie (45) (I) –381 Madison St.
Demeiner, Mrs. (m) –895 Jefferson Place
Dengler, Harry (6) (I)-123 7th St. Id'd
Derker, Frank –1010 E. 178th St. Id'd
    Theodore (3) (D)
Dersch,Ellen (41) (D) –76 1st Ave
Devine, B. M.   –2006? 8th Ave. looking for

Miss Turner (m)
Dieckhoff, Frieda (13) (D) -121 4th Ave. Brooklyn
   John (20) (D)
   Edward
Dillon, N. W. – 79 Mangin St. looking for
   Diamond, Mary (8) niece & dau. of Catherine Diamond, N. W.'s sister. Catherine fell on her knees at Mrs. Birmingham's coffin; she had insisted that the senior lady go for a "nice day out". Note- Michael Dillon of 83 Clermont Ave., Bklyn, Id'd his sister, Katherine Birmingham (72) (D) –79 Mangin St. she is also mother of N.W. Dillon.
Dillemuth, George –15th Ave & 68th St, Brooklyn–violinist in the band
Distler, Henry (m) –116 E. 14th St.
Dockdale, Mrs. George (m) -266 W. 136th St.
Doering, Martha (5) (D) -12 State ST.
Doran, Mrs. Thomas 928) (m) –477 E. 25th St.
   [no name] (14 Mo) (m)
Doerrhoefer, Fritz (7) (D) –121 Ave. A.
   Margaret (42) (I)
Duick, Mary (16) (D) –121 4th Ave., Brooklyn
Duls, Minnie (m) –103 Ave A –Milliner
   Louise
   Durker, Frank -1010 E. 178th St. Id'd
   Theodore (3) (D)
Edell, John (22) (s) looking for
   [no name] Mother (D)
   Paul (D)
Ehlig (Ellig), Conrad (Coney) (7) (8) (u) -425 (426) (433) 5th St.
   Lizzie, (6) (D)
   Margaret, (4) (D)
Eichhoff, Augusta (15) (m) -196 2nd Ave. Id'd by George Bates of the same address
   Hugo (12) (m)
Eidman, Henry –100 E. 4th St. Id'd
   Grace, sister (6) (D)

Eilar, Matilda (46) (D) –219 E. 13th St.
    Elsie (16) (D)
Elden, John (grocer at 4th St. & 1st Ave.) (n) –His condition was so bad, his friends worried for him
    [no name] Child (D)
    [no name] Child (D)
Elick, Elsie (4) (m) –433 E. 5th St.
    Lizzie (m)
    Mary (m)
    Enger, Mrs. George –1365 3rd Ave.
    Rosie (m)
    Gertrude (m)
Ensel, Mrs Louisa (28) (m) -103 E. 4th St.
    Eugene (m)
    Alfred (m)
Erdmann, Henry -100 E. 4th St. Id'd
    Grace, sister (6) (D)
Ermer, George (11) (D) -84 Stockholm St., Brooklyn
Falmeter, Lizzie 940) (D) –80 E. 1st St.
Fassner, Mrs. Johanna (m) –332 E. 5th St.
    Musie (12) (m)
    Willie (9) (m)
Feight, Mrs. Lena 926) (m) -168 1st Ave.
    Rosie (6mo) (m)
Feldhausen, George (42) (D) -50 W. 8th St.
Fellig, Peter (m) –58 Willett St.
Fickbohm, Marie (27) (m) –284 E. 7th St.
Follmer, Mrs. Mary (D) –123 1st Ave.
Frech, Henry –409 E. 54th St. Id'd
    Charles E. (4) (D)
Frey, Edna (m) #3688 -84 7th St.
    Louis (m)

Gussie (m)
Mrs. Lillian (34) (35) (D)
Frieling, Harry looking for
   Bruchard, Mrs. D.   of Hoboken (m)
   Klenen, Mrs. Fred   of Hoboken (m)
   Klenen, Mrs. Mitta   of Hoboken (m)
   Tully, Mrs. Henry S.   of Hoboken (m)
   Tully, [no name] child of Hoboken (m)
Fuller, Mrs. Annie (m) –95 2nd Ave.
Fulling, Henry F. –39 W. 110th St.
   Henry (m)
Fulmer, Joseph (m) –123 1st St. (Ave.)
Galewiski, Fredrick (D) -54 7th St.
   Henry (D)
   Anna (64) (D) –mourned by neighborhood children who left flowers on her favorite park bench
Gamberg, Henry –427 E. 9th St. Id'd
   Henry (6) (D)
Gassmann, Mrs Charles (34) (m) -128   E. 4th St.
Gates, Edward –80 1st Ave. Id'd
   Catherine (28) (D)
   Margaret (2 ½) (D)
Geisel, Emma (15) (D) –107 E. 2nd St.
Geissler, Sadie (8) (m) –439 6th St.
   [no name] (s) – small white headed boy who waited for his mother on his stoop (NYT)
   Minnie (6) (m)
Geisler, William (17 mo) (m)–201 Ave. A Id'd
   Ida (19)
   Peter (18) (D)
   Mrs. Catherine (20) (m)
   Edward (19) (m)
Geister, Ida (s) Testified at Inquest

Geluecin, (7) (D) –54 E. 7th St.
Gerdes, Mrs. Margaret Fackman (D) daughter –430 Kosciusko St., Brooklyn
Gerstenberger, James (25) (D) –Id'd by C. H. Knapp –147 W. 32nd St. –Proprietor of Central Hall, 147 W. 32nd St.
Gibbons,–225 E. 5th St.
    Mary (40) (u)
    Mary (15) (u)
    Frank (9) (u)
    Thomas, Jr. (7) (u)
    Catherine (4) (u)
Girrcler, Edith 918) (D) –20 Ave A
Glueck, Charles (8) (D) –113 St. Mark's Place -Id'd by Charles Vetter
Goerhoefer (Goerrhosfer), Fritz –121 Ave. A looking for
    Mrs. Barbara (42) (m)
    Maurice (9) (m)
    Freda (12) (m)
    Fritz (10) (m)
    Rehon, Kate (47) sister-in-law
Goetz, Margaret (2 ½) (D) –80 1st Av
Goldberg, Lena (m) –105 4th St.
    Laura (m)
Goldstrum, Helen
Granefire, Lillian (D) –998 Ave A } found locked in each others arms
Grayey, Mrs. [no name] – 84 E. 7th St. looking for
    Fred (14) (D)
    Arthur (17) (m)
Greber, Fred 914) (D) –54 7th St.
Gruning, Henry –45 E. 7th St.
    Mrs. Lena (45) (m)
Greismann, Christine (47) (I) –114 E. 4th St.
Grener, Katie (11) (D) –310 E. 25th St.
Gressler (Grissler) (Greisler), Mr. Louis–439 E. 6th St.

Anna (35) (m)
　　Lillie (7) (m)
Greve, Fred (14) (m) –54 7th St.
　　Mrs. [no name]
Grevel, Emma (15) (m) –117 2nd Ave
Grews (Growe) (Grew), Henry (60) (D) –54 E. 4th St.
　　Mrs. Catherine (60) (m)
　　Mrs. Freda (26) (m) –wife of Henry J.
　　Lillian (2) (m)
　　Henry (8) (10) (D)
　　Anna (54) (l)
Griffing (D) – no address –est. to weigh 300 lbs.
Grossarth, Charles & family were reported missing, but had not gone on the trip
Guest, Frederick –3rd Ave. & 128th St. looking for
　　Mrs Otto (m)
　　Clara 9m)
　　Walter (m)

Haas, Gertrude (46) (D) -64 E. 7th St.
　　Frederick (19) (n) – son of Rev. Geo.. only family member not aboard   (parents, 2 aunts, sister on board)
　　John A.V.   – Brother or son of Rev. Geo.–of Allentown, Pa.
　　Rev. John A. William (n) – St. Pauls Lutheran Church, W. 123rd St. – brother of Rev. Geo
　　Anna (5) (D) – Id'd by her uncle [no name]
　　Margaret (5) –Id'd by uncle
　　Mrs. William Tetemore (D) –Rev. Geo. sister-in-law –714 Bushwick Ave., Brooklyn
　　　　Edith (m)
Hach, Freda (17) (m) –541 E. 157th St.
Hagenback, Louise (14) – 102 1st Ave Id'd her 3 friends whose dying father did not go.
　　Scheele, Annie (15) (D) –14 St. Marks Pl.
　　　　Clara (D)
　　　　Vina (D)
Hagenbucher, Mamie (D) –Sherwood park, New York –(all at Geo. M. Fitzpatrick Fun.l Home, 1488 Lexington at 96th) Interments Greenwood

364

Hagenbuchter, Mrs. Kate (m) –2112 3rd Ave
    Mary (32) (D)
    Damon (m)
    Irving (m)
Margaret Molitor (D) (see Molitor)
    Arthur
    Dunn, Julia (m)
    Irwin, Fanny –a Friend
Halphausen (Halphusen) (Halthusen), John (70) (s) –138 2nd Ave. – Sexton & Pres. of the Sun.schl. at St. Marks –He & his daughters clung to paddle-wheel until picked up by the tug *Sumner*.
    Mina (12) (s)
Halpmann, Mrs. [no name] – one block over from Mangin & Houston
    Child (D)
    Child (D)
    Child (D)
    Child (D)
Hans, Otto (m) –310 E. 14th St. (see Ell)
Hanneman, Susie (54) (I) –415 E. 5th St
Hardekopf– 343 Rivington St.
    Kester, Mrs. [no name] (m)
Hardincamp, John (m) –12 E. 11th St.
    Henry (24) ld'd;
    Margaret (Mary) (11) (D), sister –June 16th was to be her 11th birthday, and a party had been planned.
    Frank (m)
    Harold (m)
Harms (Harmes) (Harnes),– 312 E. 14th St.
    [no name] father (m) from Troy NY
    [no name] (D) mother from Troy NY
Harpbrock, Mrs Henry W. (m) –2154 2nd Ave.
    [no name] child (m)
Hartung, Mildred (D),– 342 E. 21st St.

Willie (15) (m)
Laura (10) (D)
Hatterick, Mrs. John (30) (m) –420 E. 15th St.
Emil (3) (m)
Adolph (7) (m)
Hatz, Frederick -319 5th St. looking for
Anna (m)
Bertha (m)
William (m)
Hausel, Eugene 96) (m) –103 E. 4th St.
Hausen, George Heinrich (n) –167 E. 4th St.
[no name] (s) daughter   } saved by Officer Van Tassel
[no name] (s) granddaughter }
Havemeyer, William (Willie)(7) (D)–1499 1st Ave.
Anna (35) (m)
Heagy (Hagy), Barbara (17) –108 1st Ave
Hecke, Tessie (10) (D) –504 E. 15th St.
Heckert, Aniome (12) –Ave A & 5th St.
Hecklin, Kate (D) –no add.
Hedencamp, Henry-805 6th St.
Hehl, Thomas –55 1st Ave. Id'd
Gus (41) (D) –Id'd by his father
Gus (14) (D)
George (14) (D)
Emile (12) (D)
Fred 96) (D)
Andrew (41) (u)
Heidenkamp, James –Id'd by letter he was carrying
Heidkamper, Maggie (11) (D) –49 Ave A
Heil (Hel), George (18) (I) –Boston Rd. & Pelham Parkway, Bronx
Adelaide (15) (D)

366

Heins, Eddie (10) (m)– 300 Front St.
  Elsie (9) (m)
  Margarita (Marguarita) (14) (D)
  Annie (26) (D)
Heisen, Margaret (33) (D) –181 Waverley Place
  George (3) (D)
Heli (Heyl), Dora (18) (m) –9 W. 19th St.
Helm, Adam -606(?) E. 141 St St. looking for;
  Cannon, Frederick (m)
Heller (Hiller), William –29 Louis Place, Brooklyn – Sexton (Janitor) of St. Marks
  Christina (68) (D) –404 6th St.
Hendencamp (Hendkamp)(Heidencamp (Hedenkamp),–805 6th Ave.–body #289 –Carpenter & Builder, also Deacon of Evangelical
  Lutheran Church, E. Houston St.)
  Mrs. E. (s) – shock – Lb
  Margaret (50) (l)
  Frank (9) (D)
  Margaret (11) (D)
Hendersen, Barbara (30) (D) –386 W. 125th St.
Hener, Adolph (14) (u) –129 Division St.
Henkel, Lillie (8) (m) –227 7th St.
Hephman, Maurice (38) (I) –640 6th St.
Herman, Fred (13 Mo) (D) –168 1st Ave
Herman, Henry looking for
  Vitz, Mrs D. (27)   –College Pt.
  Rosie (Infant)
  Poth, Mrs. Susanna (50) (m) –75th St.
Hermann, Catherine (60) (D) –437 5th St.
Hermann, Frank –103 1st Ave. Id'd.
Hester, John –159 Varick St. looking for
  [no name] Aunt (m)
Hesterberg, Mrs. B. (m) –419 E. 5th St.

Hickman, William (3) (m) –525 E. 12th St.
  Lily (5) (m)
  Catheriene (Katie) (22) (24) (m)
Hilbert, Miss Frances (17) (s) –419 E. 5th St. –Her clothing was all but pulled off before she could jump aboard a tug. She had to borrow a man's jacket so she could go home.
Hill, Adolph T. –Brooklyn -103 Meserole St., Williamsburg -Id'd
  Christiana Richter (D) } his nieces. These bodies were claimed by
  Lydia Richter (D)   } another family and removed. They were never returned.
  Mr. William Richter (D)   Brother-in-law
  Mrs. William Richter (D)   Body #175   } Sister –Was cashier for the Queens Borough Tax Office, Lived in Long Island City
    Catherine (3) (D) Body #173 } After these three were identified, and the undertaker came to claim the bodies, they were found
    Lillian   (4) (D) Body #174 } to be missing, presumably misidentified and removed by another family. Police were called to
                                                help straighten out the matter.
Hines, Patrolman John (n) of the Mulberry St. Station sent to the scene discovered it was the ship with his family aboard
  Mrs. Margaret Wolf, Mother-in-Law (m)
  Mrs Lena Ullman, sister-in-law (m)
    [no name] son of Lena (m)
  Hines, John G. –397 E. 4th St. looking for;
    Annie (40) (D)
    Frankie (18) (13) (m)
Hirt, Mary (63) –82 W. 90th St.
Hittinger (Hettinger), Mrs. A. J. (m) –127 1st Ave.
  Elizabeth (Lizzie) (26) (D)
Hoag, Susie (Susianna) (49) (48) (D) –158 1st Ave
  Ella (13) (D)
Hoag, William (14) (D) –210 E. 14th St.
  Wilmur (Wilme) (12) (D)
  Ermina (9) (D)
Hoeff, Tilda (14) (m) –142 3rd St. –Reported by her mother
Holder, Joseph P. –34 W. 18th St. looking for
  Mary

Schaffer, Amelia –34 W. 18th St.
Holler, John (15 mo) (D) -338 6th St. –Bur. Lutheran Cemetery; Grandson of Barbara
Homan, Charles -437 E. 15th St. looking for
    Mrs., Rose (45) (m)
    Charles, Jr. (m)
Horn, Mrs. Fritz (m) –Wooster St.
    [no name] daughter (m)
Horne, Lena
Horway, Mrs. Anna (28) (D) - 313 E. 9th St.
    Cortland (27) (D)
Hotz, Lillie (9) (m) –319 5th St.
    George (5) (D)
Hubold, Margaret (76) (I) –416 E. 5th St.
Ilmar, Fritz (47) (D) –1225 Park Ave.
Jordan, John -37 3rd Ave.
    Kate
Kaas, Hermann Id'd -110 1st Ave.
Katzenberger (Hergenberger), August – 22 (223) St. Marks Pl. Id'd
Keetensch, Lizzie 930) (D) –420 E. 15th St.
Keisel (Kiessel) (Keisel), Theodore (27) (u) –266 Ave A
    Millie (6) (D)
Keisel, William (n) –worked in Office of the building Bureau, queens Borough Hall
    [Neice} (D)
Keister, Margaret (45) (D) –343 Division St.
Keppler, Minnie (19) (D) -192 1st Ave–1st Ave. btwn. 12th & 13th St.
    Lillie (16) (D)
    Irene (11) (12) (D)
    Louis (17) (D)
Keppler (Kepple), Willie (11)(s) –127 1st Ave. –Went on excursion without permission; was afraid to go home and spent the night in Harlem on a park Bench. When he jumped, he hit the water and tried to swim to mid-stream, but the current was too strong so he rolled over and floated for about ½ hour until rescued by a tug. During that time he tried to help 2 women, but they dragged him under. Then he saw his

name in the paper and went home so as not to break his mother's heart. His father gave him 50¢ for being a good swimmer.

King, Gussie (19) (D) –889 Broadway, Brooklyn
King, Katherine (46) (D) –311 E. 37th St.
Kircher, Mrs. George (s) -185 Russell St, Williamsburg, Brooklyn
    Karl (3) (D)
Klatthaar, John H. –506 E. 5th St. Id'd.
    George (56) (D)
    Joseph (14)
Klein, Tina (21) (u) –31 Ave. A –Wholesale & Retail Liquor dealer at 34 Ave. A.
    Carl (D)
    Clara (D)
Klein, Emma (6 mo.) (D) -314 6th St.
    Kate (D)
Klein, Hannah (32) (I) –444 E. 15th St.
Kleiner (Klenner), Mrs. Meta -1391 Washington Ave., Bronx
    Frederick
    Margaret (infant)
    Ethel
Klennan, Edmund (1) (u) –110 W. 129th St.
    Matilda (29) (I)
Kleinbert (Klingert), Tessie (17) (7) (u) (m) –331 (431) E. 15th (16th) St.
    Anna (m)
Klesch, Katie (6) (D) –800 E. 14th St.
Kneuster, William (12) (u) –645 (65) St. Marks Place
    Mary (46) (I)
    Charles (17) (I)
Knuessel, Mrs. (35) (m) –439 6th St.
    William (11) (m)
    Annie (9) (m)
    Nettie (8) (m)
Koch, Gussie (21) (m) –84 7th St.

Kolb, Gussie (22) (D) –517 E. 5th St. –Id'd by her sister Ida Kister
Kolinger, Mrs. Eva –Id'd by papers & bank books in her bustle
Kosel Lillian (5) (D) –266 Ave A
Krall (Kraal), Albert (n) –Expert Pearl & Ivory worker –Middle Village, Queens, Long Is. –He had been to Newport, R.I. and did not know his family Was dead until he returned on 6/21. According to the BDE, they were either among the unidentified or lost to the river.
    Mrs. Mary (26) (m)
        Julia (5) (m)
        Albert (1) (m)
Krantz, Entire family   –123 Madison Ave.
Krause, Sadie (17) (u) –158 1st Ave.
Kress, Mrs. Barbara (35) (m) –526 6th St.
Kruning, Mrs. (m) –7th St. & 2nd Ave
Kregler, Anna (Annie) (7) (D) –257 Ave. B
Lamm (Lann), Kate -303 (203) Ave C Id'd
    Amelia (Emalia), sister (40) (D) –506 6th St.
Lang, Charles A. -1843 1st Ave. Looking for rest of his family and; *
    Amelia Lang (15) (m) -68 Ave. A.
* He told Sergeant Lonergan of the emergency sub-station, E. 132 ST. that he was aboard the *Slocum* and stated that the deckhands knew there was a fire opposite 56 or 57th St., but the flames were not visible until 89th St.
Lang, Charles H. (s) –(poss. A former Coney Island Lifeguard) –all swam ashore
Lanz, Michael (n) -696 E. 145th St. looking for   Id'd 29 relatives (primarily wife's side)
    Child (m)
Laue, H. C. (n) Id'd
    Gustave, son (19) (D)
Laurie, Lona (15) (m) –111 E. 4th St.
Lebuhrl, Lizzie (9) (D) –23 Ave B
Leffler, Catherine (41) (D) –9 E. 3rd St.
    Louise (9)
Leitz, Mrs. Minnie (n) looking for
    Magdalena (7)
    Minnie (9)

    Joseph (17) – employed 457 Broome St as Office Boy who rescued many in bldg. Fire
    Amelia (abt. 17) – died clutching a chatelaine bag with 6 folded hankies
    Wolmer, Mrs. Mary, sister in law of Minnie was Id'd at Bronx, Alexander Ave. Station House
Lembach (Laubeck)(Lambeck), Albertina (33) (I) –427 (14) E. 9th St. –Ln – Family was on the hurricane deck when it collapsed, she jumped with all her children, but was the only one rescued alive
    Herman (14) (I) – was picked up by the launch *Kills* & brought to Riker's Island
    Dora (11) (D) (I) –Hung onto the paddlewheel until rescued
    Albertina (Minnie) (9) (D)
    Ernest (9) (D)
    Henry (6) (D)
    Albert (4) (3) (D)
Libbert (Lieport), Charles (12) (D) –412 E. 6th St.
    Harris (46) (u)
Liebenow, Anna Christina Wulf (76)–133 E. 125th St.
Lieberman (Liederman) (s) – 4 Smith St., White Plains looking for (he jumped with his brother, but lost hold of his hand)
    Mother (D)
    John (D) brother
    Sister (D)
    George Heinz (D) (16) (friend) –97 Ave. A.
    John Schoeneman (D) (friend)
Liebersohn Paul (s) –133 E. 125th St.
Loebinger, Henry (D) –Id'd by papers on his person
Loeffler, Louis (10) (D) -9 E. 3rd St. Id'd
Loudeman, Mrs. John (45) (D) –White Plains –Id'd by James J. Sullivan – 7 Elm St., White Plains
Lupenz, Mrs. Gertrude (46) (D) -102 Clymer St., Brooklyn
    Mildred (18) (D)
Lutz, John -148 2nd Ave
    Julius Lutz
John D. Lutz –Damerest, New Jersey
    Lusta (16) (I)
    Martha

Lyman, Samuel (8) (m) –72 Ave B
Mack, Annie (22) (D) –401 E. 5th St
    Frieda (8) (D) -Training school 414 E. 9th
Mahlstedt, Martin H. (22) (D) –629 E. 146th St., Bronx
Mamme, Mrs. Fredericks (m) -730 6th Ave.
    Charles
Mammelkamps (Mammelkampf), Lizzie (44) (D) –130 e. 4th St.
    Stella (12) (D)
    Gustav (6) (u)
Manheimer, a bookmaker came with Dr. Fleming (house physician at the Grand & Imperial Hotels) looking for
Mario, Mamie (14) (m) –121 Pitt St.
Mathews, Mrs. (D) –15_ Greenpoint, Brooklyn
Mazerath, [no name] mother (D) –330 E. 6th St. (see Schmidt)
Maurer, Julius (5) (D) – 421 9^{h} St.
    Julia (14) (m)
May, Charles -599 E. 16th St. Id'd
Mayer, Nicholas (43) –430 (130) E. 17th (107th) St. Id'd
    Louisa (Louise) (39) (D) wife
McCahon, H.(m) –504 E. 14th St.
McDonnell, Eugene & [no name] sister - 108 E. 58th St.
McGaffrey, William (14) (u) [no address] -Tossed a dazed girl over to a tug & then swam ashore. He then swam out again and saved 3 men.
Meike (Mecke), Otto –504 E. 16th St.
    Daisy (11) (D) –ID'd by father
    Tessie 99 (D)
Meinhardt, Mrs. Barbalka (D)–146 E. 4th St.
    Jacob
    John
Meirs, John (9) (u) –154 Hobart Ave, Bayone, New Jersey
Melbourne, William (D) –441 W. 46th St.
Melpen, Mrs. Frederick (m) –80 1st Ave.
Mettler, Nicholas (inf) (m) –338 E. 5th St

Mettlein, Otto - 416 5th St.
   Minnie (38) (D) wife
Miller, Bernard (Bernhardt) (34) –92 1st Ave./ 95 2nd Ave. – Tammany district leader *
   Edward (m) (3)
   Walter (m) (9) (19)
   Grover (m) (12)
   George (s) (6) -contusions of the body – Ln
* The family jumped together wearing life jackets and headed for Randall's Is., with Mary holding Edward. A delegation of Tammany representatives under the direction of Julius Harburger, the Tammany leader, accompanied the family part of the way to Lutheran Cemetery.
Miller (Muller), Elizabeth (6 Mo) (D) – ID'd by father –406 E. 6th St.
   Annie (6 Mo) (m)
Miller, Flora (17) (D) – 4 E. 46th St
Miller, Frederick C. (s) b. 1877 d. 1960, Tobacconist
   Frederick W. (s)
   Howard I. (s)
Miller (Muller), Henry (3) (D) –41 1st Ave
   Mary (7Mo) (m)
   Annie (37) (D)
   Annie (13) (D)
   George (40) (u)
   Ernest 97 (u)
Miller, Jacob (s) –officer of St. Marks Sunday School
Miller, Simon (71) (I) –111 Norfolk St.
Molitor, Adolph L. –Midland & Jerome Ave., Mt. Vernon, New York
   Fannie (m) daughter
Moller, Louisa–45 2nd Ave.
Moller, Samuel (Emil) -20 St. Marks Place
Morris, Joseph -444 6th St. looking fo;
   Klinski, Mrs. [no name] (m)
   Klinski [no name] daughter (m)
Morris, Michael (10) –69 1st Ave

374

Mueller, Walter (s) (19) (s) -95 2ⁿᵈ Ave.
  [no name] Father –He was with the family on the top deck, saw fire in time to get life preservers on all but himself. He was interviewed almost at the point of death from pneumonia caused by exposure and anxiety seeking his family in wet clothes. The surviving boys were cared for by their grandmother, Mrs. Hager who had not gone on the trip.
    Mrs. Bertrand (25) (D)
Mr. Muller, Mr. (s) –worked with the Rev. Schultz and together saved 50 children
Muller, Mr. –308 6ᵗʰ St. Id'd
    Rose (14) (D)
Muller, Anna (24) (D) -E. 4ᵗʰ St.
    Tessie (6mo) (D)
Muller, Anne (m) –2ⁿᵈ Ave & 3ʳᵈ St.
Muller, Mrs. [no name] –645 17ᵗʰ St. looking for
    Lamm, Mrs. [no name] (D) – sister, widow same address
        Frank Lamm (8) (D)
        Lillian Lamm (7) (m)
Nachman, Robert –1355 Broadway, Brooklyn looking for
    Abraham, Isaac (m) Musician in the band on the General Slocum
Nagel, Mrs. Tobias (pregnant) (D) –54 7ᵗʰ St
Nehlein, Otto -416 5ᵗʰ St. ID'd
    Minnie (41) (D)
Norman, Anna (23) (D) –402 3ʳᵈ Ave.
Norway, Mr. [no name] –313 E. 9ᵗʰ St. ID'd
    Carl (D)
Nuncle, Arthur (D) –11 7ᵗʰ St.
Oehler, Mary (55) (D) –510 6ᵗʰ St. –ID'd by son Daniel
Olfeth, Carl -339 6ᵗʰ St. Id'd
    Annie (45) (D)
Ottinger, Charles (n) – 91 7ᵗʰ St –Looking for;
    Kate, (36) (38) (33) (40) (D)
    Charles (Charley) (15) (16) (D)   } Twins

Emma (15) (14) (16)(D)
Andrew (u) -H .–one of the last off the General Slocum; taken to Hospital; checked himself out & went home
Lillian (18) (n) }
Kate (19) (n) } three older daughters of Charles were working and so not aboard
Regina (n) }
Willie (William) (13) (18) (12) (D) – H –submersion – Their mother put life preservers on all the children but Willie refused as he could swim. He pushed off and was picked up by a tug.

Pain, Annie (m) –224 E. 11th St.
Peters, Walter (n) - 50 Ave. A.
    Margaret, Wife (22) (D)
Phoma, Philip (n) -102 1st Ave
Pienning, – 45 7th St. looking for
    Carl Gruening (3) (m)
      Helen Gruening (9mo.) (m)
Poinport, Mrs. Freda –196 Guerney St., Williamsberg, Brooklyn
    Frances
    Charlotte
Pottebaum Charles G. - 61 St. Mark's Place Id'd
    Herman (5) (D)
Pullman, William H. (D) –337 E. 18th St.
    Elsie (18) (15) (n)
Quinn, George -211 8th St.
    Hattie
    Henrietta
Reichenbach (Richenback), (s) –79 (175) E. Houston St. –Put on life belt, which caught fire. Dropped baby over. Next she knew, she was saved by a negro man.
Reiss, Fred (n) -70 1st Ave. – Shoe dealer
    Lorenzo (4) (m)
Rekanski, Lena, widow (s) – 337 5th St.–Found the body of her daughter at the pier. Her daughter had talked of the trip all year and had been allowed to go with a friend from the same address. The mother was so distraught she tried suicide by jumping into the river, but was restrained and taken to Bellevue Hospital.

Wanda (10) (D)
Richter, William -6th St.
　　Lizzie
Richter, Tessie (13) (D) –90 1st Ave
Ringler, Mrs. Eva (D) –carried $30,000 in bank notes, securities and bank books
Rohme, Miss [no name] (D)–body retrieved clutching her baby. The only identification for these two was her name on some papers in her purse.
Roseman, John (11) –Ave. C. boarded without a ticket and was put off before sailing. He was reported missing by friends and had to un-report himself.
Rosenstein, Katie (27) (m) -127 1st Ave
　　Lizzie (15) (m)
Rowski, Lena –[no address] –did not know that her daughter was aboard until she ID'd the body at the morgue, she was restrained & removed to Bellevue
　　Donda (10) (D)
Rubenklau, Freda (I) –23 Einerrt St, Brooklyn
Rumpkamp, Mrs. [no name] -300 E. 4th St.
Salvation Army 4th corp. – Harlem looking for
　　Dunn, Mrs. Julia* (D)– 2112 3rd Ave.
　　　　Arthur (5) (D)
　　Buscher, Mary Hegen* (32) (D) -2112 3rd Ave. Id'd by brother
　　Mollnor, Mrs.[no name]* (D) – Mt. Vernon, NY
　　Irving, Fannie* (32) (D) – 2112 3rd Ave.
　　Alt, Father
　　Alt, Henry (14) (s), Son engineer at library & had convince the group to go
　　Hollister, Mrs. Mary *(D)
*these ladies all worked at the Astor Library
Sanders, Otto (m) –no address
Schaefer, John (D) –Military Funeral at Lutheran Cemetery-Spanish American War Veteran, Mbr. Wm.H. Hubbell Corp of Veterans, Brooklyn
Schaler, Fred (D)　} started bucket brigade on top deck
　　Charles (D)　}
　　[no name], Father (s)
Schayler, Katie (m) –322 E. 13th St.

377

Scheele, –14 St. Marks Pl.
    Emma (D) –108 1st Ave
Scheier, Fred (20) (m) –E 8th St.
Scheur, Julia (8) (D) –237 E. 10th St.
    Mrs. Margaret (24) (D)
Schiaoke, John –204 E. 5th St.
    [no name] sister-in-law (m)
Schiar (Schaier), Frederick (n) –174 E. 3rd St.
    Wife (D)
    Julia, daughter (D)
    Frederick (6mo) (D)
Schieninger, Gottlieben (D) – No address listed; his was the first will filed as a result of the disaster. He had no heirs listed but left bequests for Lena Schoenhardt, a friend –401 Sumpter St., Brooklyn Joseph Schwende, Executor, Brooklyn Robert Beyer –320 Broadway filed the will
Schier, Julia (6 Mo) (m) –124 7th St.
Schiller, George (8 Mo) (m) –45 1st Ave.
Schinde, Henry C. (35) (D) –1958 Washington Ave., Bronx
    ID'd by W. F. Mc Shane, a friend at the same address
Schluver, Harry (m) –196 Irwin St., Brooklyn
Schluver, Mr. W. (m) –462 E. 76th St.
    Mrs. (m)
Schmid, Charles L.(s) –418 E. 9th St. –ID'd the following
    Arthur (14) (D) son of Charles
    Edward (s)
Schmide, Henry -196 Guernsey St., Brooklyn
Schmidling, Charles (19) ID'd– 119 E. 7th St.
Schmidt, August (D) –1163 Greene Ave., Brooklyn
Schmidt, Sophia (D) –108 Ave B
Schmidt, Mrs. William (m) –138 E. 7th St.
Smith, Mary (46) (D) –138 E. 7th St. (Id'd by cousin, Charles Stock –142 7th St.)
Schmiiling (Schnitzerlang, Schmeling), Mrs. Eliza (Elizabeth) (36) (D) -123 Ave. A
    Annie (10) (m)

    Fred (5) (m)
Schmitt, Mrs. Annie (25) (m) -920 E. 155th (135th)St.
    Annie (10) (m)
    Mildred (12) (m)
    Beatrice (2 Mo)
Schmitt, Sophia (15) #2797 -341 E. 25th St.
Schmittberger, Joseph (n) –[no address] ID'd
Schnaff, Fred (15) (u) –645 E. 17th St.
Schneider, Augusta (3) (u) – 322 Stanhope St., Brooklyn
Schneider, Peter -326 E. 6th St. –Peter Schneider, Successor to Andrew Baldauf Mfgr. & Dealer in Cider, Est. 1858
    Mrs. T. (40) (m)
        Eva (14) (D)
        Eva (42) (D)
Schnepie, Carrie (15) (D) –54 Ave A
Schoerichs, Lotie (36) (D) –425 E. 13th St.
Schottsberg, Mrs. (22) (m) – St. Marks Place
Schrecher, Edward (n) –144 Essex St. ID'd –upon first finding his child, he threw his personal papers into the coffin crying "Here's all I have, it's no use any more" then tried to jump into the river, he was restrained.
    Elsie (9) (D)
Schreisen, Charles A. (m) -343 E. 18th St.
Schrenemann, Elsie (17) (m) –986 Holmes St., Bronx
    John (15) (m)
Schroeder, Margaret
Schruner, Bertha (16) (D) –140 1st Ave
    Lena (13) (D)
    Willie (9) (D)
Schuebbe, Mrs Mary (30) (m)–54 Ave. A
    Alfred (7) (m)
    Annie (8) (5) (m)
    Emma (23) (22 Mo) (m)
Schultz, Rev George –assistant to Pastor Haas. He worked with Mr. Muller to help 50 children to safety. He was the first witness at the Inquest.

Schultz, Rev [(J. S.) Julius G.] (s) Pastor of St. Lukes Lutheran Church –Erie, Pa. Guest of Pastor Haas
Schultz, Hermann, Butcher
Schultz, Martha (45) (D) –no address ID'd by Henry Miller (brother –202 St. Marks Place)
Schumacher, Mrs. Nicholas. (s) – 529 E. 82nd St –Someone threw her over the rail to the deck of a tug, then many people landed on her.
Schuman, Albert (D) –100 E. 8th St.
Schutte (Schiller), Annie (40) (u) –41 1st Ave.
    Henry (11) (u)
Schuttinger, Dora (m) –754 E. 149th St.
    Fred (m)
Schwartz, Louisa A. (D) –290 E. 38th St.
Schwarz, Emil -141 E. 3rd St.
Serber, Catherine (72) (D) –107 E. 84th St.
Settig, Peter, (D) saloon keeper – 120 Ave. A. –Id'd by promisory notes & cards in pockets; had loaned money to Hausman, Julius –4 1st Ave.(m)
Sibelsky, Kate (18) (D) –322 Freeman St., Greenpoint, Brooklyn
Siegwart, Harry (10) (m) –225 5th St.
Sills, Annamay (20) (D) –3 Spring St., Roundout, New York
Silver, Kate (60) (D) –207 E. 84th St.
Silverberg, Lillie (18) (m) –215 W. 13th St.
Smith, Annie (8) (D) –801 E. 147th St.   (related to Hessel, cousin of John Muth)
Smith, Henry (38) (u) –1028 Hudson St., Hoboken, New Jersey
    James (8) (u)
    August (5) (u)
Smith, George –18 Jackson St.
    Mrs. Annie (26) (D)
    Hildreth (3) (D)
Smith, Martha (18) (D) –334 E. 15th St.
    ID'd by George Schmidt,her brother
Smith, William (5) (m) –142 7th St.
Sobilinski, Mary (20) (u) –100 E. 4th St.
Sprechter (Spekter), Edward – 144 Essex St. Id'd –The father was so distraught at finding his favorite child, he threw all his valuables in her coffin on the pier and tried to jump. He was subdued until he calmed down enough to do the necessary paperwork.

Elsie (9) (D)
Spoehe, George –304 E. 6th St. ID'd
    Mrs. Susan (29) (D)
Spring, Mrs. Augusta 952) (D) –90 1st Ave.
Stahlman, Katie (14) –55 Ave A
    Lillie (10) (16) (D)
Stegel, Annie (13) (m) –1st Ave & 4th St.
Stern, Louisa (15) (m) –508 5th St.
Stick, Minnie (18) (m) –337 5th St.
    Lena (38) (D)
Stockdale, George W. –298 (296) W. 131 St.
    Mrs. Kate (52) (m)
    Edward J. (32) (m)
Straub, Mrs. Philip – 92 St. Marks Pl.
Strinz, Paul –90 1st Ave. Id'd
    Mrs. Augusta (52) (D) mother
Stuer, Clara (s) –jumped over and was picked up by a tug that put in at Randall's Island
    Millie Mannheimer (40), friend
    Lillie Mannheimer (9) (s) her niece –escaped with bruises
    Walter (11) her nephew
Suden, Edward –61 Jackson St.
Sutman, (Sudmann) Henrietta (3) –104 1st Ave.
Sullivan, James J. –7 Elm St, White Plains Id'd
Ludeman, Mrs Hannah (45) (D) – 4 Smith St., White Plains
Taylor, James (m) –153 E. 35th St. * Not a residential address, may have been given to police so that curiosity seeker could get into morgue (NYT)
    Etteler, brother (m)
Thom Suden, Margarietta 93) (D) –68 Jackson St.
Thoma, Henry (D) –90 Ave A
    Herman (4) (D)
    Emma (m)
    Stephen (D)

Tierney, Mrs. –210 E. 37th St.
Tischler, Erma (8) (D) –314 E. 9th St
Toth, 103 E. 75th St. –ID'd by Det. Sgt. McCafferty
    [no name] grandson, son of Mrs. Henry Hermans –410 E. 5th St.
Tottebaum, Charles -16 (61) St. Mark's Place Id'd
    Herman (D)
    Lizzie (48) (m)
    Henry (5) (m)
    William (9) (m)
Trapping, Lillian (26) (D) –998 Ave A –ID'd by her father
Treber, Mrs Anna 935) (D) –310 E. 25th St. –ID'd by her brother Henry
Trebing, Minnie (7) (D) –223 E. 5th St.
Trobetz, Edward (18) (I) –422 6th St.
Troell, Anna (n), wife –405 E. 5th St.
Turner, Harvey–2649 8th Ave.
Uehlien, Eddie (17) (u) -416 5th St.
Ullmann, William (14) (D)–409 5th –Porter
    Soloman
Vassmer, John (11) (D) - 333 5th St.
Vennegg, Bertha (16) -345 E. 5th St. –Ln (6/17 NYT; expected to die from fractured skull)
Vickhoff, William (25) (D) –196 2nd Ave.
Voeth (Vaeth), William, Sr. (54) –10 E. 4th St.
    William, Jr. (6) (9) (D)
Volkenberg, Albert –315 E. 17th St.
    Lucy (25) (D) –ID'd by Albert Hentze –201 1st Ave.
Von der Heiden, Alexander – Musician and teacher from Manhattan b. 1851 Germany and d. 1953, Newark, NJ.
    [wife] (D)
    [child] (D)
    [child] (D)
Wagner, Mrs. May (28) –26 Broad St. Newark
    Elizabeth (6)

May (8)
Walstenbick, Mr. (m) –2027 Madison Ave
   Mrs. (m)
   Eugene (m)
   Ernest (m)
Walter, Philip -336 6th St
Ward, Walter E. (27) (D) –Fort Lee, New Jersey
Wascher, Mary Id'd her sister in law
   Bretz, Mary (D) 304 E. 28th St.
      Baby (D) found in mother's embrace
Waurer, Mary (s) –Ln
Weaver, Regina (36) (D) –304 E. 9th St.
Wehlein, Otto –416 5th St. ID'd
   Mrs. Minnie (41) (D)
Weigwet, Carrie (6) (m) – all 4 of these from 225 5th St.
   Phoebe (8) (m)
   Henry, Sadie (12) (D)
   Givens, Maggie (12) (m)
   Wousky, Ida (14) (s) – 404 5th St.
Weil, August (15) (D) – Had broken his leg two months earlier and was in a cast.
   Charles 913) (s)
Weiner, Lena 911) (m) –707 E. 12th St.
Wiese, Mr. (D) –216 E. 11th St.
   Caroline (D) –Estate reported at $10,000
      Emilie (12) (D)
      [daughter] (D)
      [daughter] (D)
Weisl, Mrs. Caroline (50) (D) –337 E. 6th St. –ID'd by her sister
Weiss, Florence –507 E. 87th St. –Someone threw her from the deck to a tug and many other people landed on top of her.
Weiss, Mrs. Otto (s)–1235 3rd Ave. –mother of Minnie (11-13) (I)
   Lillie (47) (D)

Roth, Louisa (21) –cousin
Weis, Samuel (D) –167 Ave. A -George, William, Harry, & Louis were all orphaned as their father had died a year earlier of consumption (North)
- Sally (5) (m)
- Amelia (9) (D)
- Salome / Saloma (11) (D)
- Fred / Frederika (18) (D)
- Jake / Jacob (12) (D)
- Louis (3-10) (l)
- Harry / Henry (12) (u)

Wengert, Ethel (7) (m) –409 5th St.
Werner, Lena (11) (D) –800 E. 14th St.
Wertenberger, Margaret (22) (D) –55 1st Ave.
- Lillie (2) (D)

Westo, Lethis (14) (D) –394 E. 5th St.
Wenz, Annie Anna (22) (21) (D) –421 5th St.
- Leo (11) (m)

Whitman, Anna 960) (D) –127 1st Ave.
Wiereiter, marie 983) (D) –626 E. 12th St.
Wiese, Amelia (12–216 E. 11th St.
- [no name] Mrs. Wiese's sister-in-law

Wieser, Mrs. Caroline (D) –216 E. 9th St.
Willaim, Frank (19) (m) –119 E. 25th St.
Williams, Mrs. (m) –682 E. 148th St.
Williams, Gussie –357 E. 63rd St.
Wingerter, Peter (13) –516 5th St. (u)–He found four babies on the upper deck abandoned by their parents. He stayed with them until a tug was close and passed then over two of them. He then slid down a stanchion holding the other two, to the main deck and passed these two to a rowboat. Near him, a mother tossed her baby into the water. He dived in to save it, but someone thinking he was drowning, restrained him before he could get to the child.
Wolbern, Marvin / Martin J. (n) –1702 Dean St. Brooklyn –Ice Dealer
- Henry (1) (D)
- Hilda (D)

Wolf, Freda (20 Mo) (2) (D) –283 Himrod St, Brooklyn
Wolf, Mrs. Louise –1131 40th St, Brooklyn -Ella's daughter in law
    Albert (5) (D)
    Albert (39) (D)
    Mrs. Barbara (36/37) (D)
    Arthur (13) (u) –He lost his parents and brother. Went to live with his grandmother; his only relative. He gave up two life preservers to drowning women and went into the water without one. He was rescued by a boat. He was high on the graduation list for the 5th St. school, and will be in High school in 1905. (North)
Wolff, Mamie (24) (u) –532 5th St. (see Cordes)
Wolff, Wm.
Wolff (Wolfft), Mrs. M. (s) –420 E. 16th St. –H
Woods, Mrs. Walter E. (26) (D) –127 1st Ave.
Woolman (Woodman), Mrs. Catherine (56) (D) –2225 (2425) Jerone Ave., Bronx
    Louisa (Louise) (22)(D)
Woll, Otto Z. (1) (m) –283 Himrod St., Brooklyn
Woulrein (Woulbein), Hulda (28) (D) –1702 Dean St., Brooklyn
Wurmstich, Ella (n) –[no address] Id'd –415 or 413 5th St.
    Mrs. Louise –Ella's daughter-in-law
Yost, Edward – 393 E. 9th St. looking for
    Schoett, Christian (19) (D) brother in law & St. Marks Organist
        Josephine (m) – 98 7th St. Christian's mother
        Katie (10) (m)
        Helen (5) (m)
    Siedewande, Henry, Christian's nephew (12) (m) 184 3rd St.
    [no name given] Christian's fiancee (D)
Zetter, Mary (49) (D) – 31Beekman Place
Zimmerman, William (8) (D) –229 Bleeker St., Brooklyn
Zinger, Mrs. Susie (D) – she carried $32,000 in securities and jewelry, and a bank book which showed she was a trustee for her daughter.
Zogg, George –546 E. 143rd St.
Zuderman, Henrietta (m) –104 1st Ave.

www.ingramcontent.com/pod-product-compliance
Lightning Source LLC
Chambersburg PA
CBHW050328230426
43663CB00010B/1784